Control of Partial Differential Equations

Series in Contemporary Applied Mathematics

ISSN: 2010-2259

Editors: P. G. Ciarlet (*City University of Hong Kong*)
Tatsien Li (*Fudan University*)

The complete list of titles in the series can be found at
https://www.worldscientific.com/series/cam-bs

Series in Contemporary Applied Mathematics CAM 24

Control of Partial Differential Equations

editors

Jean-Michel Coron
Sorbonne Université, France

Tatsien Li
Fudan University, China

Zhiqiang Wang
Fudan University, China

Higher Education Press

NEW JERSEY · LONDON · SINGAPORE · BEIJING · SHANGHAI · HONG KONG · TAIPEI · CHENNAI · TOKYO

Published by

Higher Education Press Limited Company
4 Dewai Dajie, Beijing 100120, P. R. China
and
World Scientific Publishing Co. Pte. Ltd.
5 Toh Tuck Link, Singapore 596224
USA office: 27 Warren Street, Suite 401-402, Hackensack, NJ 07601
UK office: 57 Shelton Street, Covent Garden, London WC2H 9HE

Library of Congress Control Number: 2022062246

British Library Cataloguing-in-Publication Data
A catalogue record for this book is available from the British Library.

Series in Contemporary Applied Mathematics — Vol. 24
CONTROL OF PARTIAL DIFFERENTIAL EQUATIONS

ISBN 978-981-127-162-5 (hardcover)
ISBN 978-981-127-163-2 (ebook for institutions)
ISBN 978-981-127-164-9 (ebook for individuals)

Preface

This book is mainly a collection of lecture notes for the 2021 LIASFMA International Graduate School on Applied Mathematics. This graduate school consists of two seasons. Both of them are made online by Zoom and focus on the topic of "Control of Partial Differential Equations" which is a fast growing and important area in applied mathematics. The first season was held during April 6-29, 2021 and the second one was held during November 16-December 9, 2021.

This international graduate school is one of the activities organized by Laboratoire International Associé Sino-Français de Mathématiques Appliquées (LIASFMA). Established jointly by eight institutions in China and France in 2014, LIASFMA aims at providing a platform for many leading French and Chinese mathematicians to conduct in-depth researches, extensive exchanges, and student training in broad areas of applied mathematics.

The aim of the school is to increase the interest and broaden the horizons of young scholars in early stage. The graduate school consists of a series of mini-courses given by leading experts in related fields. Four 8-hour mini-courses were delivered on various aspects in control of partial differential equations, including Carleman estimates, boundary stabilization, control of stochastic equations and control of quantum systems.

Thanks to the online meeting system, the graduate school is indeed international and influential. Each season attracted about 300 registered participants from more than 20 countries. Among them, there were not only master students, PhD students and postdocs, but also many professors and researchers.

We would like to express our sincere gratitude to all the lecturers of the graduate school: Sylvain Ervedoza (CNRS & Institut de Mathématiques de Bordeaux, France), Qi Lü (Sichuan University, China), Christophe Prieur (CNRS & Gipsa-lab, Grenoble, France) and Pierre Rouchon (MINES ParisTech & INRIA, France). We are very grateful to Bopeng Rao (Université de Strasbourg, France) for offering the lecture notes on

asymptotic synchronization of dissipative systems which could have also
been a mini-course of the graduate school. The lecture notes on modelling
and control of open quantum systems by Pierre Rouchon are not included
in this book.

We shall thank Xinyue Feng for serving as contactor, thank Yubo Bai for
the efforts in webpage management and thank Mrs. Chunlian Zhou for the
help on financial issues. In the end we appreciate a lot to Fudan University
for the financial support.

Jean-Michel Coron (Sorbonne Université, France)
Tatsien Li (Fudan University, China)
Zhiqiang Wang (Fudan University, China)
December, 2021

Contents

A Short Introduction to Carleman Estimates

Jérémi Dardé

Institut de Mathématiques de Toulouse, UMR 5219, Université de Toulouse, CNRS, Toulouse, France
jeremi.darde@math.univ-toulouse.fr

Sylvain Ervedoza

Institut de Mathématiques de Bordeaux, UMR 5251, Université de Bordeaux, CNRS, Bordeaux INP, F-33400 Talence, France
sylvain.ervedoza@math.u-bordeaux.fr

Abstract. In these notes, we present a short introduction to Carleman estimates for elliptic equations, developing various proofs and points of view.

In particular, we provide Carleman estimates for the Laplace operator with linear weight functions. We also give some insights on the need of strict (pseudo-)convexity for the weight function within Carleman estimates, first when the weight function depends only on one variable, then in the general case.

We also provide some applications to unique continuation properties for the Laplace operator and to the Calderón problem appearing in Electrical Impedance Tomography.

We finally end up by mentioning several applications of these techniques, underlying their importance in the recent developments of the theory of partial differential equations.

Foreword

These notes correspond to a lecture given by Sylvain Ervedoza within the LIAFSMA International Graduate School in Applied Maths 2021, and are strongly inspired by some lectures taught in 2013 by Jérémi Dardé and Sylvain Ervedoza in the master program of Université Paul Sabatier, Toulouse, France.

These lecture notes aim at presenting some insights on the derivation of Carleman estimates and their use in some specific applications, using a self-contained approach requiring only the knowledge of basic theory for elliptic PDE and Fourier transforms.

Of course, we are fully aware that this approach will not allow to reach the many results that have been derived in the extensive literature on Carleman estimates, for which we refer to more advanced textbooks, such as the excellent books [Lebeau *et al.* (2022); Lerner (2019); Fu *et al.* (2019); Bellassoued and Yamamoto (2017)], among many others.

Still, we believe that our approach offers new perspectives on Carleman estimates and underlines how they can be derived in a somewhat pedestrian manner. In fact, this methodology is closely related to the ones in [Calderón (1958); Imanuvilov and Puel (2003)], among others.

1.0. Prerequisites

This section gathers several results which we will assume to be known next.

1.0.1. *Functional spaces*

1.0.1.1. *Lebesgue spaces*

We assume that the reader is familiar with measurable functions, integrable functions, and more specifically to the Lebesgue space $L^1(\Omega)$ for Ω an (non-empty) open set of $\mathbb{R}^d$, $d \in \mathbb{N} \setminus \{0\}$.

The Lebesgue spaces are then defined as follows:

Definition 1.1 (Lebesgue spaces). Let Ω be an (non-empty) open set of $\mathbb{R}^d$, $d \in \mathbb{N} \setminus \{0\}$ and $p \in [1, \infty)$. We set

$$L^p(\Omega) = \{f : \Omega \to \mathbb{C}, \text{ measurable, with } |f|^p \in L^1(\Omega)\},$$

$$L^\infty(\Omega) = \{f : \Omega \to \mathbb{C}, \text{ measurable,}$$

$$\text{such that there exists a constant } C \text{ such that a.e. } |f| \leqslant C\},$$

and

$$\|f\|_{L^p(\Omega)} = \left(\int_\Omega |f(x)|^p \, dx \right)^{1/p},$$

$$\|f\|_{L^\infty(\Omega)} = \inf\{C > 0 \text{ such that } |f| \leqslant C \text{ a.e.}\}.$$

The study of these Lebesgue spaces is classical and can be found for instance in Chapter 4 of [Brezis (1983)], where the following results are proved:

Proposition 1.2 (Banach spaces). *For all $p \in [1, \infty]$, the Lebesgue space $L^p(\Omega)$ is a Banach space when endowed with the norm $\| \cdot \|_{L^p(\Omega)}$.*

Proposition 1.3 (Duality). *For all $p \in (1, \infty)$, the space $(L^p(\Omega))'$ of all continuous linear forms on $L^p(\Omega)$ can be identified with $L^{p'}(\Omega)$, with p' given by*

$$\frac{1}{p} + \frac{1}{p'} = 1.$$

We also present some useful estimates:

Proposition 1.4 (Hölder's inequality). *Let p, q, $r \in [1, \infty]^3$ such that*

$$\frac{1}{p} + \frac{1}{q} = \frac{1}{r}.$$

Then, for all $f \in L^p(\Omega)$ and $g \in L^q(\Omega)$, fg belongs to $L^r(\Omega)$ and

$$\|fg\|_{L^r(\Omega)} \leqslant \|f\|_{L^p(\Omega)} \|g\|_{L^q(\Omega)}.$$

Proposition 1.5 (Young's inequality). *Let p, q, $r \in [1, \infty]^3$ with*

$$\frac{1}{p} + \frac{1}{q} = 1 + \frac{1}{r}.$$

Then for $f \in L^p(\mathbb{R}^d)$ and $g \in L^q(\mathbb{R}^d)$, the convolution of f by g, denoted by $f \star g$ and defined by

$$f \star g(x) = \int_{\mathbb{R}^d} f(x - y) g(y) \, dy, \quad x \in \mathbb{R}^d,$$

is well-defined as a function of $L^r(\mathbb{R}^d)$ and

$$\|f \star g\|_{L^r(\mathbb{R}^d)} \leqslant \|f\|_{L^p(\mathbb{R}^d)} \|g\|_{L^q(\mathbb{R}^d)}.$$

1.0.1.2. *A quick reminder on distributions*

We start by recalling the definition of distributions in a bounded domain Ω and some elementary properties, see for instance [Gel'fand and Shilov (1964)].

Definition 1.6 (Test functions). Let Ω be an (non-empty) open set of $\mathbb{R}^d$, $d \in \mathbb{N} \setminus \{0\}$. Then $\mathscr{D}(\Omega)$ denotes the set of all functions $\varphi : \Omega \to \mathbb{R}$ which are infinitely differentiable in Ω and with compact support in Ω. The set $\mathscr{D}(\Omega)$ is called the set of test functions.

Using this set $\mathscr{D}(\Omega)$, we can define the set of distributions on Ω:

Definition 1.7 (Distributions). Let Ω be an (non-empty) open set of $\mathbb{R}^d$, $d \in \mathbb{N} \setminus \{0\}$. Then a distribution u on Ω is a linear form on $\mathscr{D}(\Omega)$ such that for all compact $K \subset \Omega$, there exist $C_K > 0$ and $p_K \in \mathbb{N}$ such that for all $\varphi \in \mathscr{D}(\Omega)$ with $\mathrm{Supp}\,\varphi \subset K$, then

$$|u(\varphi)| \leqslant C_K \sup_{|\alpha| \leqslant p_k} \|D^\alpha \varphi\|_{L^\infty(K)},$$

where $\alpha \in \mathbb{N}^d$ is a multi-index, $\alpha = (\alpha_1, \cdots, \alpha_d)$, and $D^\alpha = \partial_1^{\alpha_1} \cdots \partial_d^{\alpha_d}$ and $|\alpha| = \alpha_1 + \cdots + \alpha_d$.

The set of all distributions is denoted by $\mathscr{D}'(\Omega)$.

Notation: in the following, as it is commonly done, for $u \in \mathscr{D}'(\Omega)$ and $\varphi \in \mathscr{D}(\Omega)$, we will denote $u(\varphi)$ by $\langle u, \varphi \rangle_{\mathscr{D}'(\Omega), \mathscr{D}(\Omega)}$.

A key remark is that functions can be viewed as distributions through the following identification: if $u \in L^1_{loc}(\Omega)^{\text{a}}$, then we define the corresponding distribution by, for all $\varphi \in \mathscr{D}(\Omega)$,

$$\langle u, \varphi \rangle_{\mathscr{D}'(\Omega), \mathscr{D}(\Omega)} = \int_\Omega u(x) \varphi(x) \, dx.$$

Note however that there are distributions which cannot be identified with locally integrable functions, such as, for instance, the Dirac delta distribution δ_0 on $\mathbb{R}$ defined by $\forall \varphi \in \mathscr{D}(\mathbb{R})$, $\langle \delta_0, \varphi \rangle_{\mathscr{D}'(\mathbb{R}), \mathscr{D}(\mathbb{R})} = \varphi(0)$.

Another important point is that the classical derivatives can easily be extended on distributions by the following formula:

Definition 1.8 (Distribution derivative). Let Ω be an (non-empty) open set of $\mathbb{R}^d$, $d \in \mathbb{N} \setminus \{0\}$, $u \in \mathscr{D}'(\Omega)$, and $i \in \{1, \cdots, d\}$. Then $\partial_i u$ is the distribution defined by: for all $\varphi \in \mathscr{D}(\Omega)$,

$$\langle \partial_i u, \varphi \rangle_{\mathscr{D}'(\Omega), \mathscr{D}(\Omega)} = -\langle u, \partial_i \varphi \rangle_{\mathscr{D}'(\Omega), \mathscr{D}(\Omega)},$$

where in the right hand side of this identity, $\partial_i \varphi$ is the classical derivative of φ with respect to the i-th variable.

One easily checks that Definition 1.8, together with the identification of $L^1_{loc}(\Omega)$ as a subset of $\mathscr{D}'(\Omega)$, extends the classical definition of the derivative, as a straightforward consequence of the following integration by part formula: if $u \in \mathscr{C}^1(\Omega)$ and $\varphi \in \mathscr{D}(\Omega)$,

$$-\langle u, \partial_i \varphi \rangle_{\mathscr{D}'(\Omega), \mathscr{D}(\Omega)} = -\int_\Omega u(x) \partial_i \varphi(x) \, dx = \int_\Omega \partial_i u(x) \varphi(x) \, dx.$$

---

[a] $L^1_{loc}(\Omega)$ denotes the set of functions whose restriction to any compact K of Ω belongs to $L^1(K)$.

Similarly, when $\Omega = \mathbb{R}^d$, we can define the Schwartz class:

Definition 1.9 (The Schwartz class). The Schwartz class, denoted by $\mathcal{S}(\mathbb{R}^d)$, is the set of all smooth functions $\varphi : \mathbb{R}^d \to \mathbb{R}$ such that for all multi-index α and β in $\mathbb{N}^d$,

$$\sup_{\mathbb{R}^d} |x^\alpha D^\beta \varphi| < \infty,$$

where for $\alpha = (\alpha_1, \cdots, \alpha_d) \in \mathbb{N}^d$, x^α stands for $x_1^{\alpha_1} \cdots x_d^{\alpha_d}$.

A linear form u on $\mathcal{S}(\mathbb{R}^d)$ such that there exist $p \in \mathbb{N}$ and $C > 0$, such that for all $\varphi \in \mathcal{S}(\mathbb{R}^d)$,

$$|u(\varphi)| \leqslant C \sup_{|\alpha|+|\beta| \leqslant p} \sup_{\mathbb{R}^d} |x^\alpha D^\beta \varphi|$$

is called a tempered distribution. The set of all tempered distribution is denoted by $\mathcal{S}'(\mathbb{R}^d)$.

Note that, as in the case of distribution, when $u \in \mathcal{S}'(\mathbb{R}^d)$ and $\varphi \in \mathcal{S}(\mathbb{R}^d)$, we will denote $u(\varphi)$ simply by $\langle u, \varphi \rangle$.

1.0.1.3. *Sobolev spaces*

We shall not give an extensive review of Sobolev spaces, but only some of their properties, which can be found for instance in Chapter 5 of [Evans (1998)]. In particular, we will simply focus on the Sobolev spaces which have a Hilbertian structure.

Spaces $H^k(\Omega)$, $H_0^1(\Omega)$, $H^{-1}(\Omega)$. We start with the following definition:

Definition 1.10 (The spaces $H^k(\Omega)$, $k \in \mathbb{N}$). Let Ω be an (non-empty) open set of $\mathbb{R}^d$, $d \in \mathbb{N} \setminus \{0\}$. For $k \in \mathbb{N}$, we define

$$H^k(\Omega) = \{f \in L^2(\Omega), \text{ such that for all } \alpha \in \mathbb{N}^d \text{ with } |\alpha| \leqslant k, \partial^\alpha f \in L^2(\Omega)\},$$

where for $\alpha \in \mathbb{N}^d$, $|\alpha| = \sum_{i=1}^d |\alpha_i|$ and $\partial^\alpha = \partial_1^{\alpha_1} \cdots \partial_d^{\alpha_d}$ denotes the derivative in the sense of distributions.

The set $H^k(\Omega)$ is endowed with the norm

$$\|f\|_{H^k(\Omega)} = \left(\sum_{|\alpha| \leqslant k} \|\partial^\alpha f\|_{L^2(\Omega)}^2 \right)^{1/2}. \tag{1.1}$$

We have the following proposition:

Proposition 1.11 (Basic properties of $H^k(\Omega)$). *Let Ω be an (non-empty) open set of $\mathbb{R}^d$, $d \in \mathbb{N} \setminus \{0\}$ and $k \in \mathbb{N}$.*

Then the set $H^k(\Omega)$ is a separable Hilbert space.

If Ω is bounded and its boundary $\partial\Omega$ is of class $\mathscr{C}^\infty$, then the set $\mathscr{C}^\infty(\overline{\Omega})$ is dense in $H^k(\Omega)$.

We shall also often use the space $H_0^1(\Omega)$:

Definition 1.12 (The spaces $H_0^1(\Omega)$ and $H^{-1}(\Omega)$). Let Ω be an (non-empty) open set of $\mathbb{R}^d$, $d \in \mathbb{N} \setminus \{0\}$. Then $H_0^1(\Omega)$ denotes the closure of the set $\mathscr{D}(\Omega)$ for the topology of $H^1(\Omega)$. The set $H^{-1}(\Omega)$ is defined as the topological dual of $H_0^1(\Omega)$.

Fourier transforms and fractional Sobolev spaces. It turns out that we will precisely deal with boundary data quite often. In order to do that, we will use the Fourier transform and define fractional Sobolev spaces.

Definition 1.13 (The Fourier transform). Let $f \in L^1(\mathbb{R}^d) \cap L^2(\mathbb{R}^d)$. Then the Fourier transform of f is given by

$$\mathscr{F}[f](\xi) = \frac{1}{(2\pi)^{d/2}} \int_{\mathbb{R}^d} f(x)e^{-ix\cdot\xi}\,dx, \quad \xi \in \mathbb{R}^d.$$

When no confusion occurs, we will simply denote it by $\widehat{f}$.

We have the following classical theorem:

Theorem 1.14 (Parseval's identity). *The Fourier transform $\mathscr{F}$ can be extended as an isometry of $L^2(\mathbb{R}^d)$:*

$$\int_{\mathbb{R}^d} |f(x)|^2\,dx = \int_{\mathbb{R}^d} |\widehat{f}(\xi)|^2\,d\xi.$$

Its inverse is given, for $g \in L^1(\mathbb{R}^d) \cap L^2(\mathbb{R}^d)$ by

$$\mathscr{F}^{-1}[g](x) = \frac{1}{(2\pi)^{d/2}} \int_{\mathbb{R}^d} g(\xi)e^{ix\cdot\xi}\,d\xi, \quad x \in \mathbb{R}^d.$$

Using classical properties of the Fourier transform, we easily get that the Fourier transform $\mathscr{F}$ maps $\mathcal{S}(\mathbb{R}^d)$ to itself. Accordingly, by duality, we can define the Fourier transform $\mathscr{F}$ on $\mathcal{S}'(\mathbb{R}^d)$ as follows: for $u \in \mathcal{S}'(\mathbb{R}^d)$, $\mathscr{F}u$ is the element of $\mathcal{S}'(\mathbb{R}^d)$ defined as follows: for all $\varphi \in \mathcal{S}(\mathbb{R}^d)$, $\langle \mathscr{F}u, \varphi \rangle = \langle u, \mathscr{F}^{-1}\varphi \rangle$. This extends the Fourier transform to $\mathscr{S}'(\mathbb{R}^d)$ and coincides with the usual definition of $\mathscr{F}u$ for $u \in L^1(\mathbb{R}^d) \cap L^2(\mathbb{R}^d)$ since the identity $\langle \mathscr{F}u, \varphi \rangle = \langle u, \mathscr{F}^{-1}\varphi \rangle$ obviously holds for all $\varphi \in \mathcal{S}(\mathbb{R}^d)$ when $u \in L^1(\mathbb{R}^d) \cap L^2(\mathbb{R}^d)$.

We then define fractional Sobolev spaces:

Definition 1.15 (The fractional Sobolev spaces $H^s(\mathbb{R}^d)$, $s \in \mathbb{R}$). For $s \in \mathbb{R}$, the set $H^s(\mathbb{R}^d)$ is the set of all functions $f \in L^2(\mathbb{R}^d)$ such that $\xi \mapsto (1 + |\xi|^2)^{s/2}\widehat{f}(\xi) \in L^2(\mathbb{R}^d)$. For $f \in H^s(\mathbb{R}^d)$, we set

$$\|f\|_{H^s(\mathbb{R}^d)} = \left(\int_{\mathbb{R}^d} (1 + |\xi|^2)^s |\widehat{f}(\xi)|^2 \, d\xi \right)^{1/2}.$$

If $s \in \mathbb{N}$, this norm is equivalent to the norm $H^k(\mathbb{R}^d)$ for $k = s$ given in (1.1).

For all $s \in \mathbb{R}$, the set $H^s(\mathbb{R}^d)$ is a Hilbert space, and if $s \geqslant 0$, $H^{-s}(\mathbb{R}^d)$ is the dual of $H^s(\mathbb{R}^d)$ after the identification of $L^2(\mathbb{R}^d)$ with its dual.

We are then in position to define fractional functional spaces for an open set Ω of $\mathbb{R}^d$, following for instance [Lions and Magenes (1968)].

Definition 1.16 (Definition of $H^s(\Omega)$ and $H^s(\partial\Omega)$, $s \geqslant 0$). Let Ω be a bounded subset of $\mathbb{R}^d$ whose boundary has $\mathscr{C}^\infty$ regularity. For $s \geqslant 0$, the set $H^s(\Omega)$ is defined as the restrictions to Ω of functions of $H^s(\mathbb{R}^d)$, and is endowed with the norm

$$\|u\|_{H^s(\Omega)} = \inf\{\|U\|_{H^s(\mathbb{R}^d)} \text{ for } U \in H^s(\mathbb{R}^d) \text{ with } U|_\Omega = u\}.$$

The definition of $H^s(\partial\Omega)$ for $s \geqslant 0$ is slightly more involved, and we follow the presentation of [Lions and Magenes (1968)]. We assume that Ω is a bounded subset of $\mathbb{R}^d$ whose boundary $\partial\Omega$ has $\mathscr{C}^\infty$ regularity. This means that there exists a finite family of open bounded sets $\mathcal{O}_j$, $j \in \{1, \cdots, N\}$ of $\mathbb{R}^d$ which covers $\partial\Omega$ and such that for all $j \in \{1, \cdots, N\}$, there exists a $\mathscr{C}^\infty$ mapping φ_j from $\mathcal{O}_j$ to a neighborhood $\mathcal{V}_j$ of 0 in $\mathbb{R}^d$, such that it is invertible on $\mathcal{V}_j$ as a $\mathscr{C}^\infty$ mapping, and

$$\varphi_j(\mathcal{O}_j \cap \Omega) = \mathcal{V}_j \cap \{y_d > 0\},$$
$$\varphi_j(\mathcal{O}_j \cap \partial\Omega) = \mathcal{V}_j \cap \{y_d = 0\},$$
$$\varphi_j(\mathcal{O}_j \setminus \overline{\Omega}) = \mathcal{V}_j \cap \{y_d < 0\}.$$

We then require that if $\mathcal{O}_i \cap \mathcal{O}_j \neq \emptyset$ then there exists a $\mathscr{C}^\infty$ homeomorphism $J_{i,j}$ from $\varphi_i(\mathcal{O}_i \cap \mathcal{O}_j)$ to $\varphi_j(\mathcal{O}_i \cap \mathcal{O}_j)$ with positive jacobian such that

$$\forall x \in \mathcal{O}_i \cap \mathcal{O}_j, \quad \varphi_j(x) = J_{i,j}(\varphi_i(x)).$$

We then consider a partition of unity associated to the covering $\mathcal{O}_j$, $j \in \{1, \cdots, N\}$ of $\partial\Omega$, by choosing N smooth functions α_j, $j \in \{1, \cdots, N\}$, such that for all j, α_j is compactly supported in $\mathcal{O}_j$ and $\sum_{j=1}^{N} \alpha_j = 1$ on $\partial\Omega$.

We then set

$$H^s(\partial\Omega) = \{u \in L^2(\Gamma),$$
$$\text{such that } \forall j \in \{1, \cdots, N\}, \ y' \mapsto (\alpha_j u)(\varphi_j^{-1}(y', 0)) \in H^s(\mathbb{R}^{d-1})\},$$

where the function $y' \mapsto (\alpha_j u)(\varphi_j^{-1}(y', 0))$ is extended by 0 outside $\mathcal{V}_j \cap \{y_d = 0\}$.

We also define the corresponding norm

$$\|u\|_{H^s(\partial\Omega)} = \left(\sum_{j=1}^{N} \|y' \mapsto (\alpha_j u)(\varphi_j^{-1}(y', 0))\|_{H^s(\mathbb{R}^{d-1})}^2 \right)^{1/2}.$$

Note that the definition of $H^s(\partial\Omega)$ and of $\|\cdot\|_{H^s(\partial\Omega)}$ a priori depends on the choice of the covering $\mathcal{O}_j$, $j \in \{1, \cdots, N\}$, the charts φ_j, $j \in \{1, \cdots, N\}$ and the partition of unity $\alpha_j, j \in \{1, \cdots, N\}$.

It turns out that one can prove (see [Lions and Magenes (1968)]) that the space $H^s(\partial\Omega)$ is in fact independent of all these choices, since two norms corresponding to different choices of covering, charts, and partition of unity, necessarily are equivalent.

Let us also mention that, for $s \geqslant 0$, the dual of $H^s(\partial\Omega)$ is $H^{-s}(\partial\Omega)$.

We refer to [Lions and Magenes (1968)] for a more detailed discussion on fractional Sobolev spaces.

Trace theorems. Let Ω be a smooth bounded domain. The question we address here is under which circumstances we can define the value of a function on the boundary of Ω. It turns out that this can be done only under some regularity conditions, that we recall hereafter, see [Lions and Magenes (1968)]:

Theorem 1.17. *There exists a unique continuous map γ_0 from $H^1(\Omega)$ to $L^2(\partial\Omega)$ such that for all $f \in \mathscr{C}^\infty(\overline{\Omega})$, $\gamma_0(f) = f|_{\partial\Omega}$.*

This map is called the trace map.

The trace map is continuous surjective from $H^1(\Omega)$ to $H^{1/2}(\partial\Omega)$ and its restriction $\gamma_0|_{H^2(\Omega)}$ to $H^2(\Omega)$ is surjective from $H^2(\Omega)$ to $H^{3/2}(\partial\Omega)$.

Besides, $Ker\gamma_0 = H_0^1(\Omega)$.

Similarly, to define the normal derivative of a function, one needs to add some regularity condition:

Theorem 1.18. *There exists a unique continuous map γ_1 from $H^2(\Omega)$ to $L^2(\partial\Omega)$ such that for all $f \in \mathscr{C}^\infty(\overline{\Omega})$, $\gamma_1(f) = \partial_n f|_{\partial\Omega}$, where $\partial_n f$ denotes the normal derivative of f on $\partial\Omega$.*

The map γ_1 is continuous and surjective from $H^2(\Omega)$ to $H^{1/2}(\partial\Omega)$.

For more properties, we refer to the classical textbook [Lions and Magenes (1968)].

1.0.2. *Elliptic equations*

In this section, we briefly recall the existence theory for the Laplace problem

$$\begin{cases} -\Delta u = f, & \text{in } \Omega, \\ u = g, & \text{on } \partial\Omega, \end{cases} \tag{1.2}$$

where Ω is a smooth ($\mathscr{C}^\infty$) bounded domain of $\mathbb{R}^d$, and $\Delta = \partial_{11} + \cdots + \partial_{dd}$ is the Laplace operator.

Following [Brezis (1983); Evans (1998)], we get the following result, as a consequence of Lax-Milgram theorem:

Theorem 1.19. *Let $f \in H^{-1}(\Omega)$ and $g \in H^{1/2}(\partial\Omega)$. Then there exists a unique function $u \in H^1(\Omega)$ such that*

- $\gamma_0 u = g$, *where γ_0 is the trace map defined in Theorem 1.17,*
- *for all $v \in H_0^1(\Omega)$,*

$$\int_\Omega \nabla u \cdot \nabla v \, dx = \langle f, v \rangle_{H^{-1}(\Omega), H_0^1(\Omega)}.$$

Such a function u is called a weak solution of (1.2).

Besides, there exists a constant C such that

$$\|u\|_{H^1(\Omega)} \leqslant C \left(\|f\|_{H^{-1}(\Omega)} + \|g\|_{H^{1/2}(\partial\Omega)} \right).$$

One can also prove the following regularity result, see e.g. [Evans (1998)]:

Theorem 1.20. *Let $f \in L^2(\Omega)$ and $g \in H^{3/2}(\partial\Omega)$. Then the function u provided by Theorem 1.19 satisfies $u \in H^2(\Omega)$.*

Besides, there exists a constant C such that

$$\|u\|_{H^2(\Omega)} \leqslant C \left(\|f\|_{L^2(\Omega)} + \|g\|_{H^{3/2}(\partial\Omega)} \right).$$

Let us finally remark that these theorems apply as soon as the open set Ω is smooth and bounded in one direction, since then Poincaré's estimate holds: there exists a constant $C > 0$ such that for all $u \in H_0^1(\Omega)$, $\|u\|_{L^2(\Omega)} \leqslant C \|\nabla u\|_{L^2(\Omega)}$.

In our case, we will repeatedly use these properties in the case of a vertical strip $\Omega = (0, 1) \times \mathbb{R}^{d-1}$, see the next sections.

1.1. Introduction: on the Laplace operator in a strip

1.1.1. *The Cauchy problem*

Our starting point is the so-called Cauchy problem for the Laplace operator. To be more precise, we consider a solution u of an elliptic equation in a smooth domain Ω of $\mathbb{R}^d$:

$$\Delta u = f, \quad \text{in } \Omega, \tag{1.3}$$

where $\Delta = \partial_{11} + \cdots + \partial_{dd}$ is the Laplace operator and f is assumed to be known.

We also assume that there is a non-empty open part of the boundary $\Gamma \subset \partial\Omega$ on which the Cauchy data are known, that is to say we know

$$\begin{cases} u = g_D, & \text{on } \Gamma, \\ \partial_n u = g_N, & \text{on } \Gamma. \end{cases} \tag{1.4}$$

Here, the indexes D and N respectively stand for the Dirichlet and Neumann data on Γ.

The Cauchy problem is then the following one:

Given the data f in Ω, g_D and g_N on Γ, can we determine the solution u of (1.3)–(1.4)?

Here, recall that if f is given in Ω and g_D is given on the whole boundary $\partial\Omega$, problem (1.3)–(1.4)$_1$ is well-posed and has a unique solution, see Section 1.0.2 and e.g. [Brezis (1983)] (see also Problem 1.21), hence the above question is obvious when Γ is the whole boundary, and the difficult case is the one corresponding to $\Gamma \neq \partial\Omega$.

We shall not give any precise answer to that question in this full generality, but we will give some insights on it. In order to do that, we focus on the case

$$\Omega = (0, 1) \times \mathbb{R}^{d-1} \quad (d \in \mathbb{N} \setminus \{0, 1\}), \quad \text{and} \quad \Gamma = \{0\} \times \mathbb{R}^{d-1}, \tag{1.5}$$

i.e. the case of a strip observed from one side.

In that situation, we may rewrite the problem (1.3)–(1.4) as follows:

$$\begin{cases} \partial_{11} u + \Delta' u = f, & \text{for } (x_1, x') \in (0, 1) \times \mathbb{R}^{d-1}, \\ u(0, x') = g_D(x'), & \text{for } x' \in \mathbb{R}^{d-1}, \\ \partial_1 u(0, x') = g_N(x'), & \text{for } x' \in \mathbb{R}^{d-1}, \end{cases} \tag{1.6}$$

where $\Delta' = \partial_{22} + \cdots + \partial_{dd}$ is the Laplace operator in the variable $x' = (x_2, \cdots, x_d) \in \mathbb{R}^{d-1}$.

In that setting, a natural approach consists in taking the Fourier transform of u in the x'-variable only. Therefore, for $x_1 \in [0, 1]$ and $\xi' \in \mathbb{R}^{d-1}$, we introduce

$$\widehat{u}(x_1, \xi') = \mathscr{F}_{x' \to \xi'} u(x_1, \cdot),$$
$$\widehat{f}(x_1, \xi') = \mathscr{F}_{x' \to \xi'} f(x_1, \cdot),$$
$$\widehat{g}_D(\xi') = \mathscr{F}_{x' \to \xi'} g_D(\cdot),$$
$$\widehat{g}_N(\xi') = \mathscr{F}_{x' \to \xi'} g_N(\cdot),$$

where $\mathscr{F}_{x' \to \xi'}$ is the Fourier transform in the variable x', defined for function $v \in \mathcal{S}(\mathbb{R}^{d-1})$ of $x' \in \mathbb{R}^{d-1}$ by

$$\forall \xi' \in \mathbb{R}^{d-1}, \quad (\mathscr{F}_{x' \to \xi'} v)(\xi) = \frac{1}{(2\pi)^{(d-1)/2}} \int_{\mathbb{R}^{d-1}} v(x') e^{-ix' \cdot \xi'} \, dx'.$$

Taking the partial Fourier transform $\mathscr{F}_{x' \to \xi'}$ of (1.6), we obtain

$$\begin{cases} \partial_{11} \widehat{u} - |\xi'|^2 \widehat{u} = \widehat{f}, & \text{for } (x_1, \xi') \in (0,1) \times \mathbb{R}^{d-1}, \\ \widehat{u}(0, \xi') = \widehat{g}_D(\xi'), & \text{for } \xi' \in \mathbb{R}^{d-1}, \\ \partial_1 \widehat{u}(0, \xi') = \widehat{g}_N(\xi'), & \text{for } \xi' \in \mathbb{R}^{d-1}. \end{cases} \quad (1.7)$$

At $\xi' \in \mathbb{R}^{d-1}$ fixed, these equations can be solved explicitly in x_1 and the solution $\widehat{u}(\cdot, \xi')$ is given by

$$\widehat{u}(x_1, \xi') = \cosh(|\xi'| x_1) \widehat{g}_D(\xi') + \frac{\sinh(|\xi'| x_1)}{|\xi'|} \widehat{g}_N(\xi')$$
$$+ \int_0^{x_1} \frac{\sinh(|\xi'|(x_1 - x))}{|\xi'|} \widehat{f}(x, \xi') \, dx. \quad (1.8)$$

Up to now, our arguments were mainly formal. The above computations require in particular the Fourier transform to be well-defined for $u(x_1, \cdot)$ for almost all $x_1 \in (0, 1)$. It is certainly the case if u is assumed to be for instance in $L^2((0,1) \times \mathbb{R}^{d-1})$, or even $L^2(0, 1; \mathcal{S}'(\mathbb{R}^{d-1}))$.

An important difficulty arises here. If we do not know *a priori* that u exists and wonder if one can find a solution u to the Cauchy problem (1.6), one needs to guarantee that one can take the inverse Fourier transform of formula (1.8). But, at $x_1 \in (0, 1)$ fixed, $\widehat{u}(x_1, \cdot)$ may contain growing exponentials of the form $\xi' \mapsto \exp(\alpha |\xi'|)$ with $\alpha > 0$. Such function does not belong to $\mathcal{S}'(\mathbb{R}^{d-1})$ and its inverse Fourier transform is not defined.

On the opposite, if we know from the beginning that u exists and belongs to some class where the partial Fourier transform is well-defined, then it is uniquely determined by the above formula (1.8). But even in that case, it may be difficult to derive good estimates on u via the use of formula (1.8), as we only have, for all $x_1 \in (0,1)$ and $\xi' \in \mathbb{R}^{d-1}$,

$$|\widehat{u}(x_1, \xi')| \leqslant C \exp(|\xi'|x_1) \left(|\widehat{g}_D(\xi')| + |\widehat{g}_N(\xi')| + \int_0^{x_1} |\widehat{f}(x, \xi')| \, dx \right),$$

for some constant C independent of (x_1, ξ'). This allows to derive

$$\int_0^1 \int_{\mathbb{R}^{d-1}} |\widehat{u}(x_1, \xi')|^2 \exp(-2|\xi'|x_1) \, d\xi' dx_1$$

$$\leqslant C \left(\int_{\mathbb{R}^{d-1}} |\widehat{g}_D(\xi')|^2 d\xi' + \int_{\mathbb{R}^{d-1}} |\widehat{g}_N(\xi')|^2 d\xi' + \int_0^{x_1} \int_{\mathbb{R}^{d-1}} |\widehat{f}(x, \xi')|^2 \, dx d\xi' \right)$$

$$\leqslant C \left(\|g_D\|_{L^2(\mathbb{R}^{d-1})}^2 + \|g_N\|_{L^2(\mathbb{R}^{d-1})}^2 + \|f\|_{L^2((0,1)\times\mathbb{R}^{d-1})}^2 \right).$$

Of course, the left-hand side defines a norm on u, but it is a very weak one, weaker than any norm of the form $L^2(0, 1; H^{-k}(\mathbb{R}^{d-1}))$, $k > 0$. Actually, the Cauchy problem (1.6) is the prototype of a problem which is not well-posed in the sense of Hadamard, as this norm contains an exponential degeneracy.

1.1.2. *The case of additional information*

In the above discussion, we explained that it is difficult to get estimates on u in a reasonable norm in terms of norms of the data f, g_D and g_N. The goal of this paragraph is to explain that it can be much better if we further assume that u is known on the other part of the boundary $\{1\} \times \mathbb{R}^{d-1}$. Before going further, let us note that this is expected since then the whole Dirichlet boundary conditions are known for the solution u of (1.7). Still, we believe that the computations presented below will give some interesting insights in the following.

To simplify the computations, we will assume that the solution u of (1.6) satisfies the homogeneous Dirichlet boundary conditions

$$u(1, x') = 0 \quad \text{for } x' \in \mathbb{R}^{d-1}. \tag{1.9}$$

In that case, equation (1.7) should be completed with the extra boundary conditions

$$\widehat{u}(1, \xi') = 0 \quad \text{for } \xi' \in \mathbb{R}^{d-1}. \tag{1.10}$$

Of course, formula (1.8) still holds, but it does not use the full strength of the additional boundary conditions (1.9). We shall rather use the structure of the operator

$$\partial_{11} - |\xi'|^2 = (\partial_1 + |\xi'|)(\partial_1 - |\xi'|).$$

This structure indeed suggests to introduce

$$\widehat{v}(x_1,\xi') = (\partial_1 - |\xi'|)\widehat{u}(x_1,\xi') \quad \text{for } (x_1,\xi') \in (0,1) \times \mathbb{R}^{d-1}. \quad (1.11)$$

Following, the equations (1.7)–(1.9) rewrite:

$$\begin{cases} \partial_1 \widehat{u} - |\xi'|\widehat{u} = \widehat{v}, & \text{for } (x_1,\xi') \in (0,1) \times \mathbb{R}^{d-1}, \\ \partial_1 \widehat{v} + |\xi'|\widehat{v} = \widehat{f}, & \text{for } (x_1,\xi') \in (0,1) \times \mathbb{R}^{d-1}, \\ \widehat{u}(1,\xi') = 0, & \text{for } \xi' \in \mathbb{R}^{d-1}, \\ \widehat{v}(0,\xi') = \widehat{g}_N(\xi') - |\xi'|\widehat{g}_D(\xi'), & \text{for } \xi' \in \mathbb{R}^{d-1}. \end{cases} \quad (1.12)$$

Here, we check that the equations in $\widehat{v}$ do not depend on $\widehat{u}$ and can be solved independently. The solution $\widehat{u}$ can then be computed in terms of $\widehat{v}$ from equations $(1.12)_{1,3}$. Besides, straightforward computations show

$$\widehat{v}(x_1,\xi') = \exp(-|\xi'|x_1)\widehat{v}(0,\xi') + \int_0^{x_1} \exp\left(-|\xi'|(x_1-x)\right)\widehat{f}(x,\xi')\,dx, \quad (1.13)$$

$$\widehat{u}(x_1,\xi') = -\int_{x_1}^1 \exp\left(-|\xi'|(x-x_1)\right)\widehat{v}(x,\xi')\,dx. \quad (1.14)$$

Both formulae contain only decaying exponentials (in $|\xi'|$), and we will then be able to derive estimates from them.

Indeed, we immediately get from the above formula that

$$\begin{aligned} \widehat{u}(x_1,\xi') = &-\int_{x_1}^1 \exp\left(-|\xi'|(x-x_1)\right)\exp(-|\xi'|x)\widehat{v}(0,\xi')\,dx \\ &-\int_{x_1}^1 \exp\left(-|\xi'|(x-x_1)\right)\left(\int_0^x \exp\left(-|\xi'|(x-\tilde{x})\right)\widehat{f}(\tilde{x},\xi')\,d\tilde{x}\right)dx \\ = &-\widehat{v}(0,\xi')\int_{x_1}^1 \exp\left(-|\xi'|(2x-x_1)\right)dx \quad (1.15) \\ &-\int_0^1 \widehat{f}(\tilde{x},\xi')\left(\int_{\max\{x_1,\tilde{x}\}}^1 \exp(-|\xi'|(2x-x_1-\tilde{x}))\,dx\right)d\tilde{x}. \end{aligned}$$

$$(1.16)$$

For fixed $\xi' \in \mathbb{R}^{d-1}$, we estimate the $L^2(0,1)$-norm of each term of the right hand side. For (1.15), we get for some constant C independent of ξ' that

$$\left\|\widehat{v}(0,\xi')\int_{x_1}^1 \exp\left(-|\xi'|(2x-x_1)\right)dx\right\|_{L^2(0,1)} \leqslant \frac{C}{1+|\xi'|^{3/2}}|\widehat{v}(0,\xi')|,$$

and for (1.16), we obtain

$$\left\| \int_0^1 \widehat{f}(\tilde{x}, \xi') \left(\int_{\max\{x_1, \tilde{x}\}}^1 \exp(-|\xi'|(2x - x_1 - \tilde{x})) \, dx \right) d\tilde{x} \right\|_{L^2(0,1)}$$

$$\leqslant \left\| \int_0^1 |\widehat{f}(\tilde{x}, \xi')| \left(\int_{\max\{x_1, \tilde{x}\}}^1 \exp(-|\xi'|(2x - x_1 - \tilde{x})) \, dx \right) d\tilde{x} \right\|_{L^2(0,1)}$$

$$\leqslant \frac{C}{1 + |\xi'|} \left\| \int_0^1 |\widehat{f}(\tilde{x}, \xi')| \exp(-|\xi'||x_1 - \tilde{x}|) \, d\tilde{x} \right\|_{L^2(0,1)}$$

$$\leqslant \frac{C}{1 + |\xi'|} \left\| |\widehat{f}(x_1, \xi')| 1_{x_1 \in (0,1)} \star_{x_1} \exp(-|\xi'||x_1|) 1_{x_1 \in (-1,1)} \right\|_{L^2(0,1)}$$

$$\leqslant \frac{C}{1 + |\xi'|} \left\| \widehat{f}(\cdot, \xi') \right\|_{L^2(0,1)} \left\| \exp(-|\xi'||x_1|) 1_{x_1 \in (-1,1)} \right\|_{L^1(-1,1)}$$

$$\leqslant \frac{C}{1 + |\xi'|^2} \left\| \widehat{f}(\cdot, \xi') \right\|_{L^2(0,1)},$$

for some constant C independent of ξ'. We hence obtain, for C independent of ξ', that for all $\xi' \in \mathbb{R}^{d-1}$,

$$(1 + |\xi'|^4) \|\widehat{u}(\cdot, \xi')\|_{L^2(0,1)}^2 \leqslant C(1 + |\xi'|)|\widehat{v}(0, \xi')|^2 + C \left\| \widehat{f}(\cdot, \xi') \right\|_{L^2(0,1)}^2. \tag{1.17}$$

Integrating in $\xi' \in \mathbb{R}^{d-1}$ and using Parseval's identity, we derive

$$\|u\|_{L^2(0,1;H^2(\mathbb{R}^{d-1}))}^2 \leqslant C \int_{\mathbb{R}^{d-1}} (1 + |\xi'|)|\widehat{v}(0, \xi')|^2 \, d\xi' + C \|f\|_{L^2((0,1)\times\mathbb{R}^{d-1})}^2. \tag{1.18}$$

Using the equation, $\partial_{11} u = f - \Delta' u$ belongs to $L^2((0,1) \times \mathbb{R}^{d-1})$ and can be estimated by the right hand side of (1.18). Besides, the explicit form of $\widehat{v}(0, \xi')$ in $(1.12)_4$ shows that

$$\int_{\mathbb{R}^{d-1}} (1 + |\xi'|)|\widehat{v}(0, \xi')|^2 \, d\xi' \leqslant C \|g_D\|_{H^{3/2}(\mathbb{R}^{d-1})}^2 + C \|g_N\|_{H^{1/2}(\mathbb{R}^{d-1})}^2.$$

Similar estimates can also be done on $(1 + |\xi'|)\partial_1 \widehat{u}$ in $L^2((0,1) \times \mathbb{R}^{d-1})$, thus on $\partial_1 u$ in $L^2(0,1;H^1(\mathbb{R}^{d-1}))$. We thus conclude

$$\|u\|_{H^2((0,1)\times\mathbb{R}^{d-1})}$$
$$\leqslant C \|f\|_{L^2((0,1)\times\mathbb{R}^{d-1})} + C \|g_D\|_{H^{3/2}(\mathbb{R}^{d-1})} + C \|g_N\|_{H^{1/2}(\mathbb{R}^{d-1})}. \tag{1.19}$$

Note that, since we know the whole Dirichlet data of the solution u of (1.6)–(1.9), u also satisfies the stronger estimate

$$\|u\|_{H^2((0,1)\times\mathbb{R}^{d-1})} \leqslant C \|f\|_{L^2((0,1)\times\mathbb{R}^{d-1})} + C \|g_D\|_{H^{3/2}(\mathbb{R}^{d-1})}, \tag{1.20}$$

recall Theorem 1.20, which can be proved along the same lines in this geometrical setting, see Problem 1.21.

Nevertheless, the above proof of (1.19) is important to keep in mind as it is the one that will be used and adapted in more intricate situations in the following sections.

1.1.3. *Exercises*

Problem 1.21. *Let $u \in H^1((0,1) \times \mathbb{R}^{d-1})$ be a solution of*

$$\begin{cases} \partial_{11}u + \Delta'u = f, & \text{for } (x_1, x') \in (0,1) \times \mathbb{R}^{d-1}, \\ u(0, x') = g_0(x'), & \text{for } x' \in \mathbb{R}^{d-1}, \\ u(1, x') = g_1(x'), & \text{for } x' \in \mathbb{R}^{d-1}, \end{cases} \tag{1.21}$$

for some $f \in L^2((0,1) \times \mathbb{R}^{d-1})$, $g_0, g_1 \in H^{3/2}(\mathbb{R}^{d-1})$.
Our goal is to show the following estimate:

$$\|u\|_{H^2((0,1)\times\mathbb{R}^{d-1})} \leqslant C \|f\|_{L^2((0,1)\times\mathbb{R}^{d-1})}$$
$$+ C \|g_0\|_{H^{3/2}(\mathbb{R}^{d-1})} + C \|g_1\|_{H^{3/2}(\mathbb{R}^{d-1})}. \tag{1.22}$$

1. *Taking the Fourier variable of $(1.21)_1$ in x', show that $\widehat{u}(x_1, \xi')$ satisfies the identity:*

$$\widehat{u}(x_1, \xi') = \widehat{g}_0(\xi') \frac{\sinh((1-x_1)|\xi'|)}{\sinh(|\xi'|)} + \widehat{g}_1(\xi') \frac{\sinh(x_1|\xi'|)}{\sinh(|\xi'|)}$$
$$+ \int_{x_1}^1 \frac{\sinh(|\xi|(1-x)) \sinh(x_1|\xi'|)}{\sinh(|\xi'|)} \frac{\widehat{f}(x, \xi')}{|\xi'|} \, dx$$
$$+ \int_0^{x_1} \left(-\sinh(|\xi'|(x_1 - x)) + \frac{\sinh(|\xi'|(1-x)) \sinh(x_1|\xi'|)}{\sinh(|\xi'|)} \right) \frac{\widehat{f}(x, \xi')}{|\xi'|} \, dx.$$

2. *Using similar estimates as in Section 1.1.2, show that there exists a constant C independent of ξ' such that*

$$(1+|\xi|^2) \|\widehat{u}(\cdot, \xi')\|_{L^2(0,1)} \leqslant C \left\| \widehat{f}(\cdot, \xi') \right\|_{L^2(0,1)} + C(1+|\xi|^{3/2}) \left(|\widehat{g}_0(\xi')| + |\widehat{g}_1(\xi')| \right).$$

3. *Deduce estimate (1.22).*

Problem 1.22. 1. *Show that if $u \in H^1((0,1) \times \mathbb{R}^{d-1})$ solves*

$$\begin{cases} \partial_{11}u + \Delta'u = f, & \text{for } (x_1, x') \in (0,1) \times \mathbb{R}^{d-1}, \\ u(0, x') = \partial_1 u(0, x') = 0, & \text{for } x' \in \mathbb{R}^{d-1}, \\ u(1, x') = g_1(x'), & \text{for } x' \in \mathbb{R}^{d-1}, \end{cases} \tag{1.23}$$

for some $g_1 \in H^{3/2}(\mathbb{R}^{d-1})$ and $f \in L^2((0,1) \times \mathbb{R}^{d-1})$ satisfying

$$f(x_1, x') = 0 \quad \text{for all } x_1 \in (0, 1/2), \ x' \in \mathbb{R}^{d-1},$$

then $u(x_1, x') = 0$ for all $x_1 \in (0, 1/2)$, $x' \in \mathbb{R}^{d-1}$.

2. Would that property be true if we do not assume $\partial_1 u(0, x') = 0$?

1.2. A Carleman estimate with a linear weight function and application to the Calderón problem

1.2.1. Introduction

The goal of this section is to present the most simple example of a Carleman estimate for an elliptic equation.

Similarly as in the previous section, we will again focus on the case of an elliptic equation in a strip:

$$\begin{cases} \partial_{11} u + \Delta' u = f, & \text{for } (x_1, x') \in (0, 1) \times \mathbb{R}^{d-1}, \\ u(0, x') = u(1, x') = 0, & \text{for } x' \in \mathbb{R}^{d-1}, \end{cases} \tag{1.24}$$

where $\Delta' = \partial_{22} + \cdots + \partial_{dd}$ is the Laplace operator in the variable $x' = (x_2, \cdots, x_d) \in \mathbb{R}^{d-1}$.

We further assume that u has a known normal derivative on the lateral boundary $\{x_1 = 0\}$:

$$\partial_1 u(0, x') = g_N(x'), \quad \text{for } x' \in \mathbb{R}^{d-1}. \tag{1.25}$$

Our goal is to explain how the estimate

$$\|u\|_{H^2((0,1) \times \mathbb{R}^{d-1})} \leqslant C \|f\|_{L^2((0,1) \times \mathbb{R}^{d-1})} + C \|g_N\|_{H^{1/2}(\mathbb{R}^{d-1})}, \tag{1.26}$$

proved in (1.19) can be modified by the introduction of a linear weight function[b] of the form e^{-kx_1}, where k is a free parameter assumed to satisfy $k \geqslant 1$.

To be more precise, we will show that for some constant C independent of k, for all $k \geqslant 1$, all solutions u of (1.24) with source term $f \in L^2((0, 1) \times \mathbb{R}^{d-1})$ and Neumann data $g_N \in L^2(\mathbb{R}^{d-1})$ satisfy

$$k \left\| u e^{-kx_1} \right\|_{L^2((0,1) \times \mathbb{R}^{d-1})} \leqslant C \left\| f e^{-kx_1} \right\|_{L^2((0,1) \times \mathbb{R}^{d-1})} + C k^{1/2} \|g_N\|_{L^2(\mathbb{R}^{d-1})},$$

see Theorem 1.23 for precise statements.

The freedom in the parameter $k \geqslant 1$ and the precise knowledge of the dependence of this estimate in k is what makes Carleman estimates a powerful tool, as we shall see in several situations along this course.

---

[b]In the context of Carleman estimates, the weight function usually is the function appearing in the exponential. Here, the terminology "linear weight function" comes from the fact that the function $x \mapsto kx_1$ is linear.

Indeed, it allows in particular to prove similar estimates when the elliptic equation involves lower order terms, potentials for instance. This is precisely one of the properties we shall deeply rely onto in order to show uniqueness in the Calderón problem, see Section 1.2.3.

1.2.2. A Carleman estimate

1.2.2.1. Main result

The goal of this section is to prove the following result:

Theorem 1.23. *There exists a constant $C > 0$ such that for all $k \geqslant 1$, any solution $u \in H^2 \cap H^1_0((0,1) \times \mathbb{R}^{d-1})$ of (1.24) with source term $f \in L^2((0,1) \times \mathbb{R}^{d-1})$ and Neumann data $\partial_1 u(0, \cdot) = g_N \in L^2(\mathbb{R}^{d-1})$ satisfies*

$$k^2 \left\| u e^{-kx_1} \right\|^2_{L^2((0,1) \times \mathbb{R}^{d-1})} + \left\| \nabla u e^{-kx_1} \right\|^2_{L^2((0,1) \times \mathbb{R}^{d-1})}$$
$$\leqslant Ck \left\| g_N \right\|^2_{L^2(\mathbb{R}^{d-1})} + C \left\| f e^{-kx_1} \right\|^2_{L^2((0,1) \times \mathbb{R}^{d-1})}. \tag{1.27}$$

We will give three different proofs of Theorem 1.23 in Section 1.2.2.2, each of them having its own interest, see Section 1.2.2.3.

But before going into the proof, let us emphasize that the constant C in Theorem 1.23 is independent of $k \geqslant 1$. For instance, as a straightforward corollary of Theorem 1.23, we get the following:

Corollary 1.24. *Let $q \in L^\infty((0,1) \times \mathbb{R}^{d-1})$. There exists a constant $C > 0$ such that for all $k \geqslant 1$, any solution $u \in H^2 \cap H^1_0((0,1) \times \mathbb{R}^{d-1})$ of*

$$\begin{cases} \partial_{11} u + \Delta' u + qu = f, & \text{for } (x_1, x') \in (0,1) \times \mathbb{R}^{d-1}, \\ u(0, x') = u(1, x') = 0, & \text{for } x' \in \mathbb{R}^{d-1}, \end{cases} \tag{1.28}$$

with source term $f \in L^2((0,1) \times \mathbb{R}^{d-1})$ and Neumann data $\partial_1 u(0, \cdot) = g_N \in L^2(\mathbb{R}^{d-1})$ satisfies (1.27).

Proof. If $u \in L^2((0,1) \times \mathbb{R}^{d-1})$ solves (1.28), it also solves (1.24) with source term $f - qu$. Applying Theorem 1.23, we obtain, for all $k \geqslant 1$,

$$k^2 \left\| u e^{-kx_1} \right\|^2_{L^2((0,1) \times \mathbb{R}^{d-1})} + \left\| \nabla u e^{-kx_1} \right\|^2_{L^2((0,1) \times \mathbb{R}^{d-1})}$$
$$\leqslant Ck \left\| g_N \right\|^2_{L^2(\mathbb{R}^{d-1})} + 2C \left\| f e^{-kx_1} \right\|^2_{L^2((0,1) \times \mathbb{R}^{d-1})}$$
$$+ 2C \left\| q \right\|^2_{L^\infty((0,1) \times \mathbb{R}^{d-1})} \left\| u e^{-kx_1} \right\|^2_{L^2((0,1) \times \mathbb{R}^{d-1})}.$$

Taking $k^2 \geqslant 4C \left\| q \right\|^2_{L^\infty((0,1) \times \mathbb{R}^{d-1})}$, we can absorb the last term, and we obtain

$$k^2 \left\| u e^{-kx_1} \right\|_{L^2((0,1)\times\mathbb{R}^{d-1})}^2 + \left\| \nabla u e^{-kx_1} \right\|_{L^2((0,1)\times\mathbb{R}^{d-1})}^2$$

$$\leqslant 2Ck \left\| g_N \right\|_{L^2(\mathbb{R}^{d-1})}^2 + 4C \left\| f e^{-kx_1} \right\|_{L^2((0,1)\times\mathbb{R}^{d-1})}^2,$$

for all $k \geqslant k_q = 2\sqrt{C} \left\| q \right\|_{L^\infty((0,1)\times\mathbb{R}^{d-1})}$. As the weight function is uniformly bounded for $k \in [1, k_q]$, this last estimate is also true for some constant C for $k \in [1, k_q]$. Modifying the constant if needed, we deduce the existence of a constant C such that (1.27) is true for all $k \geqslant 1$. □

Let us also remark that the Carleman estimate (1.27) involves an additional information g_N on the hyperplane $x_1 = 0$, corresponding to the set where the weight function $-kx_1$ is the largest on the domain $[0,1] \times \mathbb{R}^{d-1}$. This is a general phenomenon in Carleman estimates, as the weight function gives more weight where the additional information is available, and reflects the direction of propagation of the information.

In order to illustrate that, let us revisit the result of Problem 1.22:

Corollary 1.25. *Let* $q \in L^\infty((0,1)\times\mathbb{R}^{d-1})$, *and* $u \in H^2 \cap H_0^1((0,1)\times\mathbb{R}^{d-1})$ *be the solution of (1.28) with Neumann data* $\partial_1 u(0,\cdot) = 0$ *and source term* $f \in L^2((0,1)\times\mathbb{R}^{d-1})$ *satisfying, for some* $a \in (0,1)$,

$$f(x_1, x') = 0 \quad \text{for } x_1 \in (0,a), \ x' \in \mathbb{R}^{d-1},$$

then u *vanishes in* $(0,a) \times \mathbb{R}^{d-1}$.

Proof. Apply the Carleman estimate (1.27) to u: for all $k \geqslant 1$,

$$k \left\| u e^{-kx_1} \right\|_{L^2((0,1)\times\mathbb{R}^{d-1})} \leqslant C \left\| f e^{-kx_1} \right\|_{L^2((0,1)\times\mathbb{R}^{d-1})}. \tag{1.29}$$

Now, on one hand,

$$k \left\| u e^{-kx_1} \right\|_{L^2((0,1)\times\mathbb{R}^{d-1})} \geqslant k \left\| u e^{-kx_1} \right\|_{L^2((0,a)\times\mathbb{R}^{d-1})}$$

$$\geqslant k e^{-ka} \left\| u \right\|_{L^2((0,a)\times\mathbb{R}^{d-1})}.$$

On the other hand, using the fact that f vanishes in $(0,a) \times \mathbb{R}^{d-1}$,

$$\left\| f e^{-kx_1} \right\|_{L^2((0,1)\times\mathbb{R}^{d-1})} \leqslant e^{-ka} \left\| f \right\|_{L^2((0,1)\times\mathbb{R}^{d-1})}.$$

The estimate (1.29) then yields, for all $k \geqslant 1$,

$$k \left\| u \right\|_{L^2((0,a)\times\mathbb{R}^{d-1})} \leqslant C \left\| f \right\|_{L^2((0,1)\times\mathbb{R}^{d-1})}.$$

As k can be chosen arbitrarily large, u necessarily vanishes in $(0,a) \times \mathbb{R}^{d-1}$. □

Remark 1.26. Note that Corollary 1.25 states the same result as in Problem 1.22 but now allows non-trivial lower order terms in the elliptic equation. Note in particular that the potential q may depend on both x_1 and x' in an intricate way, as we only assumed it to be in L^∞.

1.2.2.2. *Proof of Theorem 1.23*

Strategy. Since we have to prove estimates on ue^{-kx_1} in terms of fe^{-kx_1}, it will be convenient to set

$$\begin{cases} U(x_1, x') = u(x_1, x')e^{-kx_1}, \\ F(x_1, x') = f(x_1, x')e^{-kx_1}, \end{cases} \quad \text{for } (x_1, x') \in (0,1) \times \mathbb{R}^{d-1}. \quad (1.30)$$

If u satisfies (1.24) and has Neumann data $g_N = \partial_1 u(0, \cdot)$, then U satisfies

$$\partial_{11}\left(e^{kx_1}U\right) + \Delta'\left(e^{kx_1}U\right) = Fe^{kx_1},$$

which yields to the equations

$$\begin{cases} \partial_{11}U + 2k\partial_1 U + k^2 U + \Delta'U = F, & \text{for } (x_1, x') \in (0,1) \times \mathbb{R}^{d-1}, \\ U(0, x') = U(1, x') = 0, & \text{for } x' \in \mathbb{R}^{d-1}, \\ \partial_1 U(0, x') = g_N(x'), & \text{for } x' \in \mathbb{R}^{d-1}. \end{cases} \quad (1.31)$$

Starting from there, we will present several proofs of the estimate

$$k^2 \left\| ue^{-kx_1} \right\|^2_{L^2((0,1)\times\mathbb{R}^{d-1})} \leqslant Ck \left\| g_N \right\|^2_{L^2(\mathbb{R}^{d-1})} + C \left\| fe^{-kx_1} \right\|^2_{L^2((0,1)\times\mathbb{R}^{d-1})}, \quad (1.32)$$

or equivalently

$$k^2 \left\| U \right\|^2_{L^2((0,1)\times\mathbb{R}^{d-1})} \leqslant Ck \left\| g_N \right\|^2_{L^2(\mathbb{R}^{d-1})} + C \left\| F \right\|^2_{L^2((0,1)\times\mathbb{R}^{d-1})}. \quad (1.33)$$

Approach 1: Fourier approach and explicit resolution. Similarly as in Section 1.1.1 and Section 1.1.2, we can then take the partial Fourier transform in the x'-variable:

$$\widehat{U}(x_1, \xi') = \mathscr{F}_{x' \to \xi'} U(x_1, \cdot),$$
$$\widehat{F}(x_1, \xi') = \mathscr{F}_{x' \to \xi'} F(x_1, \cdot),$$
$$\widehat{g}_N(\xi') = \mathscr{F}_{x' \to \xi'} g_N(\cdot).$$

Taking the partial Fourier transform $\mathscr{F}_{x' \to \xi'}$ of (1.31), we obtain

$$\begin{cases} \partial_{11}\widehat{U} + 2k\partial_1\widehat{U} + k^2\widehat{U} - |\xi'|^2\widehat{U} = \widehat{F}, & \text{for } (x_1, \xi') \in (0,1) \times \mathbb{R}^{d-1}, \\ \widehat{U}(0, \xi') = \widehat{U}(1, \xi') = 0, & \text{for } \xi' \in \mathbb{R}^{d-1}, \\ \partial_1\widehat{U}(0, x') = \widehat{g}_N(\xi'), & \text{for } \xi' \in \mathbb{R}^{d-1}. \end{cases}$$

$$(1.34)$$

We now consider the factorization of the operator as follows:

$$\partial_{11} + 2k\partial_1 + k^2 - |\xi'|^2 = (\partial_1 + k)^2 - |\xi'|^2 = (\partial_1 + k + |\xi'|)(\partial_1 + k - |\xi'|).$$

Following the idea of Section 1.1.2, we then introduce the function

$$\widehat{V} = (\partial_1 + k - |\xi|)\widehat{U}.$$

We are then back to the system of equations

$$\begin{cases} \partial_1 \widehat{U} + (k - |\xi'|)\widehat{U} = \widehat{V}, & \text{for } (x_1, \xi') \in (0,1) \times \mathbb{R}^{d-1}, \\ \partial_1 \widehat{V} + (k + |\xi'|)\widehat{V} = \widehat{F}, & \text{for } (x_1, \xi') \in (0,1) \times \mathbb{R}^{d-1}, \\ \widehat{U}(0, \xi') = \widehat{U}(1, \xi') = 0, & \text{for } \xi' \in \mathbb{R}^{d-1}, \\ \widehat{V}(0, \xi') = \widehat{g}_N(\xi'), & \text{for } \xi' \in \mathbb{R}^{d-1}. \end{cases} \tag{1.35}$$

For fixed $\xi' \in \mathbb{R}^{d-1}$, we can then solve this system in two steps:

(1) Compute $\widehat{V}$ in terms of $\widehat{F}$ and of $\widehat{g}_N$;
(2) Compute $\widehat{U}$ in terms of $\widehat{V}$.

The computation of $\widehat{V}$ is straightforward:

$$\widehat{V}(x_1, \xi') = \exp(-(k+|\xi'|)x_1)\widehat{g}_N(\xi') + \int_0^{x_1} \exp(-(k+|\xi'|)(x_1-x))\widehat{F}(x, \xi') \, dx. \tag{1.36}$$

The computation of $\widehat{U}$ also is straightforward, but now we have two possible formulae, depending whether we use the boundary condition at $x_1 = 0$ or at $x_1 = 1$:

$$\widehat{U}(x_1, \xi') = \int_0^{x_1} \exp(-(k - |\xi'|)(x_1 - x))\widehat{V}(x, \xi') \, dx, \tag{1.37}$$

$$\widehat{U}(x_1, \xi') = -\int_{x_1}^1 \exp(-(k - |\xi'|)(x_1 - x))\widehat{V}(x, \xi') \, dx. \tag{1.38}$$

Here, we see that we have to choose between two formulae, and we may choose the one we prefer to work with. In particular, as we want estimates as small as possible in terms of k, we will always choose the formula that does not contain exponentials of positive terms. This means that our choice will depend on the frequency parameter $|\xi'|$: we will take formula (1.37) if $|\xi'| \leqslant k$, and formula (1.38) if $|\xi'| \geqslant k$.

In what follows, the constants denoted by C are independent of the parameters ξ' and k and may change from line to line.

The case $|\xi'| \leqslant k$. In that case, combining (1.36) and (1.37), we get

$$\widehat{U}(x_1, \xi') = \widehat{g}_N(\xi') \int_0^{x_1} \exp(-(k - |\xi'|)(x_1 - x)) \exp(-(k + |\xi'|)x) \, dx$$
$$+ \int_0^{x_1} e^{-(k-|\xi'|)(x_1-x)} \left(\int_0^x e^{-(k+|\xi'|)(x-\tilde{x})} \widehat{F}(\tilde{x}, \xi') \, d\tilde{x} \right) dx$$
$$= \widehat{g}_N(\xi') \exp(-(k - |\xi'|)x_1) \int_0^{x_1} \exp(-2|\xi'|x) \, dx$$
$$+ \int_0^{x_1} \widehat{F}(\tilde{x}, \xi') e^{-k(x_1-\tilde{x})} \left(\int_{\tilde{x}}^{x_1} e^{|\xi'|(\tilde{x}+x_1-2x)} \, dx \right) d\tilde{x}.$$

On one hand, direct estimates yield

$$\left\| \widehat{g}_N(\xi') \exp(-(k - |\xi'|)x_1) \int_0^{x_1} \exp(-2|\xi'|x) \, dx \right\|_{L^2(0,1)}$$
$$\leqslant C |\widehat{g}_N(\xi')| \left(\frac{1}{1 + |\xi'|} \right) \left(\frac{1}{1 + (k - |\xi'|)^{1/2}} \right) \leqslant \frac{C}{k^{1/2}} |\widehat{g}_N(\xi')|.$$

On the other hand, writing

$$\int_0^{x_1} \widehat{F}(\tilde{x}, \xi') e^{-k(x_1-\tilde{x})} \left(\int_{\tilde{x}}^{x_1} e^{|\xi'|(\tilde{x}+x_1-2x)} dx \right) d\tilde{x} = \int_0^1 \widehat{F}(\tilde{x}, \xi') H(x_1 - \tilde{x}, \xi') d\tilde{x}$$

with

$$H(X, \xi') = 1_{X>0} e^{-(k-|\xi'|)X} \int_0^X e^{-2|\xi'|x} \, dx,$$

we obtain by Young's inequality

$$\left\| \int_0^{x_1} \widehat{F}(\tilde{x}, \xi') e^{-k(x_1-\tilde{x})} \left(\int_{\tilde{x}}^{x_1} e^{|\xi'|(\tilde{x}+x_1-2x)} \, dx \right) d\tilde{x} \right\|_{L^2(0,1)}$$
$$\leqslant C \left\| \widehat{F}(\cdot, \xi') \right\|_{L^2(0,1)} \| H(\cdot, \xi') \|_{L^1(0,1)}$$
$$\leqslant C \left\| \widehat{F}(\cdot, \xi') \right\|_{L^2(0,1)} \left(\frac{1}{1 + k - |\xi'|} \right) \left(\frac{1}{1 + |\xi'|} \right)$$
$$\leqslant \frac{C}{k} \left\| \widehat{F}(\cdot, \xi') \right\|_{L^2(0,1)}.$$

We then derive, for $\xi' \in \mathbb{R}^{d-1}$ with $|\xi'| \leqslant k$,

$$\left\| \widehat{U}(x_1, \xi') \right\|_{L^2(0,1)} \leqslant \frac{C}{k^{1/2}} |\widehat{g}_N(\xi')| + \frac{C}{k} \left\| \widehat{F}(\cdot, \xi') \right\|_{L^2(0,1)},$$

which also yields

$$k^2 \left\| \widehat{U}(x_1, \xi') \right\|_{L^2(0,1)}^2 \leqslant C k |\widehat{g}_N(\xi')|^2 + C \left\| \widehat{F}(\cdot, \xi') \right\|_{L^2(0,1)}^2. \qquad (1.39)$$

Note that, in this case, all the information comes from $x_1 = 0$ and our formulae are very close to the ones obtained in Section 1.1.1.

The case $|\xi'| \geqslant k$. In that case, combining (1.36) and (1.38), we get the following formula:

$$\widehat{U}(x_1, \xi') = -\widehat{g}_N(\xi') \int_{x_1}^1 \exp(-(k - |\xi'|)(x_1 - x)) \exp(-(k + |\xi'|)x)\, dx$$

$$- \int_{x_1}^1 e^{-(k-|\xi'|)(x_1-x)} \left(\int_0^x e^{-(k+|\xi'|)(x-\tilde{x})} \widehat{F}(\tilde{x}, \xi')\, d\tilde{x} \right) dx$$

$$= -\widehat{g}_N(\xi') e^{-(k-|\xi'|)x_1} \int_{x_1}^1 e^{-2|\xi'|x}\, dx$$

$$- \int_0^1 \widehat{F}(\tilde{x}, \xi') e^{-k(x_1-\tilde{x})} \left(\int_{\max\{\tilde{x}, x_1\}}^1 e^{|\xi'|(x_1+\tilde{x}-2x)}\, dx \right) d\tilde{x}$$

$$= -\widehat{g}_N(\xi') e^{-(k-|\xi'|)x_1} \int_{x_1}^1 e^{-2|\xi'|x}\, dx$$

$$- \int_0^1 \widehat{F}(\tilde{x}, \xi') e^{-k(x_1-\tilde{x})} e^{-|\xi'||x_1-\tilde{x}|} \left(\int_0^{1-\max\{\tilde{x}, x_1\}} e^{-2|\xi'|x}\, dx \right) d\tilde{x}.$$

The first term above can be bounded directly:

$$\left\| -\widehat{g}_N(\xi') e^{-(k-|\xi'|)x_1} \int_{x_1}^1 e^{-2|\xi'|x}\, dx \right\|_{L^2(0,1)}$$

$$\leqslant \frac{C}{1 + |\xi'|} \left\| -\widehat{g}_N(\xi') e^{-(k+|\xi'|)x_1} \right\|_{L^2(0,1)}$$

$$\leqslant \frac{C}{1 + |\xi'|} |\widehat{g}_N(\xi')| \frac{1}{(k + |\xi'|)^{1/2}} \leqslant \frac{C}{k^{3/2}} |\widehat{g}_N(\xi')|.$$

The second term can be bounded as follows:

$$\left\| -\int_0^1 \widehat{F}(\tilde{x}, \xi') e^{-k(x_1-\tilde{x})} e^{-|\xi'||x_1-\tilde{x}|} \left(\int_0^{1-\max\{\tilde{x}, x_1\}} e^{-2|\xi'|x}\, dx \right) d\tilde{x} \right\|_{L^2(0,1)}$$

$$\leqslant \left\| \int_0^1 |\widehat{F}(\tilde{x}, \xi')| e^{-k(x_1-\tilde{x})} e^{-|\xi'||x_1-\tilde{x}|} \left(\int_0^{1-\max\{\tilde{x}, x_1\}} e^{-2|\xi'|x}\, dx \right) d\tilde{x} \right\|_{L^2(0,1)}$$

$$\leqslant \left\| \int_0^1 |\widehat{F}(\tilde{x}, \xi')| e^{(k-|\xi'|)|x_1-\tilde{x}|} \left(\int_0^\infty e^{-2|\xi'|x}\, dx \right) d\tilde{x} \right\|_{L^2(0,1)}$$

$$\leqslant \frac{C}{|\xi'|} \left\| |\widehat{F}|(x_1, \xi') 1_{x_1 \in (0,1)} \star_{x_1} e^{(k-|\xi'|)|x_1|} \right\|_{L^2(0,1)}$$

$$\leqslant \frac{C}{|\xi'|} \left\| \widehat{F}(\cdot, \xi') \right\|_{L^2(0,1)} \left\| e^{(k-|\xi'|)|x_1|} \right\|_{L^1(-2,2)}$$

$$\leqslant \frac{C}{|\xi'|(1 - k + |\xi'|)} \|F(\cdot, \xi')\|_{L^2(0,1)} \leqslant \frac{C}{k} \left\| \widehat{F}(\cdot, \xi') \right\|_{L^2(0,1)}.$$

We thus obtain the same estimate as in (1.39) when $|\xi'| \geqslant k$.

Note that here, similarly as in Section 1.1.2, we use only one information at $x_1 = 0$, and one more information at $x_1 = 1$.

Conclusion of Approach 1. Since (1.39) holds for all $\xi' \in \mathbb{R}^{d-1}$, we integrate it with respect to $\xi' \in \mathbb{R}^{d-1}$ and we immediately obtain (1.33) by Parseval's identity.

Approach 2: Fourier approach and multiplier type argument. This approach starts similarly as Approach 1, up to the equations (1.35), by taking the Fourier transform in the x' variable and factorizing the operator.

Then, instead of deriving explicitly formula for $\widehat{U}$ in terms of $\widehat{F}$ and the observation $\widehat{g}_N$, we first estimate $\widehat{V}$ in terms of $\widehat{F}$ and of $\widehat{g}_N$, then $\widehat{U}$ in terms of $\widehat{V}$. These two estimates can be done using multiplier type arguments in a rather straightforward manner.

For convenience, we rewrite system (1.35):

$$\begin{cases} \partial_1 \widehat{V} + (k + |\xi'|)\widehat{V} = \widehat{F}, & \text{for } (x_1, \xi') \in (0,1) \times \mathbb{R}^{d-1}, \\ \widehat{V}(0, \xi') = \widehat{g}_N(\xi'), & \text{for } \xi' \in \mathbb{R}^{d-1}, \end{cases} \tag{1.40}$$

and

$$\begin{cases} \partial_1 \widehat{U} + (k - |\xi'|)\widehat{U} = \widehat{V}, & \text{for } (x_1, \xi') \in (0,1) \times \mathbb{R}^{d-1}, \\ \widehat{U}(0, \xi') = \widehat{U}(1, \xi') = 0, & \text{for } \xi' \in \mathbb{R}^{d-1}. \end{cases} \tag{1.41}$$

To estimate $\widehat{V}$, we take the square of each side of (1.40) and integrate by parts in x_1, and we obtain:

$$\int_0^1 \left(|\partial_1 \widehat{V}(x_1, \xi')|^2 + (k + |\xi'|)^2 |\widehat{V}(x_1, \xi')|^2 \right) dx_1 + (k + |\xi'|)|\widehat{V}(1, \xi')|^2$$

$$= \int_0^1 |\widehat{F}(x_1, \xi')|^2 dx_1 + (k + |\xi'|)|\widehat{V}(0, \xi')|^2.$$

To estimate $\widehat{U}$ from $\widehat{V}$, we also take the square of each side of (1.41) and integrate by parts in x_1:

$$\int_0^1 \left(|\partial_1 \widehat{U}(x_1, \xi')|^2 + (k - |\xi'|)^2 |\widehat{U}(x_1, \xi')|^2 \right) dx_1 = \int_0^1 |\widehat{V}(x_1, \xi')|^2 dx_1.$$

Combining these two estimates, and using Poincaré's estimate in $(0,1)$, we easily have:

$$\int_0^1 |\widehat{U}(x_1, \xi')|^2 \, dx_1 \leqslant C \int_0^1 |\partial_1 \widehat{U}(x_1, \xi')|^2 \, dx_1$$

$$\leqslant \frac{1}{(k + |\xi'|)^2} \left(\int_0^1 |\widehat{F}(x_1, \xi')|^2 \, dx_1 + (k + |\xi'|)|\widehat{V}(0, \xi')|^2 \right),$$

which of course easily implies (1.33).

Approach 3: A multiplier argument. Here, we directly start from (1.31), by multiplying the equation (1.31) by $\partial_1 U$:

$$2k \left\| \partial_1 U \right\|^2_{L^2((0,1)\times\mathbb{R}^{d-1})} = \int_{(0,1)\times\mathbb{R}^{d-1}} F \partial_1 U \, dx_1 dx'$$
$$+ \frac{1}{2} \int_{\mathbb{R}^{d-1}} |\partial_1 U(0,x')|^2 \, dx' - \frac{1}{2} \int_{\mathbb{R}^{d-1}} |\partial_1 U(1,x')|^2 \, dx',$$

so that

$$k \left\| \partial_1 U \right\|^2_{L^2((0,1)\times\mathbb{R}^{d-1})} \leqslant \frac{C}{k} \left\| F \right\|^2_{L^2((0,1)\times\mathbb{R}^{d-1})} + C \left\| g_N \right\|^2_{L^2(\mathbb{R}^{d-1})}. \quad (1.42)$$

Then, using the fact that the strip has bounded width in $x_1 \in (0,1)$ and that U satisfies homogeneous boundary conditions at $x_1 = 0$, we can use Poincaré's identity:

$$\left\| U \right\|^2_{L^2((0,1)\times\mathbb{R}^{d-1})} \leqslant C \left\| \partial_1 U \right\|^2_{L^2((0,1)\times\mathbb{R}^{d-1})}.$$

Hence we have from (1.42) that

$$k^2 \left\| U \right\|^2_{L^2((0,1)\times\mathbb{R}^{d-1})} \leqslant C \left\| F \right\|^2_{L^2((0,1)\times\mathbb{R}^{d-1})} + Ck \left\| g_N \right\|^2_{L^2(\mathbb{R}^{d-1})}. \quad (1.43)$$

Conclusion. Starting from (1.32), equivalently (1.33), we can readily estimate $\nabla u e^{-kx_1}$ using classical elliptic estimates on u. Indeed, in order to estimate $\nabla u e^{-kx_1}$, we multiply the equation (1.34) by ue^{-2kx_1}:

$$- \int_{(0,1)\times\mathbb{R}^{d-1}} |\nabla u|^2 e^{-2kx_1} \, dx_1 dx' + k^2 \int_{(0,1)\times\mathbb{R}^{d-1}} |u|^2 e^{-2kx_1} \, dx_1 dx'$$
$$= \int_{(0,1)\times\mathbb{R}^{d-1}} fu e^{-2kx_1} \, dx_1 dx',$$

so that

$$\left\| \nabla u e^{-kx_1} \right\|^2_{L^2((0,1)\times\mathbb{R}^{d-1})} \leqslant k^2 \left\| u e^{-kx_1} \right\|^2_{L^2((0,1)\times\mathbb{R}^{d-1})}$$
$$+ \left\| f e^{-kx_1} \right\|_{L^2((0,1)\times\mathbb{R}^{d-1})} \left\| u e^{-kx_1} \right\|_{L^2((0,1)\times\mathbb{R}^{d-1})}.$$

Using (1.32), we immediately derive

$$\left\| \nabla u e^{-kx_1} \right\|^2_{L^2((0,1)\times\mathbb{R}^{d-1})} \leqslant Ck \left\| g_N \right\|^2_{L^2(\mathbb{R}^{d-1})} + C \left\| f e^{-kx_1} \right\|^2_{L^2((0,1)\times\mathbb{R}^{d-1})}.$$
$$(1.44)$$

This concludes the proof of Theorem 1.23. $\qquad\square$

Problem 1.27. *Let us assume the setting of Theorem 1.23 and use the notations of the above proof.*
1. *Show that there exists $C > 0$ such that for all $\xi' \in \mathbb{R}^{d-1}$ and $k \geqslant 1$,*

$$(1+|\xi'|^2)\left\|\widehat{U}(\cdot,\xi')\right\|_{L^2(0,1)} \leqslant Ck^{1/2}(1+|\xi'|^{1/2})|\widehat{g}_N(\xi')|+Ck\left\|\widehat{F}(\cdot,\xi')\right\|_{L^2(0,1)}.$$

Hint: follow Approach 1 or Approach 2.
2. *Deduce that there exists $C > 0$ such that for all $k \geqslant 1$,*

$$\left\|ue^{-kx_1}\right\|^2_{L^2(0,1;H^2(\mathbb{R}^{d-1}))} \leqslant Ck\left\|g_N\right\|^2_{H^{1/2}(\mathbb{R}^{d-1})} + Ck^2\left\|F\right\|^2_{L^2((0,1)\times\mathbb{R}^{d-1})}.$$

3. *Deduce that for all $(i,j) \in \{2,\cdots,d\}^2$,*

$$\frac{1}{k^2}\left\|(\partial_{ij}u)e^{-kx_1}\right\|^2_{L^2((0,1)\times\mathbb{R}^{d-1})} \leqslant Ck\left\|g_N\right\|^2_{H^{1/2}(\mathbb{R}^{d-1})} + C\left\|F\right\|^2_{L^2((0,1)\times\mathbb{R}^{d-1})}.$$

4. *Estimate $\partial_{11}\widehat{U}$ and show the same estimate as above for $\partial_{11}u\, e^{-kx_1}$.*
5. *Estimate $(1 + |\xi'|)\partial_1\widehat{U}$, show that the same estimate also holds for $\partial_{1i}u\, e^{-kx_1}$ with $i \in \{1,\cdots,d\}$.*
6. *Prove that there exists C independent of k such that for all $k \geqslant 1$,*

$$\frac{1}{k^2}\left\|ue^{-kx_1}\right\|^2_{H^2((0,1)\times\mathbb{R}^{d-1})} \leqslant Ck\left\|g_N\right\|^2_{H^{1/2}(\mathbb{R}^{d-1})} + C\left\|F\right\|^2_{L^2((0,1)\times\mathbb{R}^{d-1})}.$$

$$(1.45)$$

Problem 1.28. *Discuss the optimality of the powers of k in the Carleman estimate (1.27).*

1.2.2.3. *Advantages of each approach*

In this section, we discuss the advantages of each approach developed in the proof of Theorem 1.23.

When reading the various proofs, it seems that the easiest one is Approach 3. Indeed, this approach, based on pure multiplier arguments, has the advantage of requiring integration by parts only. In particular, one can easily get convinced that it can be adapted quite easily to more general domains (see next section, in which the domain Ω will not need to be a strip orthogonal to the gradient of the weight function), and to general elliptic operator with Lipschitz coefficient (at least for strictly pseudo-convex functions, see Section 1.3). For this reason, it clearly appears that this is the most robust and efficient approach in general.

However, Approach 3 does not allow to be very precise on the boundary terms. In particular, for a source term in $L^2(\Omega)$ and homogeneous Dirichlet boundary conditions, solutions u of (1.24) belong to $H^2(\Omega)$, and it would be much more natural to consider the observation in the space $H^{1/2}(\partial\Omega)$. This method does not really allow to do this.

Approach 2, which relies on performing a Fourier transform in the transverse variable, factorizing the operator, and doing multiplier type estimates on (1.40)–(1.41), is much more precise when it comes to the boundary conditions, since all the estimates are done frequency by frequency, and thus allows to consider the boundary conditions in appropriate Sobolev spaces. We refer for instance to [Imanuvilov and Puel (2003)] where such computations are done for strictly pseudo-convex weight functions (see Section 1.3 for the definition) and non-homogeneous Dirichlet boundary conditions, or in the case of elliptic operators with discontinuous conductivities to the recent works [Le Rousseau and Robbiano (2010); Le Rousseau and Lerner (2013)] (again, for strictly pseudo-convex weight functions).

Still, to use it, one needs to get the possibility of factorizing the operator, which leads to some technicalities in general settings (general bounded domains, or general weight functions), that can be solved using appropriate semi-classical techniques, or, as we will do in the next section, proving it by hand by suitable localizing process and change of coordinates.

Approach 1, finally, might seem even more restrictive, since the computations are already quite involved, even in that specific case of a strip with a weight depending linearly on x_1. However, this is a good point of view when deriving L^p Carleman estimates, since Parseval's identity holds only in Hilbertian setting and has no substitute in the L^p setting. Thus, Approach 1, which provides a "good" (meaning, yielding to nice estimates, at least in L^2) formula for the solution of (1.31) is useful in this setting. Then, to derive good L^p estimates, the strategy consists in going back in the original variables and performs a detailed analysis of the kernel function, using Hardy-Littlewood-Sobolev theory, see for instance the textbook [Sogge (1993)] and the articles [Kenig *et al.* (1987); Dos Santos Ferreira (2005); Koch and Tataru (2001)] and the references therein.

1.2.2.4. *More general setting*

The goal of this section is to generalize Theorem 1.23 to more general domains.

Theorem 1.29. *Let Ω be a smooth ($\mathscr{C}^2$) bounded domain of $\mathbb{R}^d$.*

There exists a constant $C > 0$ such that for all $k \in \mathbb{R}^d$ with $|k| \geqslant 1$, any solution $u \in H_0^1(\Omega)$ of

$$\begin{cases} \Delta u = f, & \text{for } x \in \Omega, \\ u(x) = 0, & \text{for } x \in \partial\Omega, \end{cases} \tag{1.46}$$

with source term $f \in L^2(\Omega)$ *satisfies*

$$|k|^2 \left\| u e^{-k \cdot x} \right\|_{L^2(\Omega)}^2 + \left\| \nabla u e^{-k \cdot x} \right\|_{L^2(\Omega)}^2$$

$$\leqslant C|k| \left\| \partial_n u e^{-k \cdot x} \right\|_{L^2(\Gamma_k)}^2 + C \left\| f e^{-k \cdot x} \right\|_{L^2(\Omega)}^2, \tag{1.47}$$

where $\Gamma_k = \{ x \in \partial\Omega \mid k \cdot n_x < 0 \}$, *where* n_x *is the outward pointing normal vector at* x.

Proof. Setting $U = u e^{-k \cdot x}$ and $F = f e^{-k \cdot x}$, one easily checks that U satisfies

$$\begin{cases} \Delta U + 2k \cdot \nabla U + |k|^2 U = F, & \text{for } x \in \Omega, \\ U(x) = 0, & \text{for } x \in \partial\Omega. \end{cases} \tag{1.48}$$

We then follow the strategy developed in Section 1.2.2.2 based on Approach 3. Multiplying the equation by U, we first derive

$$\| \nabla U \|_{L^2(\Omega)}^2 \leqslant C|k|^2 \| U \|_{L^2(\Omega)}^2 + C \| F \|_{L^2(\Omega)}^2. \tag{1.49}$$

Multiplying (1.48) by $k \cdot \nabla U$, we derive

$$\int_\Omega F k \cdot \nabla U \, dx = 2 \int_\Omega |k \cdot \nabla U|^2 \, dx + \int_{\partial\Omega} \partial_n U k \cdot \nabla U \, d\sigma - \frac{1}{2} \int_{\partial\Omega} k \cdot n_x |\nabla U|^2 \, d\sigma$$

$$= 2 \int_\Omega |k \cdot \nabla U|^2 \, dx + \frac{1}{2} \int_{\partial\Omega} k \cdot n_x |\partial_n U|^2 \, d\sigma,$$

where we used that, as $U = 0$ on $\partial\Omega$, $\nabla U = (\partial_n U) n_x$ on $\partial\Omega$.

Similarly as in (1.42), this yields for some constant C independent of $k \in \mathbb{R}^d$ that

$$\int_\Omega |k \cdot \nabla U|^2 \, dx \leqslant \int_\Omega |F|^2 \, dx + |k| \int_{\Gamma_k} |\partial_n U|^2 \, d\sigma.$$

Using the fact that Ω is bounded in all directions, thus in the direction $k/|k|$ for all $k \in \mathbb{R}^d \setminus \{0\}$, we can use Poincaré's inequality and we obtain a constant C independent of k such that for all $k \in \mathbb{R}^d$,

$$|k|^2 \| U \|_{L^2(\Omega)}^2 \leqslant C \| F \|_{L^2(\Omega)}^2 + |k| \| \partial_n U \|_{L^2(\Gamma_k)}^2.$$

Combined with (1.49), we thus derive

$$|k|^2 \| U \|_{L^2(\Omega)}^2 + \| \nabla U \|_{L^2(\Omega)}^2 \leqslant C \| F \|_{L^2(\Omega)}^2 + |k| \| \partial_n U \|_{L^2(\Gamma_k)}^2. \tag{1.50}$$

Recalling that $u e^{-k \cdot x} = U$ and $\nabla u e^{-k \cdot x} = \nabla U + kU$, we immediately conclude estimate (1.47). $\qquad\qquad\square$

Remark 1.30. An H^2 estimate on U similar to (1.45) can also be proved: indeed, from the equation (1.48),

$$\frac{1}{|k|^2} \|\Delta U\|_{L^2(\Omega)}^2 \leqslant C|k|^2 \|U\|_{L^2(\Omega)}^2 + C \|\nabla U\|_{L^2(\Omega)}^2 + C \|F\|_{L^2(\Omega)}^2 .$$

Since U vanishes on the boundary, elliptic regularity results give

$$|k|^2 \|U\|_{L^2(\Omega)}^2 + \|\nabla U\|_{L^2(\Omega)}^2 + \frac{1}{|k|^2} \|U\|_{H^2(\Omega)}^2 \leqslant C \|F\|_{L^2(\Omega)}^2 + |k| \|\partial_n U\|_{L^2(\Gamma_k)}^2 .$$

$$(1.51)$$

1.2.3. *The Calderón problem*

1.2.3.1. *Setting and main result*

The Calderón problem is also known in the physical literature as the Electrical Impedance Tomography (EIT in short). This corresponds to a medical imaging technique which consists in the recovery of the conductivity of a tissue (or a material) by applying currents on the surface on the body and measuring the electrical potentials on the surface of the body. Actually, to make the setting we describe hereafter slightly simpler, we impose the voltages and measure the corresponding currents.

To be more precise, let Ω be a bounded domain of $\mathbb{R}^d$, and consider the elliptic problem

$$\begin{cases} \text{div} \, (\sigma \nabla u) = 0, & \text{for } x \in \Omega, \\ u(x) = g_D(x), & \text{for } x \in \partial\Omega. \end{cases} \quad (1.52)$$

Here, $\sigma = \sigma(x)$ is a scalar function modeling the conductivity of the material.

The Dirichlet boundary data g_D is a voltage imposed on the boundary of the object. In the EIT process, it is assumed that for all voltage g_D, we can measure the current $(\sigma \nabla u) \cdot n_x$ on the whole boundary $\partial\Omega$. This is the so-called Voltage-to-Current map.

The question known as the Calderón problem is then the following:

Can we determine the conductivity of the material σ from the knowledge on the Voltage-to-Current map?

Before going further, let us give some more precisions on the mathematical setting under consideration.

First, we assume that the conductivity σ belongs to the class:

$$\sigma \in C^0(\overline{\Omega}), \, \exists C_* > 0, \text{ such that } \forall x \in \Omega, \, \frac{1}{C_*} \leqslant \sigma(x) \leqslant C_*. \quad (1.53)$$

Under these conditions, the elliptic problem (1.52) is well-posed for Dirichlet data $g_D \in H^{1/2}(\partial\Omega)$ (see [Brezis (1983)]), and the solution u of (1.52) belongs to $H^1(\Omega)$.

We are then able to define $(\sigma\nabla u) \cdot n_x$ as an element of $H^{-1/2}(\partial\Omega)$ by the following formula: for all $g \in H^{1/2}(\partial\Omega)$, there exists $v_g \in H^1(\Omega)$ such that $v_g = g$ on $\partial\Omega$, and we define

$$\langle (\sigma\nabla u) \cdot n_x, g \rangle_{H^{-1/2}(\partial\Omega), H^{1/2}(\partial\Omega)} = \int_\Omega \sigma\nabla u \cdot \nabla v_g \, dx. \qquad (1.54)$$

Remark 1.31. More generally, for a function $\vec{v} \in (L^2(\Omega))^d$ such that div $(\vec{v}) \in L^2(\Omega)$, we can define $\vec{v} \cdot n_x \in H^{-1/2}(\partial\Omega)$ by the formula: for all $g \in H^{1/2}(\partial\Omega)$,

$$\langle \vec{v} \cdot n_x, g \rangle_{H^{-1/2}(\partial\Omega), H^{1/2}(\partial\Omega)} = \int_\Omega \text{div } (\vec{v})w_g + \int_\Omega \vec{v} \cdot \nabla w_g \, dx,$$

where w_g is any function of $H^1(\Omega)$ such that $w_g = g$ on $\partial\Omega$.

Following, the Voltage-to-Current map is defined as follows:

$$\Lambda_\sigma : \begin{array}{l} H^{1/2}(\partial\Omega) \to H^{-1/2}(\partial\Omega), \\ g_D \mapsto (\sigma\nabla u) \cdot n_x, \text{ where } u \text{ solves (1.52)}. \end{array} \qquad (1.55)$$

Calderón's problem now consists in studying the map

$$\Lambda : \sigma \mapsto \Lambda_\sigma.$$

But let us emphasize that Calderón's problem contains many different sub-problems:

- Uniqueness/Identifiability: If $\Lambda_{\sigma_1} = \Lambda_{\sigma_2}$, can we deduce $\sigma_1 = \sigma_2$? This corresponds to the injectivity of the map Λ.
- Stability: If $\Lambda_{\sigma_1} - \Lambda_{\sigma_2}$ is small (in suitable norms), can we deduce that $\sigma_1 - \sigma_2$ is small (in suitable norms)? This corresponds to the continuity of the inverse of Λ.
- Reconstruction: given Λ_σ, can we compute σ? This corresponds to the construction of the inverse of Λ. Note that there, this also includes questions like design of numerical schemes to recover σ, being able to propose a good numerical algorithm even in the event of imperfect measurements (noise), ...

In the following, we shall only focus on the uniqueness question, and we shall not address the problem in its full complexity. The interested reader may have a look on [Uhlmann (1999); Salo (2008); Choulli (2009)].

Remark that the map Λ is non-linear. Of course, this is one of the main difficulties of that problem.

Besides that, we shall assume that

- σ is known on the boundary.
- $\sqrt{\sigma}$ belongs to $W^{2,\infty}(\Omega)$.

These conditions allow us to perform the so-called Liouville's transform: setting $v = \sigma^{1/2}u$, u solves (1.52) if and only if v solves

$$\begin{cases} \Delta v + qv = 0, & \text{for } x \in \Omega, \\ v(x) = h_D(x), & \text{for } x \in \partial\Omega, \end{cases} \tag{1.56}$$

with

$$h_D = \sigma^{1/2}g_D, \quad q = -\frac{\Delta(\sigma^{1/2})}{\sigma^{1/2}}.$$

As σ is known on the boundary, this potential $q \in L^\infty(\Omega)$ determines uniquely the conductivity σ.

It is then natural to introduce the map

$$\Lambda^q : \begin{array}{l} H^{1/2}(\partial\Omega) \to H^{-1/2}(\partial\Omega), \\ h_D \mapsto \partial_n v, \text{ where } v \text{ solves (1.56)}. \end{array} \tag{1.57}$$

To simplify notations, we shall further assume that there is no non-trivial solutions of (1.56) when $h_D = 0$. Otherwise, the above map should be defined as a set-valued map (or set of Cauchy data). In the following, we simply omit this issue for sake of simplicity (also note that, if the potential q corresponds to some conductivity σ satisfying (1.53), there is no non-trivial solutions of (1.56) when $h_D = 0$).

In particular, if we restrict ourselves to classes of conductivity σ for which (1.53) are satisfied, σ is known on the boundary, and $\sqrt{\sigma} \in W^{2,\infty}$, the injectivity of the map $\sigma \mapsto \Lambda_\sigma$ is equivalent to the injectivity of the map

$$\Lambda : q \in L^\infty(\Omega) \mapsto \Lambda^q.$$

Our goal is to prove the following result:

Theorem 1.32. *Assume that the dimension is greater than 3, i.e. $d \geqslant 3$. If q_1, q_2 belong to $L^\infty(\Omega)$ and $\Lambda^{q_1} = \Lambda^{q_2}$, then $q_1 = q_2$.*

The proof is done in the following sections.

Let us emphasize that our result focuses on the case of a dimension $d \geqslant 3$. This is due to the use of the *Complex Geometric Optics* in the proof below. The 2d case can also be handled, but using a completely different construction based on complex analysis (see [Astala and Päivärinta (2006)]).

1.2.3.2. *Preliminaries*

Let us begin with the following remark:

Proposition 1.33. *Let $q \in L^\infty(\Omega)$. Then the map Λ^q is self-adjoint.*

Proof. Let $q \in L^\infty(\Omega)$ and, for h_1 and h_2 in $H^{1/2}(\partial\Omega)$, set v_1 and v_2 as the respective solutions of

$$\begin{cases} \Delta v_1 + qv_1 = 0, & \text{for } x \in \Omega, \\ v_1(x) = h_1(x), & \text{for } x \in \partial\Omega, \end{cases} \qquad \begin{cases} \Delta v_2 + qv_2 = 0, & \text{for } x \in \Omega, \\ v_2(x) = h_2(x), & \text{for } x \in \partial\Omega. \end{cases}$$

According to the definition of the normal derivative on the boundary,

$$\begin{aligned} \langle \Lambda^q h_1, h_2 \rangle_{H^{-1/2}(\partial\Omega), H^{1/2}(\partial\Omega)} &= \langle \partial_n v_1, v_2 \rangle_{H^{-1/2}(\partial\Omega), H^{1/2}(\partial\Omega)} \\ &= \int_\Omega \Delta v_1 v_2 \, dx + \int_\Omega \nabla v_1 \cdot \nabla v_2 \, dx \\ &= -\int_\Omega q v_1 v_2 \, dx + \int_\Omega \nabla v_1 \cdot \nabla v_2 \, dx \\ &= \langle h_1, \Lambda^q h_2 \rangle_{H^{1/2}(\partial\Omega), H^{-1/2}(\partial\Omega)}, \end{aligned}$$

and this concludes the proof of Proposition 1.33. □

We then show the following polarization formula:

Proposition 1.34. *Let q_1 and q_2 in $L^\infty(\Omega)$. Then for all h_1, h_2 in $H^{1/2}(\partial\Omega)$, the solutions v_1, v_2 of*

$$\begin{cases} \Delta v_1 + q_1 v_1 = 0, & \text{for } x \in \Omega, \\ v_1(x) = h_1(x), & \text{for } x \in \partial\Omega, \end{cases} \qquad \begin{cases} \Delta v_2 + q_2 v_2 = 0, & \text{for } x \in \Omega, \\ v_2(x) = h_2(x), & \text{for } x \in \partial\Omega, \end{cases} \tag{1.58}$$

satisfy

$$\langle (\Lambda^{q_1} - \Lambda^{q_2}) h_1, h_2 \rangle_{H^{-1/2}(\partial\Omega), H^{1/2}(\partial\Omega)} = \int_\Omega (q_2 - q_1) v_1 v_2 \, dx. \tag{1.59}$$

Proof. We compute

$$\begin{aligned} \langle \Lambda^{q_1} h_1, h_2 \rangle_{H^{-1/2}(\partial\Omega), H^{1/2}(\partial\Omega)} &= \langle \partial_n v_1, v_2 \rangle_{H^{-1/2}(\partial\Omega), H^{1/2}(\partial\Omega)} \\ &= \int_\Omega \Delta v_1 v_2 \, dx + \int_\Omega \nabla v_1 \cdot \nabla v_2 \, dx \\ &= -\int_\Omega q_1 v_1 v_2 \, dx + \int_\Omega \nabla v_1 \cdot \nabla v_2 \, dx. \end{aligned}$$

Similar computations yield

$$\langle h_1, \Lambda^{q_2} h_2 \rangle_{H^{1/2}(\partial\Omega), H^{-1/2}(\partial\Omega)} = -\int_\Omega q_2 v_1 v_2 \, dx + \int_\Omega \nabla v_1 \cdot \nabla v_2 \, dx.$$

As Λ^{q_2} is self-adjoint from Proposition 1.33,

$$\langle h_1, \Lambda^{q_2} h_2\rangle_{H^{1/2}(\partial\Omega), H^{-1/2}(\partial\Omega)} = \langle \Lambda^{q_2} h_1, h_2\rangle_{H^{-1/2}(\partial\Omega), H^{1/2}(\partial\Omega)},$$

and the subtraction of the two above identities immediately yield (1.59). □

The polarization formula is very important as it gives a simple way to describe the condition $\Lambda^{q_1} = \Lambda^{q_2}$. In particular, we see that, if we are able to show that the set described by the products $v_1 v_2$ is dense in $L^1(\Omega)$, then $q_2 = q_1$. This is where the Complex Geometric Optics solutions, constructed from the Carleman estimate (1.47), come into play, see afterwards.

1.2.3.3. *Complex Geometric Optics solutions*

Let us begin by a brief description of the main idea. The first basic remark is that, if ρ denotes a constant vector,

$$\Delta e^{\rho \cdot x} = \rho \cdot \rho\, e^{\rho \cdot x}.$$

In particular, we see that, if $\rho \cdot \rho = 0$, $e^{\rho \cdot x}$ is an harmonic solution. This suggests to take complex vectors $\rho \in \mathbb{C}^d$ with

$$\rho = a + \mathrm{i} b, \quad a, b \in \mathbb{R}^d, \quad a \cdot b = 0, \quad |a| = |b|, \tag{1.60}$$

which implies that $\rho \cdot \rho = |a|^2 - |b|^2 + 2\mathrm{i} a \cdot b = 0$. As $e^{\rho \cdot x}$ is an harmonic function and a potential function is a low-order perturbation of the Laplace operator, we expect it to "almost" solve

$$(\Delta + q)e^{\rho \cdot x} = 0$$

at high-frequency ($|\rho| \gg 1$).

To give a precise meaning to these insights, we show the following result:

Theorem 1.35. *Let $q \in L^\infty(\Omega)$. There exists $C > 0$, such that for all $\rho = a + \mathrm{i} b \in \mathbb{C}^d$ as in (1.60) with $|a| \geqslant 1$, there exists a solution $v_\rho \in L^2(\Omega)$ of*

$$\Delta v_\rho + q v_\rho = 0 \text{ in } \Omega, \tag{1.61}$$

that can be written as

$$v_\rho(x) = e^{\rho \cdot x} + e^{a \cdot x} r_\rho(x), \tag{1.62}$$

with r_ρ satisfying the estimate

$$|a| \, \|r_\rho\|_{L^2(\Omega)} \leqslant C. \tag{1.63}$$

Proof. First remark that v_ρ as in (1.62) solves (1.61) if and only if r_ρ solves

$$e^{-a\cdot x}(\Delta + q)(e^{a\cdot x} r_\rho) = -q e^{ib\cdot x} \text{ in } \Omega.$$

We set $\tilde{q} = -q e^{ib\cdot x}$. The question is then reduced to show that there exists $r_\rho \in L^2(\Omega)$ solution of

$$e^{-a\cdot x}(\Delta + q)(e^{a\cdot x} r_\rho) = \tilde{q} \text{ in } \Omega, \tag{1.64}$$

and satisfying estimate (1.63). Though it is completely straightforward that one can choose $r_\rho \in L^2(\Omega)$ solving (1.64) (for instance by setting $r_\rho(x) = e^{-a\cdot x} R$, where R solves $(\Delta + q)R = \tilde{q} e^{a\cdot x}$ in Ω with $R = 0$ on the boundary $\partial\Omega$), the difficult part is to show that one can choose r_ρ solution of (1.64) such that it moreover satisfies the estimate (1.63).

The function r_ρ solves (1.64) if and only if for all $w \in \mathscr{D}(\Omega)$,

$$\int_\Omega r_\rho \left(e^{a\cdot x}(\Delta + q)(e^{-a\cdot x} w)\right) dx = \int_\Omega \tilde{q} w \, dx, \tag{1.65}$$

or, by density, for all $w \in H_0^2(\Omega)$.

It follows that the set of r_ρ solving (1.64) is an affine space with vector space

$$\{e^{a\cdot x}(\Delta + q)(e^{-a\cdot x} w), \ w \in H_0^2(\Omega)\}^{\perp_{L^2(\Omega)}}.$$

Therefore, the solution r_ρ of (1.64) of smallest norm should belong to the set

$$\overline{\{e^{a\cdot x}(\Delta + q)(e^{-a\cdot x} w), \ w \in H_0^2(\Omega)\}}^{L^2(\Omega)}.$$

Note that

$$e^{a\cdot x}(\Delta + q)(e^{-a\cdot x} w) = \Delta w - 2a \cdot \nabla w + |a|^2 w + q w.$$

Hence, applying Carleman estimate (1.51) with $k = -a$ and $U = w$, we get, for all $w \in H_0^2(\Omega)$,

$$|a|^2 \|w\|_{L^2(\Omega)}^2 + \frac{1}{|a|^2} \|w\|_{H^2(\Omega)}^2 \leqslant C \left\|e^{a\cdot x}(\Delta + q)(e^{-a\cdot x} w)\right\|_{L^2(\Omega)}^2, \tag{1.66}$$

for some constant C independent of a (cf Corollary 1.24 and its proof: this is the place where we use the fact that $q \in L^\infty(\Omega)$). Following, the set

$$\{e^{a\cdot x}(\Delta + q)(e^{-a\cdot x} w), \ w \in H_0^2(\Omega)\}$$

is closed in $L^2(\Omega)$. Indeed, if (w_n) is a sequence of $H_0^2(\Omega)$ for which $(e^{a\cdot x}(\Delta + q)e^{-a\cdot x} w_n)$ is a Cauchy sequence in $L^2(\Omega)$, then (1.66) implies that (w_n) is a Cauchy sequence in $H^2(\Omega)$.

Thus there exists $W \in H_0^2(\Omega)$ such that

$$r_\rho = e^{a \cdot x}(\Delta + q)(e^{-a \cdot x}W) \tag{1.67}$$

satisfies (1.65), i.e. r_ρ solves (1.64), and is the solution of (1.64) of minimal $L^2(\Omega)$-norm.

Now, to derive estimates on r_ρ, we apply (1.65) to W: that way, we obtain

$$\int_\Omega |e^{a \cdot x}(\Delta + q)(e^{-a \cdot x}W)|^2 \, dx = \int_\Omega \tilde{q}W \, dx \leqslant \|\tilde{q}\|_{L^2} \|W\|_{L^2}.$$

Using (1.66), we check that

$$\left(\int_\Omega |e^{a \cdot x}(\Delta + q)(e^{-a \cdot x}W)|^2 \, dx \right)^{1/2} \leqslant \frac{C \|\tilde{q}\|_{L^\infty}}{|a|} \leqslant \frac{C \|q\|_{L^\infty}}{|a|}.$$

Using the definition of r_ρ in (1.67), we conclude (1.63). $\qquad\square$

1.2.3.4. *Proof of Theorem 1.32*

Let q_1 and q_2 in $L^\infty(\Omega)$ be such that $\Lambda^{q_1} = \Lambda^{q_2}$. According to Proposition 1.34, this implies that

$$\int_\Omega (q_2 - q_1)v_1 v_2 = 0 \tag{1.68}$$

for all v_1, v_2 solutions of (1.58).

The idea is to use the CGO solutions v_ρ given by Theorem 1.35 in order to construct, for all $\xi \in \mathbb{R}^d$, solutions v_1 and v_2 corresponding respectively to potentials q_1 and q_2 according to (1.58), so that

$$v_1 v_2 \simeq e^{-i\xi \cdot x}.$$

We will not be able to do that exactly, but we will manage to get sequence of solutions $(v_{1,n})$, $(v_{2,n})$ such that the product $(v_{1,n}v_{2,n})$ converges to $e^{-i\xi \cdot x}$.

Indeed, let $\xi \in \mathbb{R}^d$. Consider α, β in $\mathbb{R}^d$ of unit norm such that

$$\alpha \cdot \beta = \beta \cdot \xi = \alpha \cdot \xi = 0. \tag{1.69}$$

Note that this requires $d \geqslant 3$.

For each $n \in \mathbb{N}$ satisfying $n \geqslant |\xi|$, there exists $\gamma_n > 0$ such that

$$\rho_{n,1} = n\alpha + \mathbf{i}\left(\gamma_n \beta - \frac{\xi}{2}\right), \quad \rho_{n,2} = -n\alpha - \mathbf{i}\left(\gamma_n \beta + \frac{\xi}{2}\right)$$

both satisfy (1.60) (take $\gamma_n = \sqrt{n^2 - |\xi|^2/4}$). Note that this step requires the orthogonality conditions (1.69), and that by construction,

$$\rho_{n,1} + \rho_{n,2} = -\mathbf{i}\xi.$$

According to Theorem 1.35, one can construct

$$v_1 = e^{\rho_{n,1}\cdot x} + e^{n\alpha\cdot x}r_{n,1}(x), \quad v_2 = e^{\rho_{n,2}\cdot x} + e^{-n\alpha\cdot x}r_{n,2}(x),$$

solutions of (1.58) corresponding respectively to potentials q_1 and q_2, with

$$\|r_{n,1}\|_{L^2(\Omega)} + \|r_{n,2}\|_{L^2(\Omega)} \leqslant \frac{C}{n}. \tag{1.70}$$

Using the polarization formula (1.68), we obtain

$$\int_\Omega (q_2 - q_1)e^{-i\xi\cdot x}\,dx$$
$$= -\int_\Omega (q_2 - q_1)\left(e^{-i(\gamma_n\beta - \xi/2)\cdot x}r_{n,1}(x) + e^{i(\gamma_n\beta + \xi/2)\cdot x}r_{n,2}(x) + r_{n,1}r_{n,2}\right)\,dx,$$

and thus we have

$$|\mathscr{F}((q_2 - q_1)1_\Omega)(\xi)|$$
$$\leqslant C\|q_2 - q_1\|_{L^\infty(\Omega)}\left(\|r_{n,1}\|_{L^2(\Omega)} + \|r_{n,2}\|_{L^2(\Omega)} + \|r_{n,1}\|_{L^2(\Omega)}\|r_{n,2}\|_{L^2(\Omega)}\right).$$

Using (1.70) and letting $n \to \infty$, we deduce that the Fourier transform of $(q_2 - q_1)1_\Omega$ vanishes identically, hence $q_2 = q_1$ on Ω. This concludes the proof of Theorem 1.32. $\qquad\square$

Problem 1.36. *The goal of this problem is to study the stability of the map* $q \mapsto \Lambda^q$ *when* $d \geqslant 3$.
Let $m > 0$ *and let* $L^\infty_{\leqslant m}(\Omega) = \{q \in L^\infty(\Omega)|\ \|q\|_{L^\infty(\Omega)} \leqslant m\}$.
1. *Show that there exist* C *and* $a_0 \geqslant 1$ *depending only on* m *such that for all* $q \in L^\infty_{\leqslant m}(\Omega)$, *for all* $\rho = a + ib \in \mathbb{C}^d$ *as in (1.60) with* $|a| \geqslant a_0$, *there exists a solution* $v_\rho \in H^1(\Omega)$ *of (1.61) that can be written as (1.62) with* r_ρ *satisfying the estimate (1.63).*
2. *Show that for all* $\eta \in \mathscr{C}^\infty_c(\Omega)$, *the above constructed* r_ρ *satisfy, for some constant* C *depending on* m *and independent of* ρ *and* $q \in L^\infty_{\leqslant m}(\Omega)$, *that*

$$\|\eta\nabla r_\rho\|_{L^2(\Omega)} \leqslant C.$$

Hint: use (1.64).
3. *Let* $\mathcal{O}$ *be a bounded open set such that* $\overline{\Omega} \subset \mathcal{O}$. *Extending the potentials by 0 outside* Ω *and applying the above estimates to derive solutions* v_ρ *on* $\mathcal{O}$, *show the following result: there exist* C *and* $a_0 \geqslant 1$ *depending only on* m *such that for all* $q \in L^\infty_{\leqslant m}(\Omega)$, *for all* $\rho = a + ib \in \mathbb{C}^d$ *as in (1.60) with* $|a| \geqslant a_0$, *there exists a solution* $v_\rho \in H^1(\Omega)$ *of (1.61) that can be written as (1.62) with* r_ρ *satisfying the estimate (1.63) and*

$$\|\nabla r_\rho\|_{L^2(\Omega)} \leqslant C.$$

4. *Deduce from these estimates that for some constant C independent of ρ and $q \in L^\infty_{\leqslant m}(\Omega)$,*

$$\|v_\rho\|_{H^1(\Omega)} \leqslant Ce^{C|a|}(1 + |a|).$$

5. *We recall that for $v \in H^1(\Omega)$ satisfying $\Delta v \in L^2(\Omega)$, we have $v|_{\partial\Omega} \in H^{1/2}(\partial\Omega)$ and $\partial_n v|_{\partial\Omega} \in H^{-1/2}(\partial\Omega)$ with*

$$\|v|_{\partial\Omega}\|_{H^{1/2}(\partial\Omega)} + \|\partial_n v|_{\partial\Omega}\|_{H^{-1/2}(\partial\Omega)} \leqslant C\left(\|v\|_{H^1(\Omega)} + \|\Delta v\|_{L^2(\Omega)}\right).$$

Using the polarization formula (1.59) and following the above proof, show that there exists $C > 0$ and n_0 such that for all $\xi \in \mathbb{R}^d$, and $n \in \mathbb{N}$ larger than $\max\{n_0, |\xi|\}$,

$$\begin{aligned}
|\mathscr{F}((q_1 - q_2)1_\Omega)(\xi)| &\leqslant C \|\Lambda^{q_1} - \Lambda^{q_2}\|_{\mathfrak{L}(H^{1/2}(\partial\Omega), H^{-1/2}(\partial\Omega))} e^{Cn} \\
&\quad + \frac{C}{n} \|q_1 - q_2\|_{L^\infty(\Omega)}.
\end{aligned}$$

6. *If $\|\Lambda^{q_1} - \Lambda^{q_2}\|_{\mathfrak{L}(H^{1/2}(\partial\Omega), H^{-1/2}(\partial\Omega))}$ is small enough, deduce that for all $\xi \in \mathbb{R}^d$ with $|\xi| < \log(1/\|\Lambda^{q_1} - \Lambda^{q_2}\|_{\mathfrak{L}(H^{1/2}(\partial\Omega), H^{-1/2}(\partial\Omega))})$,*

$$|\mathscr{F}((q_1 - q_2)1_\Omega)(\xi)| \leqslant \frac{C}{\log\left(\dfrac{1}{\|\Lambda^{q_1} - \Lambda^{q_2}\|_{\mathfrak{L}(H^{1/2}(\partial\Omega), H^{-1/2}(\partial\Omega))}}\right)}.$$

7. *Recalling that $q_1 - q_2 \in L^2(\Omega)$, and extending $q_1 - q_2$ by 0 outside Ω, for $k > 0$, deduce an estimate on $\|q_1 - q_2\|_{H^{-k}(\mathbb{R}^d)}$.*
Hint: estimate

$$\int_{\mathbb{R}^d} |\mathscr{F}((q_1 - q_2)1_\Omega)(\xi)|^2 (1 + |\xi|^2)^{-k} \, d\xi$$

and divide the integral in the parts $|\xi| \leqslant \rho$ and $|\xi| > \rho$ and optimize in ρ.

1.2.3.5. *Further comments*

In this section, we focused on the Calderón problem for a scalar conductivity σ. But this question also makes sense for conductivity σ taking value in the set of positive definite matrices: physically, this corresponds to anisotropic materials.

In the case of anisotropic conductivities, L. Tartar proposed a simple construction to show that uniqueness cannot hold. Indeed, given any $\mathscr{C}^2$ diffeomorphism Ψ on Ω with $\Psi = Id$ on the boundary $\partial\Omega$, then

$$\Lambda_{\tilde\sigma} = \Lambda_\sigma, \quad \text{where } \tilde\sigma(y) = \left(\frac{D\Psi^T(\Psi^{-1}(y)) \times \sigma(\Psi^{-1}(y)) \times D\Psi(\Psi^{-1}(y))}{|\det(D\Psi(\Psi^{-1}(y)))|}\right).$$

This shows in particular that one cannot distinguish between σ and $\tilde{\sigma}$ when allowing anisotropic conductivities.

Actually, this counterexample is the basis of several recent works on invisibility, see for instance the article [Uhlmann (2009)], where one possible idea is to construct diffeomorphisms that approximate the singular transformation between $B(0,2) \setminus \{0\}$ and $B(0,2) \setminus B(0,1)$ given by

$$\Psi(x) = \frac{x}{|x|}(1+|x|).$$

In such case, the material inside the ball $B(0,1)$ will be invisible from the boundary.

1.3. Carleman estimates with general weights in a strip

1.3.1. *Introduction*

In Corollary 1.25, we saw that if $u \in H^2 \cap H_0^1((0,1) \times \mathbb{R}^{d-1})$ solves

$$\Delta u + qu = 0 \text{ in } (0,1) \times \mathbb{R}^{d-1}, \tag{1.71}$$

for some $q \in L^\infty(\Omega)$, satisfies

$$u(0,x') = \partial_1 u(0,x') = 0, \text{ for all } x' \in \mathbb{R}^{d-1}, \tag{1.72}$$

and $u(1,x') = 0$ for all $x' \in \mathbb{R}^{d-1}$, then u vanishes everywhere.

In this section, our goal is to generalize this result to more intricate situations. In order to do that, we will produce new Carleman estimates corresponding to more general weight functions, still considering the case of a strip $\Omega = (0,1) \times \mathbb{R}^{d-1}$ with an additional observation on $\Gamma = \{0\} \times \mathbb{R}^{d-1}$.

1.3.2. *General Carleman weights in a strip for a weight function depending on x_1*

1.3.2.1. *Goal*

The goal of this section is to understand if the Carleman estimate obtained in Theorem 1.23 in the case of a strip can be generalized to more general weight functions, but depending on x_1 only.

To be more precise, we want to know under which conditions on $\varphi = \varphi(x_1)$ we can guarantee an estimate of the form

$$s^{\theta_1} \|ue^{s\varphi}\|_{L^2((0,1)\times\mathbb{R}^{d-1})}$$
$$\leqslant C \|\Delta u e^{s\varphi}\|_{L^2((0,1)\times\mathbb{R}^{d-1})} + Cs^{\theta_2} \left\|\partial_n u(0,\cdot)e^{s\varphi(0)}\right\|_{L^2(\mathbb{R}^{d-1})}, \tag{1.73}$$

for all solution u of (1.24), with some constants C and θ_1, θ_2 independent of u and s, and valid for all $s \geqslant 1$.

1.3.2.2. *Analysis*

In this section, we explain what would be the conditions on φ to develop the same type of proof as in Section 1.2.2.

In order to do that, for u satisfying

$$\begin{cases} \partial_{11}u + \Delta'u = f, & \text{for } (x_1, x') \in (0,1) \times \mathbb{R}^{d-1}, \\ u(0, x') = u(1, x') = 0, & \text{for } x' \in \mathbb{R}^{d-1}, \\ \partial_1 u(0, x') = g_N(x'), & \text{for } x' \in \mathbb{R}^{d-1}, \end{cases} \quad (1.74)$$

we set

$$\begin{aligned} U(x_1, x') &= u(x_1, x')e^{s\varphi(x_1)}, \\ F(x_1, x') &= f(x_1, x')e^{s\varphi(x_1)}, \\ G_N(x') &= g_N(x')e^{s\varphi(0)}. \end{aligned} \quad (1.75)$$

Easy computations show that U satisfies the equation

$$\begin{cases} \partial_{11}U - 2s\partial_1\varphi\partial_1 U + (s^2|\partial_1\varphi|^2 - s\partial_{11}\varphi)U + \Delta'U = F, \\ \qquad\qquad\qquad\qquad \text{for } (x_1, x') \in (0,1) \times \mathbb{R}^{d-1}, \\ U(0, x') = U(1, x') = 0, \qquad \text{for } x' \in \mathbb{R}^{d-1}, \\ \partial_1 U(0, x') = G_N(x'), \qquad \text{for } x' \in \mathbb{R}^{d-1}. \end{cases} \quad (1.76)$$

Similarly as before, we can then take the partial Fourier transform in the x'-variable:

$$\begin{aligned} \widehat{U}(x_1, \xi') &= \mathscr{F}_{x' \to \xi'} U(x_1, \cdot), \\ \widehat{F}(x_1, \xi') &= \mathscr{F}_{x' \to \xi'} F(x_1, \cdot), \\ \widehat{G}_N(\xi') &= \mathscr{F}_{x' \to \xi'} G_N(\cdot). \end{aligned}$$

Using these notations, $\widehat{U}(x_1, \xi')$ satisfies

$$\begin{cases} \partial_{11}\widehat{U} - 2s\partial_1\varphi\partial_1\widehat{U} + (s^2|\partial_1\varphi|^2 - s\partial_{11}\varphi)U - |\xi'|^2\widehat{U} = \widehat{F}, \\ \qquad\qquad\qquad\qquad \text{for } (x_1, \xi') \in (0,1) \times \mathbb{R}^{d-1}, \\ \widehat{U}(0, \xi') = \widehat{U}(1, \xi') = 0, \qquad \text{for } \xi' \in \mathbb{R}^{d-1}, \\ \partial_1 \widehat{U}(0, x') = \widehat{G}_N(\xi'), \qquad \text{for } \xi' \in \mathbb{R}^{d-1}. \end{cases} \quad (1.77)$$

Here again, the key is that this operator can be factorized into

$$(\partial_1 - s\partial_1\varphi + |\xi'|)(\partial_1 - s\partial_1\varphi - |\xi'|),$$

or

$$(\partial_1 - s\partial_1\varphi - |\xi'|)(\partial_1 - s\partial_1\varphi + |\xi'|).$$

Remark then that the information on the first derivative can only come from the left (recall $(1.77)_3$).

Considering for an instant the resolution of $\partial_1 y + a(x_1)y = h(x_1)$ for $x_1 \in (0,1)$ with the boundary condition $y(0) = b$, we get

$$y(x_1) = y(0)\exp(-A(x_1)) + \int_0^{x_1} \exp(A(\tilde{x}_1) - A(x_1))h(\tilde{x}_1)d\tilde{x}_1,$$

where A is such that $A' = a$ and $A(0) = 0$. Therefore, to get that each exponential in this formula is smaller than 1, one needs to impose that for all $\tilde{x}_1 < x_1$, $A(\tilde{x}_1) \leqslant A(x_1)$, that is A increasing, or equivalently $a \geqslant 0$.

Accordingly, to transfer the information from the left to the right in our problem, we should have for all ξ', that, either

$$(\forall x_1 \in (0,1), \ -s\partial_1\varphi - |\xi'| \geqslant 0) \quad \text{or} \quad (\forall x_1 \in (0,1), \ -s\partial_1\varphi + |\xi'| \geqslant 0).$$

Taking $\xi' = 0$, we will thus require the condition

$$\partial_1\varphi(x_1) \leqslant 0, \quad \text{for all } x_1 \in (0,1). \tag{1.78}$$

Using this condition, it is natural to set

$$\widehat{V}(x_1, \xi') = (\partial_1 - s\partial_1\varphi - |\xi'|)\widehat{U}(x_1, \xi'),$$

so that we have

$$\begin{cases} \partial_1\widehat{U} + (-s\partial_1\varphi - |\xi'|)\widehat{U} = \widehat{V}, & \text{for } (x_1, \xi') \in (0,1) \times \mathbb{R}^{d-1}, \\ \partial_1\widehat{V} + (-s\partial_1\varphi + |\xi'|)\widehat{V} = \widehat{F}, & \text{for } (x_1, \xi') \in (0,1) \times \mathbb{R}^{d-1}, \\ \widehat{U}(0, \xi') = \widehat{U}(1, \xi') = 0, & \text{for } \xi' \in \mathbb{R}^{d-1}, \\ \widehat{V}(0, \xi') = \widehat{G}_N(\xi'), & \text{for } \xi' \in \mathbb{R}^{d-1}. \end{cases} \tag{1.79}$$

Similarly as in Section 1.2.2.2, we first solve the equation in $\widehat{V}$, which is well-posed:

$$\widehat{V}(x_1, \xi') = e^{s\varphi(x_1) - |\xi'|x_1}\widehat{G}_N(\xi') + \int_0^{x_1} e^{-s\varphi(x) + |\xi'|x}e^{s\varphi(x_1) - |\xi'|x_1}\widehat{F}(x, \xi')\,dx.$$

One should then solve the equation on $\widehat{U}$. We see there that the solution $\widehat{U}$ can be expressed as an integral of exponentials involving only negative coefficients if for all $\xi' \in \mathbb{R}^{d-1}$, there exists $x_\xi \in [0,1]$ such that

$$-s\partial_1\varphi(x_\xi) - |\xi'| = 0,$$

and

$$\begin{cases} \forall x_1 \in [0, x_\xi], & -s\partial_1\varphi(x_1) - |\xi'| \geqslant 0, \\ \forall x_1 \in [x_\xi, 1], & -s\partial_1\varphi(x_1) - |\xi'| \leqslant 0. \end{cases}$$

In that case indeed, for $x_1 \leqslant x_\xi$, we solve the equation $(1.79)_{(1)}$ from the left, whereas for $x_1 \geqslant x_\xi$, we solve the equation $(1.79)_{(1)}$ from the right.

As for all X, there exists $\xi' \in \mathbb{R}^d$ such that $-s\partial_1\varphi(X) - |\xi'| = 0$, this latter condition is satisfied if

$$\forall X \in (0,1), \quad \begin{cases} \forall x_1 < X, & -s\partial_1\varphi(x_1) \geqslant -s\partial_1\varphi(X), \\ \forall x_1 > X, & -s\partial_1\varphi(x_1) \leqslant -s\partial_1\varphi(X), \end{cases}$$

i.e. if $\partial_1\varphi$ is non-decreasing, i.e.

$$\partial_{11}\varphi \geqslant 0. \tag{1.80}$$

Under this convexity assumption, for all $\xi' \in \mathbb{R}^{d-1}$, we have the following cases:

- Low-frequency case: for all $x_1 \in [0,1]$, $-s\partial_1\varphi(x_1) - |\xi'| > 0$. In that case, we have

$$\widehat{U}(x_1,\xi') = \int_0^{x_1} e^{s\varphi(x_1)+|\xi'|x_1} e^{-s\varphi(x)-|\xi'|x} \widehat{V}(x,\xi')\,dx. \tag{1.81}$$

- High-frequency case: for all $x_1 \in [0,1]$, $-s\partial_1\varphi(x_1) - |\xi'| < 0$. In that case, we have

$$\widehat{U}(x_1,\xi') = -\int_{x_1}^1 e^{s\varphi(x_1)+|\xi'|x_1} e^{-s\varphi(x)-|\xi'|x} \widehat{V}(x,\xi')\,dx. \tag{1.82}$$

- Intermediate range: there exists $x_\xi \in [0,1]$ such that $-s\partial_1\varphi(x_\xi) - |\xi'| = 0$. Then, for $x_1 < x_\xi$, we use formula (1.81), whereas for $x_1 > x_\xi$, we use formula (1.82).

One may then estimate the solutions $\widehat{U}(\cdot,\xi')$ in $L^2(0,1)$ for fixed $\xi' \in \mathbb{R}^{d-1}$ as in Section 1.2.2, distinguishing between the above cases.

Roughly speaking, the low frequency and high frequency cases will follow the same path as in Section 1.2.2. In order to deal with these cases, we will use that φ is strictly decreasing, i.e. the assumption

$$\sup_{x_1 \in (0,1)} \partial_1\varphi(x_1) < -\alpha,$$

for some $\alpha > 0$.

The new case corresponding to the intermediate range of frequency will require a deeper analysis. First, to make it simpler, it is usually assumed that φ is strictly convex instead of (1.80), i.e. we will assume

$$\inf_{x_1 \in (0,1)} \partial_{11}\varphi(x_1) \geqslant \beta,$$

for some $\beta > 0$. In particular, this condition implies that in the intermediate range of frequency, there exists a *unique* x_ξ such that $-s\partial_1\varphi(x_\xi) - |\xi'| = 0$. One then needs to get estimates on the weight function $s\varphi(x_1) + |\xi'|x_1$, which can be deduced by the facts that $\Psi(x_1, \xi') = s\varphi(x_1) + |\xi'| x_1$ satisfies $\Psi(0, \xi') = 0$, $\partial_1 \Psi(x_\xi, \xi') = 0$, and $\partial_{11} \Psi \geqslant s\beta$. We will not give more details about that explicit approach as the computations rapidly become heavy.

Still, Approach 2 in the proof of Theorem 1.23 can be performed here assuming that

$$\exists \alpha > 0 \text{ and } \beta > 0, \text{ such that } \sup_{x_1 \in (0,1)} \partial_1 \varphi(x_1) \leqslant -\alpha, \text{ and}$$

$$\inf_{x_1 \in (0,1)} \partial_{11} \varphi(x_1) \geqslant \beta. \tag{1.83}$$

Indeed, one can perform multiplier type arguments on the system (1.79), by estimating $\widehat{V}$ in terms of $\widehat{F}$ and $\widehat{G}_N$, and on $\widehat{U}$ in terms of $\widehat{V}$.

Indeed, since $\widehat{V}$ satisfies

$$\begin{cases} \partial_1 \widehat{V} + (-s\partial_1 \varphi + |\xi'|)\widehat{V} = \widehat{F}, & \text{for } (x_1, \xi') \in (0,1) \times \mathbb{R}^{d-1}, \\ \widehat{V}(0, \xi') = \widehat{G}_N(\xi'), & \text{for } \xi' \in \mathbb{R}^{d-1}, \end{cases} \tag{1.84}$$

taking the square of the equation and integrating in x_1, we obtain:

$$\int_0^1 \left(|\partial_1 \widehat{V}(x_1, \xi')|^2 + \left((-s\partial_1 \varphi(x_1) + |\xi'|)^2 + s\partial_{11}\varphi \right) |\widehat{V}(x_1, \xi')|^2 \right) dx_1$$

$$+(-s\partial_1\varphi+|\xi'|)|\widehat{V}(1,\xi')|^2 = \int_0^1 |\widehat{F}(x_1,\xi')|^2 \, dx_1 + (-s\partial_1\varphi+|\xi'|)|\widehat{G}_N(\xi')|^2,$$

hence, due to (1.83), for some constant $C > 0$ independent of s and ξ',

$$\int_0^1 \left(|\partial_1 \widehat{V}(x_1, \xi')|^2 + (s + |\xi'|)^2 |\widehat{V}(x_1, \xi')|^2 \right) dx_1$$

$$\leqslant C \int_0^1 |\widehat{F}(x_1, \xi')|^2 \, dx_1 + C(s + |\xi'|)|\widehat{G}_N(\xi')|^2. \tag{1.85}$$

To estimate $\widehat{U}$, we also take the square of both sides of

$$\begin{cases} \partial_1 \widehat{U} + (-s\partial_1 \varphi - |\xi'|)\widehat{U} = \widehat{V}, & \text{for } (x_1, \xi') \in (0,1) \times \mathbb{R}^{d-1}, \\ \widehat{U}(0, \xi') = \widehat{U}(1, \xi') = 0, & \text{for } \xi' \in \mathbb{R}^{d-1}, \end{cases} \tag{1.86}$$

and integrate in the x_1 variable:

$$\int_0^1 \left(|\partial_1 \widehat{U}(x_1, \xi')|^2 + \left((-s\partial_1 \varphi(x_1) - |\xi'|)^2 + s\partial_{11}\varphi \right) |\widehat{U}(x_1, \xi')|^2 \right) dx_1$$

$$= \int_0^1 |\widehat{V}(x_1, \xi')|^2 \, dx_1. \tag{1.87}$$

Here, we emphasize that the term $(-s\partial_1\varphi(x_1) - |\xi'|)^2$ may vanish, but this does not matter as the other term $s\partial_{11}\varphi$ is always strictly positive according to (1.83). Therefore, we get that for some constant C independent of s and ξ',

$$\int_0^1 \left(|\partial_1 \widehat{U}(x_1,\xi')|^2 + s|\widehat{U}(x_1,\xi')|^2\right) dx_1 \leqslant C \int_0^1 |\widehat{V}(x_1,\xi')|^2 dx_1. \quad (1.88)$$

Combining the estimates on $\widehat{U}$ and on $\widehat{V}$, we easily deduce that there exists a constant C independent of s and ξ', such that

$$s(s+|\xi'|)^2 \int_0^1 |\widehat{U}(x_1,\xi')|^2 dx_1 \leqslant C \int_0^1 |\widehat{F}(x_1,\xi')|^2 dx_1 + C(s+|\xi'|)|\widehat{G}_N(\xi')|^2.$$

Accordingly,

$$s^3 \int_0^1 |\widehat{U}(x_1,\xi')|^2 dx_1$$
$$\leqslant C\frac{s^2}{(s+|\xi'|)^2} \int_0^1 |\widehat{F}(x_1,\xi')|^2 dx_1 + C\frac{s^2}{(s+|\xi'|)}|\widehat{G}_N(\xi')|^2$$
$$\leqslant C \int_0^1 |\widehat{F}(x_1,\xi')|^2 dx_1 + Cs|\widehat{G}_N(\xi')|^2.$$

Integrating in ξ' and using Parseval's identity, we obtain:

$$s^3 \int_\Omega |U(x)|^2 dx \leqslant C \int_\Omega |F(x)|^2 dx + Cs \|G_N\|^2_{L^2(\Gamma)}. \quad (1.89)$$

1.3.2.3. *A Carleman estimate for strictly convex weight functions*

Based on the analysis done in the previous section, we derive the following result:

Theorem 1.37. *Let* $\varphi = \varphi(x_1)$ *be a* $\mathscr{C}^4([0,1])$ *function such that there exist* $\alpha > 0$ *and* $\beta > 0$ *for which*

$$\sup_{x_1\in(0,1)} \partial_1\varphi(x_1) \leqslant -\alpha, \quad \text{and} \quad \inf_{x_1\in(0,1)} \partial_{11}\varphi(x_1) \geqslant \beta. \quad (1.90)$$

Then there exists a constant $C > 0$ *such that for all* $u \in H^2 \cap H^1_0((0,1) \times \mathbb{R}^{d-1})$ *satisfying* (1.74) *with source term* $f \in L^2((0,1) \times \mathbb{R}^{d-1})$, *and for all* $s \geqslant 1$,

$$s^3 \|e^{s\varphi}u\|^2_{L^2((0,1)\times\mathbb{R}^{d-1})} + s \|e^{s\varphi}\nabla u\|^2_{L^2((0,1)\times\mathbb{R}^{d-1})}$$
$$\leqslant C\left(\|e^{s\varphi}f\|^2_{L^2((0,1)\times\mathbb{R}^{d-1})} + s\left\|e^{s\varphi(0)}\partial_1 u(0,\cdot)\right\|^2_{L^2(\mathbb{R}^{d-1})}\right). \quad (1.91)$$

Remark 1.38. Note that the fact that the powers of the parameter s in (1.91) are strictly greater than the powers of the parameter $|k|$ in Theorem 1.29. This allows to handle first-order potentials W and zero-order potentials q, while Corollary 1.25 based on Theorem 1.29 only yields unique continuation for zero-order potentials q.

In fact, strictly speaking, Carleman estimates usually refer to cases in which the strict convexity condition in (1.90) is satisfied, so that the linear weight does not truly correspond to a Carleman weight function. To be precisely, using the wording of [Kenig *et al.* (2007)], such linear weight is usually referred to as a *limiting* Carleman weight.

Proof. We present below a proof of Theorem 1.37 which relies on similar arguments as the Approach 3 presented for the proof of Theorem 1.23.

Let $u \in H^2 \cap H_0^1((0,1) \times \mathbb{R}^{d-1})$ satisfying (1.74) with source term $f \in L^2((0,1) \times \mathbb{R}^{d-1})$. Let U and F be as in (1.75). Then U satisfies (1.76).

Multiply the equation (1.76) by $s\partial_1\varphi\partial_1 U$:

$$\iint_{(0,1)\times\mathbb{R}^{d-1}} Fs\partial_1\varphi\partial_1 U dx$$

$$= \iint_{(0,1)\times\mathbb{R}^{d-1}} \left(\partial_{11}U - 2s\partial_1\varphi\partial_1 U + (s^2|\partial_1\varphi|^2 - s\partial_{11}\varphi)U + \Delta'U\right) s\partial_1\varphi\partial_1 U dx$$

$$= \frac{s}{2}\int_{\mathbb{R}^{d-1}} \partial_1\varphi|\nabla U|^2 dx' \Big|_{x_1=0}^{x_1=1} + \frac{s}{2}\iint_{(0,1)\times\mathbb{R}^{d-1}} \partial_{11}\varphi|\nabla'U|^2 dx$$

$$- \iint_{(0,1)\times\mathbb{R}^{d-1}} \left(\frac{s}{2}\partial_{11}\varphi|\partial_1 U|^2 + 2s^2(\partial_1\varphi)^2|\partial_1 U|^2 + \frac{3s^3}{2}\partial_{11}\varphi(\partial_1\varphi)^2|U|^2\right) dx$$

$$+ \iint_{(0,1)\times\mathbb{R}^{d-1}} \frac{s^2}{2}((\partial_{11}\varphi)^2 + \partial_1\varphi\partial_{111}\varphi)|U|^2 dx.$$

Using the conditions (1.90) and using the bound

$$\left|\iint_{(0,1)\times\mathbb{R}^{d-1}} Fs\partial_1\varphi\partial_1 U dx\right| \leq \frac{1}{4}\iint_{(0,1)\times\mathbb{R}^{d-1}} F^2 dx + s^2\iint_{(0,1)\times\mathbb{R}^{d-1}} (\partial_1\varphi)^2(\partial_1 U)^2 dx,$$

we have proved:

$$\iint_{(0,1)\times\mathbb{R}^{d-1}} \left(\frac{s}{2}\partial_{11}\varphi|\partial_1 U|^2 + s^2(\partial_1\varphi)^2|\partial_1 U|^2 + \frac{3s^3}{2}\partial_{11}\varphi(\partial_1\varphi)^2|U|^2\right) dx$$

$$\leq \frac{s}{2}\iint_{(0,1)\times\mathbb{R}^{d-1}} \partial_{11}\varphi|\nabla'U|^2 dx + Cs\int_{\mathbb{R}^{d-1}} |\nabla U(0,x')|^2 dx'$$

$$+ C\iint_{(0,1)\times\mathbb{R}^{d-1}} |F|^2 dx + Cs^2\iint_{(0,1)\times\mathbb{R}^{d-1}} |U|^2 dx. \tag{1.92}$$

In order to estimate

$$\frac{s}{2} \iint_{(0,1)\times\mathbb{R}^{d-1}} \partial_{11}\varphi |\nabla' U|^2 dx, \qquad (1.93)$$

we multiply the equation (1.76) by $s\partial_{11}\varphi U$:

$$\iint_{(0,1)\times\mathbb{R}^{d-1}} F s\partial_{11}\varphi U dx$$

$$= \iint_{(0,1)\times\mathbb{R}^{d-1}} (\partial_{11}U - 2s\partial_1\varphi\partial_1 U + (s^2|\partial_1\varphi|^2 - s\partial_{11}\varphi)U + \Delta' U)\, s\partial_{11}\varphi U dx$$

$$= -s \iint_{(0,1)\times\mathbb{R}^{d-1}} \partial_{11}\varphi |\nabla U|^2 dx + \frac{s}{2} \iint_{(0,1)\times\mathbb{R}^{d-1}} \partial_1^{(4)}\varphi |U|^2 dx$$

$$+ s^2 \iint_{(0,1)\times\mathbb{R}^{d-1}} \partial_1(\partial_1\varphi\partial_{11}\varphi)|U|^2 dx + s^3 \iint_{(0,1)\times\mathbb{R}^{d-1}} \partial_{11}\varphi(\partial_1\varphi)^2 |U|^2 dx$$

$$- s^2 \iint_{(0,1)\times\mathbb{R}^{d-1}} (\partial_{11}\varphi)^2 |U|^2 dx.$$

Accordingly,

$$s \iint_{(0,1)\times\mathbb{R}^{d-1}} \partial_{11}\varphi |\nabla U|^2 dx \leqslant s^3 \iint_{(0,1)\times\mathbb{R}^{d-1}} \partial_{11}\varphi(\partial_1\varphi)^2 |U|^2 dx$$

$$+ C \iint_{(0,1)\times\mathbb{R}^{d-1}} |F|^2 dx + C s^2 \iint_{(0,1)\times\mathbb{R}^{d-1}} |U|^2 dx.$$
$$(1.94)$$

Combined with (1.92) and the fact that $\nabla U(0,\cdot) = \partial_1 U(0,\cdot)\vec{e}_1$ on $\{x_1 = 0\}$ due to the boundary condition $U(0,\cdot) = 0$ on $\{x_1 = 0\}$, we easily deduce the existence of a constant $C > 0$ such that

$$\iint_{(0,1)\times\mathbb{R}^{d-1}} \left(\frac{s}{2}\partial_{11}\varphi|\partial_1 U|^2 + s^2(\partial_1\varphi)^2|\partial_1 U|^2 \right.$$

$$\left. + \frac{s^3}{2}\partial_{11}\varphi(\partial_1\varphi)^2|U|^2 + \frac{s}{2}\partial_{11}\varphi|\nabla U|^2 \right) dx$$

$$\leqslant C \|F\|_{L^2((0,1)\times\mathbb{R}^{d-1})}^2 + Cs \|\partial_1 U(0,\cdot)\|_{L^2(\mathbb{R}^{d-1})}^2 + Cs^2 \|U\|_{L^2((0,1)\times\mathbb{R}^{d-1})}^2.$$

Using the assumptions (1.90) on φ, we deduce the existence of a constant $C > 0$ and of $s_0 \geqslant 1$ large enough such that for all $s \geqslant s_0$,

$$\iint_{(0,1)\times\mathbb{R}^{d-1}} \left(s|\nabla U|^2 + s^2|\partial_1 U|^2 + s^3|U|^2 \right) dx$$

$$\leqslant C \|F\|_{L^2((0,1)\times\mathbb{R}^{d-1})}^2 + Cs \|\partial_1 U(0,\cdot)\|_{L^2(\mathbb{R}^{d-1})}^2.$$

We then write $ue^{s\varphi} = U$ and $\nabla u e^{s\varphi} = \nabla(Ue^{-s\varphi})e^{s\varphi} = \nabla U - s\nabla\varphi U$. Straightforward estimates then yield (1.91) for $s \geqslant s_0$. As the weight function is bounded, modifying the constant if needed, the Carleman estimate (1.91) holds for all $s \geqslant 1$. $\qquad\square$

Remark 1.39. Note that a direct approach consists in multiplying the equation of U by

$$2s\partial_1\varphi\partial_1 U + 2s\partial_{11}\varphi U.$$

But the two above terms do not play the same role. Here $s\partial_1\varphi\partial_1 U$ corresponds to a multiplier chosen for analyzing the propagation in the x_1 variable. The other term, $s\partial_{11}\varphi U$, rather corresponds to an energy method to compensate the "bad" term (1.93) coming from the multiplier $s\partial_1\varphi\partial_1 U$. Note in particular that $s\partial_{11}\varphi U$ is a weaker order term in U than $s\partial_1\varphi\partial_1 U$.

Remark 1.40. Another approach would consist in writing (1.76) under the form

$$AU + SU = F \text{ for } (x_1, x') \in (0,1) \times \mathbb{R}^{d-1},$$

with

$$AU = -2s\partial_1\varphi\partial_1 U - s\partial_{11}\varphi U,$$
$$SU = \Delta U + s^2|\partial_1\varphi|^2 U,$$

corresponding respectively to the skew-adjoint and symmetric parts of the conjugated operator. Then we can write:

$$\int_\Omega |F|^2 \, dx = \int_\Omega (|AU|^2 + |SU|^2) \, dx + 2\int_\Omega AU \, SU \, dx,$$

and computations will yield that

$$\int_\Omega AU \, SU \, dx = 2s^3 \int_\Omega \partial_{11}\varphi(\partial_1\varphi)^2 |U|^2 \, dx - \frac{s}{2}\int_\Omega \partial_{1111}\varphi|U|^2 \, dx$$
$$- s\int_{\partial\Omega} \partial_1\varphi\vec{n}\cdot\vec{e}_1|\partial_1 U|^2 \, d\sigma.$$

According to assumptions (1.90), we get some positive constants c_* and C such that for all s large enough,

$$\int_\Omega AU \, SU \, dx \geqslant c_* s^3 \int_\Omega |U|^2 \, dx - Cs\int_{\mathbb{R}^{d-1}} |\partial_1 U(0, x')|^2 \, dx'.$$

This gives another proof of Theorem 1.37 using multiplier type arguments, also in the spirit of the third approach developed for the proof of Theorem 1.23.

Problem 1.41 (*Difficult*). *Prove the same estimate as in* (1.91) *using Approach 1 in Section* 1.3.2.2.

1.3.2.4. *A Carleman estimate for coefficients depending on x_1*

One may wonder under which general settings the above strategy can be applied. It is quite clear that it can be extended when the coefficients of the tangential derivatives possibly depend on x_1, while the coefficients in the derivatives with respect to x_1 should not depend on x_1.

We present below a class of coefficients for which we can perform the above strategy. As we will see later, even if this looks like a rather restrictive case, we will be able to use these computations in the next section in a much more general setting.

We introduce constants $\lambda_{j,k} = \lambda_{k,j}$ defined for $j, k \in \{2, \cdots, d\}$, and we let $X_1 > 0$ be such that there exists $c > 0$ such that for all $\xi' \in \mathbb{R}^{d-1}$ and $x_1 \in [-X_1, X_1]$,

$$|\xi'|^2 - x_1 \sum_{j,k=2}^{d} \lambda_{j,k} \xi'_j \xi'_k \geqslant c|\xi'|^2. \tag{1.95}$$

We also assume the following condition

$$\exists c_* > 0, \forall \xi' \in \mathbb{R}^{d-1}, \quad \sum_{j,k=2}^{d} \lambda_{j,k} \xi'_j \xi'_k \geqslant c_*|\xi'|^2. \tag{1.96}$$

We then let

$$X_0 \in [-X_1, 0], \quad \Omega_x = (X_0, X_1) \times \mathbb{R}^{d-1} \tag{1.97}$$

and, for $s > 0$, and $G \in L^2(\Omega_x)$, we consider the equation:

$$\begin{cases} \Delta w - x_1 \sum_{j,k=2}^{d} \lambda_{j,k} \partial_j \partial_k w - 2s\partial_1 w + s^2 w = G, & \text{in } \Omega_x, \\ w(X_0, x') = w(X_1, x') = 0, & \text{for } x' \in \mathbb{R}^{d-1}, \end{cases} \tag{1.98}$$

and we assume that $\partial_1 w$ is known on $\{x_1 = X_1\}$ and satisfies:

$$\partial_1 w(X_1, x') = 0, \quad \text{for } x' \in \mathbb{R}^{d-1}. \tag{1.99}$$

Remark 1.42. Setting $\tilde{w}(x) = w(x)e^{-sx_1}$, one easily checks that w solves (1.98) with $s > 0$ and $G \in L^2(\Omega_x)$ if and only if $\tilde{w}$ solves

$$\begin{cases} \Delta \tilde{w} - x_1 \sum_{j,k=2}^{d} \lambda_{j,k} \partial_j \partial_k \tilde{w} = \tilde{G}, & \text{in } \Omega_x, \\ \tilde{w}(X_0, x') = \tilde{w}(X_1, x') = 0, & \text{for } x' \in \mathbb{R}^{d-1}, \end{cases} \tag{1.100}$$

with $\tilde{G}(x) = G(x)e^{-sx_1}$. Accordingly, equation (1.98) should be seen as the conjugated operator of (1.100).

We then claim the following result:

Lemma 1.43. *Let* $(\lambda_{j,k})$, X_1 *satisfying assumption* (1.95) *with* c *and* (1.96) *with* c_*, *and consider the geometry given by* (1.97).

Then there exist positive constants C *and* $s_0 \geqslant 1$, *depending only on* c *and* c_*, *such that for all* $s \geqslant s_0$ *and* w *satisfying* (1.98) *and* (1.99),

$$s^3 \|w\|^2_{L^2(\Omega_x)} + s \|\nabla w\|^2_{L^2(\Omega_x)} + \frac{1}{s} \|D^2 w\|^2_{L^2(\Omega_x)} \leqslant C \|G\|^2_{L^2(\Omega_x)}. \quad (1.101)$$

Proof. Taking the Fourier transform in the transverse variable, we obtain

$$L_- L_+ \widehat{w} = \widehat{H},$$

where

$$L_- = \partial_1 - s - \sqrt{|\xi'|^2 - x_1 \sum_{j,k=2}^{d} \lambda_{j,k} \xi_j \xi_k}, \qquad x_1 \in [X_0, X_1], \ \xi' \in \mathbb{R}^{d-1},$$

$$L_+ = \partial_1 - s + \sqrt{|\xi'|^2 - x_1 \sum_{j,k=2}^{d} \lambda_{j,k} \xi_j \xi_k}, \qquad x_1 \in [X_0, X_1], \ \xi' \in \mathbb{R}^{d-1},$$

$$\widehat{H} = \widehat{G} - \partial_1 \left(\sqrt{|\xi'|^2 - x_1 \sum_{j,k=2}^{d} \lambda_{j,k} \xi_j \xi_k} \right) \widehat{w}, \qquad x_1 \in [X_0, X_1], \ \xi' \in \mathbb{R}^{d-1}.$$

We first prove that

$$\widehat{z} = L_+ \widehat{w}, \quad (1.102)$$

satisfies, for some constant C independent of s and $\xi' \in \mathbb{R}^{d-1}$,

$$(s + |\xi'|)^2 \|\widehat{z}(\cdot, \xi')\|^2_{L^2(X_0, X_1)} + \|\partial_1 \widehat{z}(\cdot, \xi')\|^2_{L^2(X_0, X_1)} \leqslant C \left\| \widehat{H}(\cdot, \xi') \right\|^2_{L^2(X_0, X_1)}. \quad (1.103)$$

Indeed, for $\xi' \in \mathbb{R}^{d-1}$,

$$L_- \widehat{z} = \widehat{H} \text{ in } (X_0, X_1), \qquad \widehat{z}(X_0, \xi') = 0. \quad (1.104)$$

Hence, taking then the L^2 norm of both sides of this identity, we get

$$\left\| \widehat{H}(\cdot, \xi') \right\|^2_{L^2(X_0, X_1)}$$

$$= \|\partial_1 \widehat{z}(\cdot, \xi')\|^2_{L^2(X_0, X_1)} + \left\| \left(s + \sqrt{|\xi'|^2 - x_1 \sum_{j,k=2}^{d} \lambda_{j,k} \xi_j \xi_k} \right) \widehat{z}(\cdot, \xi') \right\|^2_{L^2(X_0, X_1)}$$

$$+ \int_{X_0}^{X_1} \partial_1 \left(\sqrt{|\xi'|^2 - x_1 \sum_{j,k=2}^{d} \lambda_{j,k}\xi_j\xi_k} \right) |\widehat{z}(x_1,\xi')|^2 \, dx_1$$

$$= \|\partial_1\widehat{z}(\cdot,\xi')\|_{L^2(X_0,X_1)}^2 + \left\| \left(s + \sqrt{|\xi'|^2 - x_1 \sum_{j,k=2}^{d} \lambda_{j,k}\xi_j\xi_k} \right) \widehat{z}(\cdot,\xi') \right\|_{L^2(X_0,X_1)}^2$$

$$- \int_{X_0}^{X_1} \frac{\sum_{j,k=2}^{d} \lambda_{j,k}\xi_j\xi_k}{\sqrt{|\xi'|^2 - x_1 \sum_{j,k=2}^{d} \lambda_{j,k}\xi_j\xi_k}} |\widehat{z}(x_1,\xi')|^2 \, dx_1.$$

Using then the assumptions (1.95) and (1.96), we obtain, for some $c > 0$ and $C > 0$ and independent of s and ξ',

$$\left\| \widehat{H}(\cdot,\xi') \right\|_{L^2(X_0,X_1)}^2 \geq \|\partial_1\widehat{z}(\cdot,\xi')\|_{L^2(X_0,X_1)}^2 + c(s+|\xi'|)^2 \|\widehat{z}(\cdot,\xi')\|_{L^2(X_0,X_1)}^2$$

$$- C|\xi'| \|\widehat{z}(\cdot,\xi')\|_{L^2(X_0,X_1)}^2.$$

We then easily deduce the estimate (1.103) from this last estimate by taking s large enough.

The estimate of $\widehat{w}$ is more involved. Taking the $L^2(X_0,X_1)$ norm of both sides of (1.102) at ξ' fixed, we obtain

$$\|\widehat{z}(\cdot,\xi')\|_{L^2(X_0,X_1)}^2$$

$$= \|\partial_1\widehat{w}(\cdot,\xi')\|_{L^2(X_0,X_1)}^2 + \left\| \left(s - \sqrt{|\xi'|^2 - x_1 \sum_{j,k=2}^{d} \lambda_{j,k}\xi_j\xi_k} \right) \widehat{w}(\cdot,\xi') \right\|_{L^2(X_0,X_1)}^2$$

$$- \int_{X_0}^{X_1} \partial_1 \left(\sqrt{|\xi'|^2 - x_1 \sum_{j,k=2}^{d} \lambda_{j,k}\xi_j\xi_k} \right) |\widehat{w}(x_1,\xi')|^2 \, dx_1$$

$$= \|\partial_1\widehat{w}(\cdot,\xi')\|_{L^2(X_0,X_1)}^2 + \left\| \left(s - \sqrt{|\xi'|^2 - x_1 \sum_{j,k=2}^{d} \lambda_{j,k}\xi_j\xi_k} \right) \widehat{w}(\cdot,\xi') \right\|_{L^2(X_0,X_1)}^2$$

$$+ \int_{X_0}^{X_1} \frac{\sum_{j,k=2}^{d} \lambda_{j,k}\xi_j\xi_k}{\sqrt{|\xi'|^2 - x_1 \sum_{j,k=2}^{d} \lambda_{j,k}\xi_j\xi_k}} |\widehat{w}(x_1,\xi')|^2 \, dx_1.$$

In particular, in view of (1.96), the last identity provides, for some $C > 0$ depending only on c and c_*,

$$\|\widehat{z}(\cdot,\xi')\|^2_{L^2(X_0,X_1)}$$

$$\geqslant \|\partial_1\widehat{w}(\cdot,\xi')\|^2_{L^2(X_0,X_1)} + \left\|\left(s - \sqrt{|\xi'|^2 - x_1 \sum_{j,k=2}^{d} \lambda_{j,k}\xi_j\xi_k}\right)\widehat{w}(\cdot,\xi')\right\|^2_{L^2(X_0,X_1)}$$

$$+ C\int_{X_0}^{X_1} |\xi'| |w(x_1,\xi')|^2 \, dx_1.$$

It follows that we always have, for some C strictly positive independent of s and ξ',

$$\|\widehat{z}(\cdot,\xi')\|^2_{L^2(X_0,X_1)} \geqslant \|\partial_1\widehat{w}(\cdot,\xi')\|^2_{L^2(X_0,X_1)} + C(s+|\xi'|)\|\widehat{w}(\cdot,\xi')\|^2_{L^2(X_0,X_1)}.$$

We then easily conclude that

$$(s+|\xi'|)^2 \|\partial_1\widehat{w}(\cdot,\xi')\|^2_{L^2(X_0,X_1)} + (s+|\xi'|)^3 \|\widehat{w}(\cdot,\xi')\|^2_{L^2(X_0,X_1)}$$

$$\leqslant C\left\|\widehat{H}\right\|^2_{L^2(X_0,X_1)} \leqslant C\left\|\widehat{G}\right\|^2_{L^2(X_0,X_1)} + C|\xi'|^2 \|\widehat{w}(\cdot,\xi')\|^2_{L^2(X_0,X_1)}.$$

In particular, for s large enough, we get

$$(s+|\xi'|)^2 \|\partial_1\widehat{w}(\cdot,\xi')\|^2_{L^2(X_0,X_1)} + (s+|\xi'|)^3 \|\widehat{w}(\cdot,\xi')\|^2_{L^2(X_0,X_1)}$$

$$\leqslant C\left\|\widehat{G}\right\|^2_{L^2(X_0,X_1)}.$$

We can then integrate in $\xi' \in \mathbb{R}^{d-1}$, use Parseval's identity and obtain

$$s^3 \|w\|^2_{L^2(\Omega_x)} + s^2 \|\partial_1 w\|^2_{L^2(\Omega_x)} + s \|\nabla' w\|^2_{L^2(\Omega_x)} + \|\partial_1 \nabla' w\|^2_{L^2(\Omega_x)}$$

$$\leqslant C \|G\|^2_{L^2(\Omega_x)}.$$

From the equation satisfied by w, we easily check that

$$\left\|\Delta w - x_1 \sum_{j,k=2}^{d} \lambda_{j,k}\partial_j\partial_k w\right\|_{L^2(\Omega_x)} \leqslant C s^{1/2} \|G\|_{L^2(\Omega_x)}.$$

Besides, the operator $\Delta - x_1 \sum_{j,k=2}^{d} \lambda_{j,k}\partial_j\partial_k$ is elliptic due to the assumption (1.95), so the standard elliptic regularity results give

$$\frac{1}{s^{1/2}} \|w\|_{H^2(\Omega_x)} \leqslant C \|G\|_{L^2(\Omega_x)}.$$

This concludes the proof of Lemma 1.43.   □

Remark 1.44. In fact, the estimates on w in the direction of $\vec{e}_1$ are better than in the other directions. We have indeed proved that there exists $C > 0$ such that for all $s \geqslant s_0$,

$$s^2 \|\partial_1 w\|^2_{L^2(\Omega_x)} + \|\partial_1 \nabla' w\|^2_{L^2(\Omega_x)} \leqslant C \|G\|^2_{L^2(\Omega_x)}.$$

As we also have the estimate

$$\|\partial_1 \widehat{z}\|^2_{L^2(\Omega_x)} \leqslant C \left\|\widehat{G}\right\|^2_{L^2(\Omega_x)},$$

recalling the definition of $\widehat{z}$, we easily check that we have in fact

$$s^2 \|\partial_1 w\|^2_{L^2(\Omega_x)} + \|\partial_1 \nabla w\|^2_{L^2(\Omega_x)} \leqslant C \|G\|^2_{L^2(\Omega_x)}. \tag{1.105}$$

1.4. Global Carleman estimates

1.4.1. *Elliptic Carleman estimate under Hörmander's strict pseudo-convexity condition: the distributed case*

The goal of this section is to prove a Carleman estimate for solutions of Laplace equation with homogeneous Dirichlet boundary conditions. More precisely, we consider a smooth bounded domain Ω of $\mathbb{R}^d$, and u such that

$$\begin{cases} \Delta u = f, & \text{in } \Omega, \\ u = 0, & \text{on } \partial\Omega. \end{cases} \tag{1.106}$$

Then, we have the following result:

Theorem 1.45. *Let Ω be a smooth bounded domain. Let ω be an open subset of Ω with $\overline{\omega} \subset \Omega$. Let $\varphi \in \mathscr{C}^4(\overline{\Omega})$ be such that there exist $\alpha, \beta > 0$ such that*

$$\inf_{x \in \overline{\Omega}\setminus\omega} |\nabla\varphi(x)| > \alpha \tag{1.107}$$

and

$$\forall x \in \overline{\Omega} \setminus \omega, \ \forall \xi \in \mathbb{R}^d, |\nabla\varphi(x)| = |\xi| \ and \ \nabla\varphi(x) \cdot \xi = 0$$
$$\Rightarrow D^2\varphi_x(\nabla\varphi(x), \nabla\varphi(x)) + D^2\varphi_x(\xi, \xi) \geqslant \beta|\nabla\varphi(x)|^2, \tag{1.108}$$

and

$$\forall x \in \partial\Omega, \quad \varphi(x) = 0, \quad and \quad \partial_n\varphi(x) < 0. \tag{1.109}$$

Then for all $\varepsilon > 0$, there exist $C, s_0 > 0$ such that for all $s \geqslant s_0$, for all $u \in L^2(\Omega)$ solution of (1.106) with source term $f \in L^2(\Omega)$,

$$s^3\|e^{s\varphi}u\|^2_{L^2(\Omega)} + s\|e^{s\varphi}\nabla u\|^2_{L^2(\Omega)} \leqslant C\big(\|e^{s\varphi}f\|^2_{L^2(\Omega)} + s^3\|e^{s\varphi}u\|^2_{L^2(\omega_\varepsilon)}\big), \tag{1.110}$$

where $\omega_\varepsilon = \{x \in \Omega, d(x,\omega) \leqslant \varepsilon\}$.

Remark 1.46. Assumption (1.108) is often referred as Hörmander's strict pseudo-convexity condition.

Proof. Here, we provide a proof of Theorem 1.45 using Fourier techniques as we did earlier. As before, we start by setting

$$w = e^{s\varphi}u, \quad F = e^{s\varphi}f \qquad (1.111)$$

which verifies

$$\begin{cases} \Delta w - 2s\nabla\varphi \cdot \nabla w + s^2|\nabla\varphi|^2 w - s\Delta\varphi w = F, & \text{in } \Omega, \\ w = 0, & \text{on } \partial\Omega. \end{cases} \qquad (1.112)$$

In order to prove Theorem 1.45, we will prove it locally. Namely, for $x_0 \in \overline{\Omega} \setminus \omega$, we introduce $\eta_{x_0}(x)$ a cut-off function and

$$w_{x_0}(x) = \eta_{x_0}(x)w(x), \qquad (1.113)$$

which solves

$$\begin{cases} \Delta w_{x_0} - 2s\nabla\varphi \cdot \nabla w_{x_0} + s^2|\nabla\varphi|^2 w_{x_0} = F_{x_0}, & \text{in } \Omega, \\ w_{x_0} = 0, & \text{on } \partial\Omega, \end{cases} \qquad (1.114)$$

where F_{x_0} is defined by

$$F_{x_0} = \eta_{x_0}F + s\Delta\varphi w_{x_0} + [\Delta, \eta_{x_0}]w - 2s[\nabla\varphi \cdot \nabla, \eta_{x_0}]w, \qquad (1.115)$$

where for operators A and B, $[A, B]$ denotes the operator $AB - BA$, i.e. the commutator of A and B: in the above cases, $[\Delta, \eta_{x_0}]w = 2\nabla\eta_{x_0} \cdot \nabla w + \Delta\eta_{x_0}w$ and $[\nabla\varphi \cdot \nabla, \eta_{x_0}]w = \nabla\varphi \cdot \nabla\eta_{x_0}w$.

In particular, all our proof is based on the following Lemma, whose proof will be postponed to Section 1.4.2:

Lemma 1.47. *There exist constants $C > 0$ and $s_0 > 0$ such that for all $x_0 \in \overline{\Omega} \setminus \omega$, and for all $F_{x_0} \in L^2(\Omega)$ and w_{x_0} satisfying (1.114), we have*

$$s^3\|w_{x_0}\|^2_{L^2(\Omega)} + s\|\nabla w_{x_0}\|^2_{L^2(\Omega)} + \frac{1}{s}\|w_{x_0}\|^2_{H^2(\Omega)} \leqslant C\|F_{x_0}\|^2_{L^2(\Omega)} \qquad (1.116)$$

for all w_{x_0} supported in $B(x_0, s^{-1/3}) \cap \overline{\Omega}$.

We will thus choose

$$\eta_{x_0}(x) = \eta^0(s^{1/3}|x - x_0|)$$

with $\text{Supp}\,\eta^0 \subset [0, 1)$ and $\eta^0(\rho) = 1$ for $\rho \in [0, 1/2)$. Therefore, we obtain, for C and s independent of x_0,

$$s^3\|w_{x_0}\|^2_{L^2(\Omega)} + s\|\nabla w_{x_0}\|^2_{L^2(\Omega)} \leqslant C\|\eta_{x_0}F\|^2_{L^2(\Omega)} + C\||\Delta\eta_{x_0}|w\|^2_{L^2(\Omega)}$$
$$+ C\||\nabla\eta_{x_0}|\nabla w\|^2_{L^2(\Omega)} + Cs^2\||\nabla\eta_{x_0}|w\|^2_{L^2(\Omega)},$$

and then,

$$s^3\|\eta_{x_0}w\|^2_{L^2(\Omega)}+s\|\eta_{x_0}\nabla w\|^2_{L^2(\Omega)}\leqslant C\|\eta_{x_0}F\|^2_{L^2(\Omega)}+C\||\Delta\eta_{x_0}|w\|^2_{L^2(\Omega)}$$
$$+C\||\nabla\eta_{x_0}|\nabla w\|^2_{L^2(\Omega)}+Cs^2\||\nabla\eta_{x_0}|w\|^2_{L^2(\Omega)}.$$

Now, integrating in $x_0 \in \overline{\Omega} \setminus \omega$ and using Fubini's identity, we get

$$s^3\int_\Omega \rho_0(x)|w(x)|^2\,dx + s\int_\Omega \rho_0(x)|\nabla w(x)|^2\,dx \leqslant C\int_\Omega \rho_0(x)|F(x)|^2\,dx$$
$$+C\int_\Omega (\rho_{r,1}(x)+s^2\rho_{r,2}(x))|w(x)|^2\,dx + C\int_\Omega \rho_{r,2}(x)|\nabla w(x)|^2\,dx,$$

where the weights ρ_0, $\rho_{r,i}$ are defined as follows:

$$\rho_0(x) = \int_{\Omega\setminus\omega} |\eta_{x_0}(x)|^2\,dx_0,$$

$$\rho_{r,1}(x) = \int_{\Omega\setminus\omega} |\Delta\eta_{x_0}(x)|^2\,dx_0,$$

$$\rho_{r,2}(x) = \int_{\Omega\setminus\omega} |\nabla\eta_{x_0}(x)|^2\,dx_0.$$

Now let $\varepsilon > 0$. We check that:

$$\forall x \in \Omega, \quad |\rho_0(x)| \leqslant Cs^{-d/3};$$
$$\forall x \in \Omega, \quad |\rho_{r,1}(x)| \leqslant Cs^{4/3-d/3};$$
$$\forall x \in \Omega, \quad |\rho_{r,2}(x)| \leqslant Cs^{2/3-d/3};$$
$$\text{if } s \geqslant \frac{1}{\min\{\varepsilon^3, d(\omega,\Omega)^3\}}, \, \forall x \in \Omega\setminus\omega_\varepsilon, \, \rho_0(x) \geqslant \frac{s^{-d/3}}{3}\|\eta^0\|^2_{L^2}.$$

Thus, for s large enough,

$$s^3\int_{\Omega\setminus\omega_\varepsilon}|w(x)|^2\,dx + s\int_{\Omega\setminus\omega_\varepsilon}|\nabla w(x)|^2\,dx \leqslant C\int_\Omega |F(x)|^2\,dx$$
$$+Cs^{8/3}\int_{\omega_\varepsilon}|w(x)|^2\,dx + Cs^{2/3}\int_{\omega_\varepsilon}|\nabla w(x)|^2\,dx. \quad (1.117)$$

Adding then

$$s^3\int_{\omega_\varepsilon}|w(x)|^2\,dx + s\int_{\omega_\varepsilon}|\nabla w(x)|^2\,dx$$

to both sides of (1.117), we get:

$$s^3\int_\Omega |w(x)|^2\,dx + s\int_\Omega |\nabla w(x)|^2\,dx$$
$$\leqslant C\int_\Omega |F(x)|^2\,dx + Cs^3\int_{\omega_\varepsilon}|w(x)|^2\,dx + Cs\int_{\omega_\varepsilon}|\nabla w(x)|^2\,dx.$$

To remove the term in ∇w in the right hand side, we take a smooth cut-off function η_ω taking value 1 in ω_ε and vanishing in $\overline{\Omega} \setminus \omega_{2\varepsilon}$ and write[c]

$$s \int_{\omega_\varepsilon} |\nabla w(x)|^2 \, dx \leqslant s \int_{\omega_{2\varepsilon}} \eta_\omega(x) |\nabla w(x)|^2 \, dx$$

$$\leqslant -s \int_{\omega_{2\varepsilon}} w \nabla \eta_\omega \cdot \nabla w - s \int_{\omega_{2\varepsilon}} \eta_\omega w \Delta w$$

$$\leqslant \frac{s}{2} \int_{\omega_{2\varepsilon}} \Delta \eta_\omega |w|^2$$

$$- s \int_{\omega_{2\varepsilon}} \eta_\omega w (F + 2s\nabla\varphi \cdot \nabla w - s^2 |\nabla\varphi|^2 w + s\Delta\varphi w)$$

$$\leqslant \frac{s}{2} \int_{\omega_{2\varepsilon}} \Delta \eta_\omega |w|^2 - s \int_{\omega_{2\varepsilon}} \eta_\omega w F + s^2 \int_{\omega_{2\varepsilon}} \operatorname{div}(\eta_\omega \nabla\varphi) |w|^2$$

$$+ s^3 \int_{\omega_{2\varepsilon}} \eta_\omega |\nabla\varphi|^2 |w|^2 - s^2 \int_{\omega_{2\varepsilon}} \eta_\omega \Delta\varphi |w|^2$$

$$\leqslant C s^3 \int_{\omega_{2\varepsilon}} |w|^2 + C \int_\Omega |F|^2.$$

This concludes the proof of Theorem 1.45 up to the proof of Lemma 1.47 since $\varepsilon > 0$ is arbitrary. $\qquad\square$

Remark 1.48. The above argument relies on the local character of Carleman estimates: a global Carleman estimate, such as (1.110), can be deduced by gluing local Carleman estimates such as (1.116).

1.4.2. *Proof of Lemma* 1.47

In order to prove the local Carleman estimate stated in Lemma 1.47, we first rewrite the problem (1.114) in suitable coordinates, which will turn out to be adapted to prove the local Carleman estimate (1.116).

1.4.2.1. *A suitable change of variable*

We fix $x_0 \in \overline{\Omega} \setminus \omega$ and introduce $L_1 \in \mathbb{R}^d$ and $A_1 \in \mathbb{R}^{d \times d}$ as follows:

$$L_1 = \nabla\varphi(x_0) \in \mathbb{R}^d, \quad A_1 = D^2\varphi(x_0) \in \mathbb{R}^{d \times d}. \qquad (1.118)$$

We then introduce $d - 1$ vectors $(L_i)_{i \in \{2,\cdots,d\}}$ such that the family $(L_1, (L_i)_{i \in \{2,\cdots,d\}})$ form an orthogonal basis of $\mathbb{R}^d$, and such that for all $i \in \{2, \cdots, d\}$,

$$|L_i| = |L_1|.$$

---

[c]This estimate is also known as Cacciopoli's inequality.

For $i \in \{2, \cdots, d\}$, we then introduce self-adjoint matrices A_i such that

(i) $A_i L_1 = -A_1 L_i$;
(ii) for $j \in \{2, \cdots, d\}$, $A_i L_i \cdot L_j = A_1 L_1 \cdot L_j$;
(iii) for $(j, k) \in \{2, \cdots, d\}^2$ with i, j, k two by two distinct, $A_i L_j \cdot L_k = 0$;
(iv) for $j \in \{2, \cdots, d\}$ with $j \neq i$, $A_i L_j \cdot L_j = -A_1 L_1 \cdot L_i$.

It is then easy to check that each matrix A_i is fully determined. Indeed, writing the matrix A_i in the base of $(L_1, (L_i)_{i \in \{2, \cdots, d\}})$, the first condition imposes the first column of A_i, and thus its first line as A_i is symmetric. The second condition imposes the lines 2 to d of the i-th column of A_i, and thus also its i line. The two last conditions then impose the lines 2 to d of the j-th line of A_i for $j \notin \{1, i\}$. It remains to check that A_i is symmetric. This is a consequence of the third condition.

We shall then introduce the following coordinates for x in a neighborhood of x_0:

$$y_1(x) = \varphi(x) - \varphi(x_0), \tag{1.119}$$

$$\text{for } j \in \{2, \cdots, d\}, \; y_j(x) = L_j \cdot (x - x_0) + \frac{1}{2} A_j(x - x_0) \cdot (x - x_0). \tag{1.120}$$

By construction, there exists a neighborhood, whose size depends on the $\mathscr{C}^2$ norm of φ only, such that $x \mapsto y(x)$ is a local diffeomorphism between a neighborhood of x_0 (in Ω) and a neighborhood of 0 (in $y(\Omega)$). In particular, for s large enough, we can ensure that the ball of center x_0 and radius $s^{-1/3}$ (intersected with Ω if needed) is included in a set on which $x \mapsto y(x)$ is a diffeomorphism, and its image is included in a ball $B(0, C s^{-1/3})$. Therefore, for w_{x_0} solving (1.114), we set

$$\tilde{w}(y) = w_{x_0}(x) \text{ for } y = y(x), \quad \tilde{F}(y) = F_{x_0}(x) \text{ for } y = y(x). \tag{1.121}$$

Explicit computations then give that $\tilde{w}$ satisfies

$$\sum_{j,k=1}^{d} b_{j,k}(x) \partial_{y_j} \partial_{y_k} \tilde{w}(y(x)) + \nabla_y \tilde{w}(y(x)) \cdot \Delta_x y(x)$$

$$- 2s \sum_{j=1}^{d} c_j \partial_{y_j} \tilde{w}(y(x)) + s^2 |\nabla \varphi(x)|^2 \tilde{w}(y(x)) = \tilde{F}(y(x)), \quad \text{in } \Omega, \tag{1.122}$$

where

$$b_{j,k}(x) = \nabla y_j(x) \cdot \nabla y_k(x), \quad \text{and} \quad c_j(x) = \nabla \varphi(x) \cdot \nabla y_j(x).$$

We then remark that $c_j(x) = b_{j,1}(x)$ and that $b_{j,k}(x) = b_{k,j}(x)$ for all x. We now briefly analyze the coefficients $b_{j,k}$. By construction of the coordinates y_j, we easily check that

$$b_{j,k}(x_0) = |L_1|^2 \delta_{j,k},$$

$$\partial_\ell b_{j,k}(x_0) = \sum_{i=1}^d (\partial_\ell \partial_i y_j(x_0) \partial_i y_k(x_0) + \partial_i y_j(x_0) \partial_\ell \partial_i y_k(x_0))$$

$$= \sum_{i=1}^d ((A_j e_i \cdot e_\ell)(L_k \cdot e_i) + (L_j \cdot e_i)(A_k e_i \cdot e_\ell))$$

$$= A_j e_\ell \cdot L_k + A_k e_\ell \cdot L_j = (A_j L_k + A_k L_j, e_\ell),$$

so that we have in particular that

$$L_\ell \cdot \nabla b_{j,k}(x_0) = (A_j L_k + A_k L_j) \cdot L_\ell,$$

and, corresponding to $j = k = 1$,

$$|\nabla \varphi(x_0)|^2 = |L_1|^2, \quad \text{and} \quad \partial_\ell(|\nabla \varphi(x_0)|^2) = 2 A_1 L_1 \cdot e_\ell.$$

We can thus analyze $b_{j,k}/|\nabla \varphi|^2$ close to $x = x_0$:

$$\frac{b_{j,k}}{|\nabla \varphi(x)|^2}(x_0) = \delta_{j,k},$$

$$L_\ell \cdot \nabla \left(\frac{b_{j,k}}{|\nabla \varphi(x)|^2} \right)(x_0) = \frac{1}{|L_1|^2} \left((A_j L_k + A_k L_j) \cdot L_\ell - 2\delta_{j,k} A_1 L_1 \cdot L_\ell \right).$$

$$(1.123)$$

In particular, for all $j, k \in \{1, \cdots, d\}^2$,

$$L_\ell \cdot \nabla \left(\frac{b_{j,k}}{|\nabla \varphi(x)|^2} \right)(x_0) = 0 \quad \text{when } \ell \in \{2, \cdots, d\}.$$

To check this property, we shall consider many different cases and the properties (i)–(iv): for $j, k \geqslant 2$ with j, k, ℓ two by two distinct, this relies on (iii); for $j, k \geqslant 2$ with $k = j \neq \ell$, this relies on (ii); for $j, k \geqslant 2$ with $j = \ell \neq k$, this relies on (iv) and (ii); for $j = k = 1$, this is obvious; for $j = 1$ and $k \geqslant 2$, this relies on (i).

In particular, doing a Taylor expansion of $\frac{b_{j,k}}{|\nabla \varphi(x)|^2}$ close to $x = x_0$, we get for all $j, k \in \{1, \cdots, d\}$,

$$\frac{b_{j,k}}{|\nabla \varphi(x)|^2} = \delta_{j,k} + \left\langle \nabla \left(\frac{b_{j,k}}{|\nabla \varphi(x)|^2} \right)(x_0), (x - x_0) \right\rangle + \mathcal{O}(|x - x_0|^2)$$

$$= \delta_{j,k} + \frac{1}{|L_1|^2} \left\langle \nabla \left(\frac{b_{j,k}}{|\nabla \varphi(x)|^2} \right)(x_0), L_1 \right\rangle \langle L_1, (x - x_0) \rangle + \mathcal{O}(|x - x_0|^2)$$

$$= \delta_{j,k} + \frac{1}{|L_1|^2} \left\langle \nabla \left(\frac{b_{j,k}}{|\nabla \varphi(x)|^2} \right)(x_0), L_1 \right\rangle y_1(x) + \mathcal{O}(|y|^2),$$

where we used the Taylor expansion of y_1 close to x_0, $y_1(x) = \langle L_1, (x - x_0)\rangle + \mathcal{O}(|x - x_0|^2)$ and the fact that y is a local diffeomorphism from a neighborhood of x_0 to a neighborhood of 0.

We thus compute the value of

$$\left\langle \nabla\left(\frac{b_{j,k}}{|\nabla\varphi(x)|^2}\right)(x_0), L_1\right\rangle$$

corresponding to $\ell = 1$ in (1.123):

- If $j = k \geqslant 2$, by (i), we get

$$L_1 \cdot \nabla\left(\frac{b_{j,j}}{|\nabla\varphi(x)|^2}\right)(x_0) = -\frac{2}{|L_1|^2}(A_1 L_j \cdot L_j + A_1 L_1 \cdot L_1).$$

- If $j = k = 1$,

$$L_1 \cdot \nabla\left(\frac{b_{1,1}}{|\nabla\varphi(x)|^2}\right)(x_0) = 0.$$

- If $j \neq k$ and $j, k \geqslant 2$, by (i)

$$L_1 \cdot \nabla\left(\frac{b_{j,k}}{|\nabla\varphi(x)|^2}\right)(x_0) = -\frac{2}{|L_1|^2}A_1 L_j \cdot L_k.$$

- If j or k equals 1, and $j \neq k$, again using (i),

$$L_1 \cdot \nabla\left(\frac{b_{j,k}}{|\nabla\varphi(x)|^2}\right)(x_0) = 0.$$

Consequently, as a consequence of Taylor expansion of $b_{j,k}/|\nabla\varphi(x)|^2$ close to $x = x_0$, for w supported in $B(x_0, s^{-1/3})$

$$\left|\sum_{j,k=1}^{d} \frac{b_{j,k}(x)}{|\nabla\varphi(x)|^2}\partial_{y_j}\partial_{y_k}\tilde{w}(y(x))\right.$$

$$-\left(\partial_{11}\tilde{w}(y(x)) + \sum_{j=2}^{d}(1 - \frac{2}{|L_1|^4}(A_1 L_1 \cdot L_1)y_1(x))\partial_{jj}\tilde{w}(y(x))\right.$$

$$\left.\left.-\frac{2}{|L_1|^4}\sum_{j,k=2}^{d} y_1(x)(A_1 L_j \cdot L_k)\partial_j\partial_k\tilde{w}(y(x))\right)\right| \leqslant Cs^{-2/3}\|w\|_{H^2}. \quad (1.124)$$

We thus obtain that in $B(0, Cs^{-1/3})$ (intersected with $y(\Omega)$ if needed) $\tilde{w}$ satisfies

$$\partial_{11}\tilde{w} + \sum_{j=2}^{d}\left(1 - \frac{2}{|L_1|^4}(A_1 L_1 \cdot L_1)y_1\right)\partial_{jj}\tilde{w}$$

$$-y_1\sum_{j,k=2}^{d}\frac{2}{|L_1|^4}(A_1 L_j \cdot L_k)\partial_j\partial_k\tilde{w} - 2s\partial_1\tilde{w} + s^2\tilde{w} = \tilde{G}, \quad (1.125)$$

where

$$\left\| \tilde{G}(y) - \frac{1}{|\nabla\varphi(x(y))|^2}(\tilde{F}(y) - \nabla_y\tilde{w}(y) \cdot \Delta_x y(x(y))) \right\|_{L^2} \leqslant Cs^{-2/3} \|\tilde{w}\|_{H^2}.$$
(1.126)

If the ball $B(0, Cs^{-1/3})$ intersects the boundary of $y(\Omega)$, then we simply recall that the weight function φ has been chosen such that $\varphi = 0$ on the boundary. In particular, the boundary is locally parameterized by $y_1(x) = Y_0$, for a suitable $Y_0 \leqslant 0$, and Ω can be locally defined by $y_1 > Y_0$. Thus, in this case, the equation (1.125) of $\tilde{w}$ should be completed with

$$\tilde{w}(Y_0, y') = 0, \quad \text{for } y' \in \mathbb{R}^{d-1}.$$
(1.127)

Remark 1.49. To better understand this change of variable, it is interesting to focus on the case of a weight function depending only on the x_1 variable. Then, after some computations left as an exercise, the above strategy amounts to transform equation (1.76) into equation (1.98), and the strict convexity of the weight function φ in x_1 is then equivalent to (1.96).

1.4.2.2. *Proof of Lemma 1.47*

For $(j, k) \in \{2, \cdots, d\}^2$, we set

$$\lambda_{j,j} = \frac{2}{|L_1|^4}(A_1 L_j \cdot L_j + A_1 L_1 \cdot L_1) \text{ and } \lambda_{j,k} = \frac{2}{|L_1|^4}(A_1 L_j \cdot L_k).$$

In particular, for all $\xi' \in \mathbb{R}^{d-1}$,

$$\begin{aligned}
\sum_{j,k=2}^{d} \lambda_{j,k}\xi'_j\xi'_k &= \frac{2}{|L_1|^4}\left(A_1 L_1 \cdot L_1|\xi'|^2 + A_1\left(\sum_{j=2}^{d}\xi_j L_j\right) \cdot \left(\sum_{j=2}^{d}\xi_j L_j\right)\right) \\
&= \frac{2|\xi'|^2}{|\nabla\varphi(x_0)|^4}\left(D^2\varphi_{x_0}(\nabla\varphi(x_0), \nabla\varphi(x_0)) + D^2\varphi_{x_0}(\zeta', \zeta')\right),
\end{aligned}$$

where $\zeta' = \frac{1}{|\xi'|}\sum_{j=2}^{d}\xi_j L_j$ is of modulus $|\nabla\varphi(x_0)|$ and satisfies $\zeta' \cdot \nabla\varphi(x_0) = 0$. Thus, condition (1.108) at a point x_0 is in fact equivalent to the condition (1.96). Condition (1.95) necessarily holds if we restrict ourselves to a neighborhood of $y_1 = 0$ small enough depending only on the coefficients of $D^2\varphi(x_0)$. Since we will take $\tilde{w}_{x_0}$ localized in a neighborhood of $y = 0$ when taking s large enough, condition (1.95) automatically holds for s large enough.

We can thus apply Lemma 1.43 to each $\tilde{w}_{x_0}$: there exist $C > 0$ and $s_0 > 0$ independent of x_0 such that for all $s \geqslant s_0$ and all w_{x_0} supported in $B(x_0, s^{-1/3}) \cap \Omega$ (recall that then $\tilde{w}_{x_0}$ is supported in $y(B(x_0, s^{-1/3}) \cap \Omega))$,

$$s^3 \|\tilde{w}_{x_0}\|^2_{L^2(\Omega_y)} + s\|\nabla \tilde{w}_{x_0}\|^2_{L^2(\Omega_y)} + \frac{1}{s}\|\tilde{w}_{x_0}\|^2_{H^2(\Omega_y)} \leqslant C\|\tilde{G}_{x_0}\|^2_{L^2(\Omega_y)}.$$

Using the estimate (1.126) and the fact that w_{x_0} is supported in $B(0, s^{-1/3})$, we get for s large enough,

$$s^3 \|\tilde{w}_{x_0}\|^2_{L^2(\Omega_y)} + s\|\nabla \tilde{w}_{x_0}\|^2_{L^2(\Omega_y)} + \frac{1}{s}\|\tilde{w}_{x_0}\|^2_{H^2(\Omega_y)}$$
$$\leqslant C\|\tilde{F}_{x_0}\|^2_{L^2(\Omega_y)} + C\|\nabla \tilde{w}_{x_0}\|^2_{L^2(\Omega_y)}.$$

Taking s large enough, the last term can be absorbed and we obtain

$$s^3 \|\tilde{w}_{x_0}\|^2_{L^2(\Omega_y)} + s\|\nabla \tilde{w}_{x_0}\|^2_{L^2(\Omega_y)} + \frac{1}{s}\|\tilde{w}_{x_0}\|^2_{H^2(\Omega_y)} \leqslant C\|\tilde{F}_{x_0}\|^2_{L^2(\Omega_y)}.$$

We then undo the change of variable, and conclude immediately the proof of Lemma 1.47.

Remark 1.50. We can add a non-homogeneous boundary condition in $H^{3/2}(\partial\Omega)$ and do the same thing.

We can also get a similar estimate with a source term F lying in $H^{-1}(\Omega)$ and a non-homogeneous boundary condition in $H^{1/2}(\partial\Omega)$, see [Imanuvilov and Puel (2003)].

1.4.3. *Elliptic Carleman estimate under Hörmander's strict pseudo-convexity condition: the boundary case*

We next derive a similar estimate when the observation is on the boundary.

Again, we consider a smooth bounded domain Ω of $\mathbb{R}^d$, and u a solution of (1.106).

Then, we have the following result:

Theorem 1.51. *Let* $\varphi \in \mathscr{C}^4(\overline{\Omega})$ *be such that there exist* $\alpha, \beta > 0$ *such that*

$$\inf_{x \in \overline{\Omega}} |\nabla\varphi(x)| \geqslant \alpha > 0 \tag{1.128}$$

and

$$\forall x \in \Omega, \forall \xi \in \mathbb{R}^d \text{ with } |\xi| = |\nabla\varphi(x)| \text{ and } \xi \cdot \nabla\varphi(x) = 0,$$
$$D^2\varphi_x(\nabla\varphi(x), \nabla\varphi(x)) + D^2\varphi_x(\xi, \xi) \geqslant \beta|\nabla\varphi(x)|^2. \tag{1.129}$$

Then there exist $C, s_0 > 0$ such that for all $s \geqslant s_0$, for all $u \in L^2(\Omega)$ solution of (1.106) with source term $f \in L^2(\Omega)$,

$$s^3 \|e^{s\varphi} u\|^2_{L^2(\Omega)} + s\|e^{s\varphi} \nabla u\|^2_{L^2(\Omega)} \leqslant C\big(\|e^{s\varphi} f\|_{L^2(\Omega)} + s\|e^{s\varphi} \partial_\nu u\|^2_{L^2(\Gamma_\varphi)}\big),$$

$$(1.130)$$

where

$$\Gamma_\varphi = \{x \in \partial\Omega, \ \partial_n \varphi(x) > 0\}.$$

Remark 1.52. Note that by standard elliptic regularity result, if $u \in L^2(\Omega)$ is solution of (1.106) with source term $f \in L^2(\Omega)$, then u belongs to $H^2(\Omega)$, and therefore expression (1.130) makes sense.

Remark 1.53. Note that the linear weight function $x \mapsto \vec{e} \cdot x$ for a unitary vector $\vec{e}$ does not satisfy the convexity condition (1.129). This explains why the powers of the parameter $|k|$ in Theorem 1.29 are not the same as the powers of the parameter s in Theorem 1.51.

Actually, it is easy to check that it satisfies the degenerate convexity condition (1.129) with $\beta = 0$.

When condition (1.108) is saturated, that is when

$$\forall x \in \Omega, \forall \xi \in \mathbb{R}^d \text{ with } |\xi| = |\nabla\varphi(x)| \text{ and } \xi \cdot \nabla\varphi(x) = 0,$$
$$D^2\varphi_x(\nabla\varphi(x), \nabla\varphi(x)) + D^2\varphi_x(\xi, \xi) = 0,$$

the weight function belongs to the so-called *Limiting Carleman Weights*, see e.g. [Dos Santos Ferreira *et al.* (2009); Kenig *et al.* (2007)], where it is used to solve the Calderón problem in dimension $d \geqslant 3$ with partial knowledge of the Dirichlet-to-Neumann map.

Remark 1.54. Actually, Ω does not necessarily need to be bounded. It is sufficient to assume φ and all its derivative to be bounded on $\overline{\Omega}$ to obtain the same result.

As usual, to prove Theorem 1.51, we work on the conjugated variable and source term

$$w = e^{s\varphi} u, \quad F = e^{s\varphi} f$$

which verifies

$$\begin{cases} \Delta w + s^2 |\nabla\varphi|^2 w - 2s\,\nabla\varphi \cdot \nabla w - s\Delta\varphi\,w = F, & \text{in } \Omega, \\ w = 0, & \text{on } \partial\Omega. \end{cases} \quad (1.131)$$

Then Theorem 1.51 is implied by the following Carleman estimate on w:

Theorem 1.55. *Let $\varphi \in \mathscr{C}^4(\overline{\Omega})$ verify (1.128) and (1.129). Then there exist $s_0, C > 0$ such that for all $s \geqslant s_0$ and all $w \in L^2(\Omega)$ satisfying (1.131) with source term $F \in L^2(\Omega)$, we have*

$$s^3\|w\|_{L^2(\Omega)}^2 + s^2\|\xi_2\|_{L^2(\Omega)}^2 + s\|\nabla w\|_{L^2(\Omega)}^2$$
$$\leqslant C\big(\|F\|_{L^2(\Omega)}^2 + s\|\partial_n w\|_{L^2(\Gamma_\varphi)}^2\big), \tag{1.132}$$

where

$$\xi_2 = (\nabla w \cdot \eta)\,\eta,$$

with, for all $x \in \Omega$,

$$\eta(x) = \frac{\nabla\varphi(x)}{|\nabla\varphi(x)|}.$$

In Theorem 1.55, the term $s^2\|\xi_2\|_{L^2(\Omega)}^2$ emphasizes that we obtain a better estimate on the component of the gradient *in the direction of $\nabla\varphi$* than on the total gradient.

We now focus on the proof of Theorem 1.55, which we will provide using a multiplier approach. From now on, we consider $\varphi \in \mathscr{C}^4(\overline{\Omega})$ verifying (1.128) and (1.129).

1.4.3.1. *Some basic estimates*

For any $w \in H_0^1(\Omega)$, we define

$$P_S w = \Delta w + s^2|\nabla\varphi|^2 w$$

and

$$P_A w = -2s\,\nabla\varphi \cdot \nabla w - s\Delta\varphi\,w.$$

The operators P_S and P_A are respectively symmetric and skew-adjoint, as for any $w, \tilde{w} \in H_0^1(\Omega)$,

$$\langle P_S w, \tilde{w}\rangle_{H^{-1}(\Omega), H_0^1(\Omega)} = \langle w, P_S \tilde{w}\rangle_{H_0^1(\Omega), H^{-1}(\Omega)}$$

and

$$\langle P_A w, \tilde{w}\rangle_{H^{-1}(\Omega), H_0^1(\Omega)} = -\langle w, P_A \tilde{w}\rangle_{H_0^1(\Omega), H^{-1}(\Omega)}.$$

We start with some basic estimates that shall be useful in the following.

Proposition 1.56. *There exists a constant $C > 0$ such that for any $w \in H^2(\Omega) \cap H_0^1(\Omega)$ and $s \geqslant 1$,*

$$s^2\|\xi_2\|_{L^2(\Omega)}^2 \leqslant C\big(\|P_A w\|_{L^2(\Omega)}^2 + s^2\|w\|_{L^2(\Omega)}^2\big), \tag{1.133}$$

$$\|\xi_1\|^2_{L^2(\Omega)} \leqslant C\left(s^2\|w\|^2_{L^2(\Omega)} + \frac{1}{s^2}\left(\|P_S w\|^2_{L^2(\Omega)} + \|P_A w\|^2_{L^2(\Omega)}\right)\right), \quad (1.134)$$

and

$$s\int_\Omega |D^2\varphi(\xi_1,\xi_2)|\,dx$$

$$\leqslant C\left(s^{5/2}\|w\|^2_{L^2(\Omega)} + \frac{1}{\sqrt{s}}\left(\|P_S w\|^2_{L^2(\Omega)} + \|P_A w\|^2_{L^2(\Omega)}\right)\right), \quad (1.135)$$

where

$$\nabla w = \xi_1 + \xi_2, \quad \text{with } \xi_2 = (\nabla w \cdot \eta)\eta, \ \eta(x) = \frac{\nabla\varphi(x)}{|\nabla\varphi(x)|}, \ \xi_1 \cdot \xi_2 = 0. \quad (1.136)$$

Proof. Estimate (1.133) is direct as

$$|\xi_2| = \frac{1}{|\nabla\varphi|}|\nabla\varphi \cdot \nabla w|$$

and

$$2s\,\nabla\varphi \cdot \nabla w = -P_A w - s\nabla\varphi\, w.$$

To prove estimate (1.134), we note that

$$\int_\Omega P_S w\, w\, dx = -\int_\Omega |\nabla w|^2\, dx + s^2 \int_\Omega |\nabla\varphi|^2 |w|^2\, dx.$$

As $|\nabla w|^2 = |\xi_1|^2 + |\xi_2|^2$, we obtain for some constant $C > 0$,

$$\int_\Omega |\xi_1|^2\, dx \leqslant C\left(s^2 \int_\Omega |w|^2 + \left|\int_\Omega P_S w\, w\, dx\right|\right).$$

Estimate (1.134) then follows from

$$\left|\int_\Omega P_S w\, w\, dx\right| \leqslant \frac{1}{2s^2}\int_\Omega |P_S w|^2\, dx + \frac{s^2}{2}\int_\Omega |w|^2\, dx.$$

Finally, estimate (1.135) simply comes from the inequality

$$s\int_\Omega |D^2\varphi(\xi_1,\xi_2)|\,dx \leqslant C\sqrt{s}\int_\Omega |\xi_1|^2\, dx + C\, s^{3/2}\int_\Omega |\xi_2|^2\, dx,$$

and estimates (1.134) and (1.133). $\qquad\square$

Lemma 1.57. *For any $w \in H^2(\Omega) \cap H^1_0(\Omega)$ and any $s \geqslant 1$,*

$$\int_\Omega P_S w\, P_A w\, dx = 2s^3 \int_\Omega D^2\varphi(\nabla\varphi,\nabla\varphi)|w|^2\, dx - \frac{s}{2}\int_\Omega \Delta^2\varphi\,|w|^2\, dx$$

$$+ 2s\int_\Omega D^2\varphi(\nabla w,\nabla w)\, dx - s\int_\Omega \partial_n\varphi\,|\partial_n w|^2 ds(x).$$

Proof. By definition, for any smooth function w,

$$\int_\Omega P_S w \, P_A w \, dx = \int_\Omega \left(\Delta w + s^2 |\nabla\varphi|^2 w\right)(-2\,s\,\nabla\varphi\nabla w - s\nabla\varphi w)\,dx,$$

so we have to compute each term appearing in the product.

First term: we have

$$\int_\Omega \Delta w(-2\,s\,\nabla\varphi\nabla w)\,dx = -2\,s\int_{\partial\Omega}(\partial_n w)(\nabla\varphi\cdot\nabla w)\,ds(x)$$

$$+2\,s\int_\Omega \nabla w\cdot\nabla(\nabla\varphi\cdot\nabla w)\,dx.$$

On the boundary of Ω, as $w = 0$, we have $\nabla w = (\nabla w \cdot n)n$, hence

$$-2\,s\int_{\partial\Omega}(\partial_n w)(\nabla\varphi\cdot\nabla w)\,ds(x) = -2\,s\int_{\partial\Omega}\partial_n\varphi|\partial_n w|^2\,ds(x).$$

On the other hand, an easy computation shows that

$$\nabla w\cdot\nabla(\nabla\varphi\cdot\nabla w) = D^2\varphi(\nabla w, \nabla w) + \frac{1}{2}\nabla\varphi\cdot\nabla(|\nabla w|^2),$$

which leads to

$$2\,s\int_\Omega \nabla w\cdot\nabla(\nabla\varphi\cdot\nabla w)\,dx$$

$$= 2\,s\int_\Omega D^2\varphi(\nabla w, \nabla w)\,dx + s\int_\Omega \nabla\varphi\cdot\nabla(|\nabla w|^2)\,dx$$

$$= 2\,s\int_\Omega D^2\varphi(\nabla w, \nabla w)\,dx + s\int_{\partial\Omega}\partial_n\varphi\,|\partial_n w|^2\,ds(x) - s\int_\Omega \Delta\varphi\,|w|^2\,dx,$$

where we have again used that on $\partial\Omega$, $\nabla w = (\nabla w\cdot n)n$. So we finally obtain

$$\int_\Omega \Delta w(-2\,s\,\nabla\varphi\nabla w)\,dx = 2\,s\int_\Omega D^2\varphi(\nabla w, \nabla w)\,dx - s\int_\Omega \Delta\varphi\,|w|^2\,dx$$

$$- s\int_{\partial\Omega}\partial_n\varphi\,|\partial_n w|^2\,ds(x). \qquad (1.137)$$

Second term: as $w = 0$ on $\partial\Omega$, we have

$$-s\int_\Omega \Delta w\,\Delta\varphi\,w\,dx = s\int_\Omega \Delta\varphi|\nabla w|^2\,dx + s\int_\Omega \nabla w\cdot\nabla(\Delta\varphi)w\,dx$$

$$= s\int_\Omega \Delta\varphi|\nabla w|^2\,dx + \frac{s}{2}\int_\Omega \nabla(|w|^2)\cdot\nabla(\Delta\varphi)\,dx$$

$$= s\int_\Omega \Delta\varphi|\nabla w|^2\,dx - \frac{s}{2}\int_\Omega \Delta^2\varphi\,|w|^2\,dx. \qquad (1.138)$$

Third term: first we note that

$$\nabla\cdot\left(\nabla\varphi\,|\nabla\varphi|^2\right) = 2D^2\varphi(\nabla\varphi, \nabla\varphi) + |\nabla\varphi|^2\,\Delta\varphi.$$

Therefore, we obtain

$$
- 2s^3 \int_\Omega |\nabla\varphi|^2 w\, \nabla\varphi \cdot \nabla w\, ds
$$

$$
= - s^3 \int_\Omega |\nabla\varphi|^2\, \nabla\varphi \cdot \nabla(|w|^2)\, dx
$$

$$
= s^3 \int_\Omega \nabla \cdot \left(\nabla\varphi\, |\nabla\varphi|^2\right) |w|^2\, dx
$$

$$
= 2s^3 \int_\Omega D^2\varphi(\nabla\varphi, \nabla\varphi)|w|^2\, dx + s^3 \int_\Omega |\nabla\varphi|^2\, \Delta\varphi\, |w|^2\, dx. \tag{1.139}
$$

Fourth term: nothing to do here, the fourth term simply reads

$$
-s^3 \int_\Omega |\nabla\varphi|^2\, \Delta\varphi\, |w|^2\, dx. \tag{1.140}
$$

Adding equations (1.137), (1.138), (1.139) and (1.140), we obtain the desired result. $\qquad\square$

Lemma 1.58. *For any $w \in H^2(\Omega)$, we have*

$$
D^2\varphi(\nabla w, \nabla w) \geqslant |\xi_2|^2 \frac{D^2\varphi(\nabla\varphi, \nabla\varphi)}{|\nabla\varphi|^2} + \beta|\xi_1|^2
$$

$$
- |\xi_1|^2 \frac{D^2\varphi(\nabla\varphi, \nabla\varphi)}{|\nabla\varphi|^2} + 2D^2\varphi(\xi_1, \xi_2), \tag{1.141}
$$

with $\nabla w = \xi_1 + \xi_2$ as in (1.136).

Proof. It is clear that

$$
D^2\varphi(\nabla w, \nabla w) = D^2\varphi(\xi_1, \xi_1) + 2\, D^2\varphi(\xi_1, \xi_2) + D^2\varphi(\xi_2, \xi_2).
$$

By construction,

$$
\tilde\xi_1 = |\nabla\varphi| \frac{\xi_1}{|\xi_1|}
$$

verifies

$$
|\tilde\xi_1| = |\nabla\varphi| \text{ and } \tilde\xi_1 \cdot \nabla\varphi = 0.
$$

Therefore, Hörmander's pseudo-convexity condition (1.129) implies that

$$
D^2\varphi(\tilde\xi_1, \tilde\xi_1) \geqslant \beta|\nabla\varphi|^2 - D^2\varphi(\nabla\varphi, \nabla\varphi).
$$

An easy computation shows that

$$
D^2\varphi(\xi_2, \xi_2) = |\xi_2|^2 \frac{D^2\varphi(\nabla\varphi, \nabla\varphi)}{|\nabla\varphi|^2},
$$

which ends the proof. $\qquad\square$

Lemma 1.59. *For any $w \in H^2(\Omega) \cap H_0^1(\Omega)$, we have*

$$s^3 \int_\Omega D^2\varphi(\nabla\varphi, \nabla\varphi)|w|^2 \, dx - s \int_\Omega \frac{D^2\varphi(\nabla\varphi, \nabla\varphi)}{|\nabla\varphi|^2} |\nabla w|^2 \, dx$$
$$= s \int_\Omega P_S w \, w \, \frac{D^2\varphi(\nabla\varphi, \nabla\varphi)}{|\nabla\varphi|^2} \, dx - \frac{s}{2} \int_\Omega \Delta\left(\frac{D^2\varphi(\nabla\varphi, \nabla\varphi)}{|\nabla\varphi|^2}\right) |w|^2 \, dx. \tag{1.142}$$

Proof. We have

$$s \int_\Omega P_S w \, w \, \frac{D^2\varphi(\nabla\varphi, \nabla\varphi)}{|\nabla\varphi|^2} \, dx$$
$$= s^3 \int_\Omega D^2\varphi(\nabla\varphi, \nabla\varphi)|w|^2 \, dx + s \int_\Omega \Delta w \, \frac{D^2\varphi(\nabla\varphi, \nabla\varphi)}{|\nabla\varphi|^2} \, w \, dx$$
$$= s^3 \int_\Omega D^2\varphi(\nabla\varphi, \nabla\varphi)|w|^2 \, dx - s \int_\Omega \frac{D^2\varphi(\nabla\varphi, \nabla\varphi)}{|\nabla\varphi|^2} |\nabla w|^2 \, dx$$
$$- s \int_\Omega \nabla\left(\frac{D^2\varphi(\nabla\varphi, \nabla\varphi)}{|\nabla\varphi|^2}\right) \cdot \nabla w \, w \, dx.$$

As $\nabla w \, w = \frac{1}{2}\nabla\left(|w|^2\right)$, another integration by parts of the last term gives the result. $\qquad\square$

1.4.3.2. *Proof of theorem 1.55*

To prove Theorem 1.55, we first note that w solution of (1.131) verifies

$$P_S w + P_A w = F \in L^2(\Omega),$$

with implies in particular $w \in H^2(\Omega)$ (see Remark 1.52), hence $P_S w \in L^2(\Omega)$ and $P_A w \in L^2(\Omega)$. Therefore, it is readily seen that

$$\|P_S w\|_{L^2(\Omega)}^2 + 2 \, (P_S w, P_A w)_{L^2(\Omega)} + \|P_A w\|_{L^2(\Omega)}^2 \leqslant \|F\|_{L^2(\Omega)}^2. \tag{1.143}$$

Then, Theorem 1.55 is a direct consequence of the following result, whose proof is postponed afterwards:

Proposition 1.60. *There exist $s_0, C > 0$ such that for any w solution of (1.131) and $s \geqslant s_0$,*

$$s^3 \|w\|_{L^2(\Omega)}^2 \leqslant C\left(\|P_S w\|_{L^2(\Omega)}^2 + (P_S w, P_A w)_{L^2(\Omega)} + \|P_A w\|_{L^2(\Omega)}^2 + s\|\partial_n w\|_{\Gamma_\varphi}^2\right). \tag{1.144}$$

Proof of Theorem 1.55. Proposition 1.60 and equation (1.143) implies that for any $s \geqslant s_0$,

$$s^3 \|w\|_{L^2(\Omega)}^2 \leqslant C \left(\|F\|_{L^2(\Omega)}^2 + s\|\partial_n w\|_{\Gamma_\varphi}^2 \right).$$

Then, if we write

$$\nabla w = \xi_1 + \xi_2$$

with

$$\xi_2 = (\nabla w \cdot \nu)\nu, \ \nu(x) = \frac{\nabla\varphi(x)}{|\nabla\varphi(x)|}, \ \xi_1 \cdot \xi_2 = 0.$$

Proposition 1.56 implies that for any $s \geqslant \max(s_0, 1)$,

$$s^2 \|\xi_2\|_{L^2(\Omega)}^2 \leqslant C \left(\|P_A w\|_{L^2(\Omega)}^2 + s^3 \|w\|_{L^2(\Omega)}^2 \right),$$

and

$$s\|\xi_1\|_{L^2(\Omega)}^2 \leqslant C \left(s^3 \|w\|_{L^2(\Omega)}^2 + \|P_S w\|_{L^2(\Omega)}^2 + \|P_A w\|_{L^2(\Omega)}^2 \right),$$

which, combined again with Proposition 1.60 and equation (1.143), gives (1.132). $\qquad\qquad\square$

Proof of Proposition 1.60. From Lemma 1.57, we know that

$$\int_\Omega P_S w \, P_A w \, dx = 2\,s^3 \int_\Omega D^2\varphi(\nabla\varphi, \nabla\varphi)|w|^2 \, dx - \frac{s}{2} \int_\Omega \Delta^2\varphi \, |w|^2 \, dx$$
$$+ 2\,s \int_\Omega D^2\varphi(\nabla w, \nabla w) \, dx - s \int_{\partial\Omega} \partial_n\varphi \, |\partial_n w|^2 ds(x),$$

which immediately implies

$$(P_S w, P_A w)_{L^2(\Omega)} + s \int_{\Gamma_\varphi} \partial_n\varphi \, |\partial_n w|^2 ds(x)$$
$$\geqslant 2\,s^3 \int_\Omega D^2\varphi(\nabla\varphi, \nabla\varphi)|w|^2 \, dx - \frac{s}{2} \int_\Omega \Delta^2\varphi \, |w|^2 \, dx + 2\,s \int_\Omega D^2\varphi(\nabla w, \nabla w) \, dx.$$
$$(1.145)$$

Now, from Lemma 1.58 we get

$$2\,s \int_\Omega D^2\varphi(\nabla w, \nabla w) \, dx \geqslant 2\,s \int_\Omega G(\varphi)|\xi_2|^2 \, dx + 2\,s\,\beta \int_\Omega |\xi_1|^2 \, dx$$
$$- 2\,s \int_\Omega G(\varphi)|\xi_1|^2 \, dx + 4\,s \int_\Omega D^2\varphi(\xi_1, \xi_2) \, dx,$$
$$(1.146)$$

where we have introduced $G(\varphi) = \dfrac{D^2\varphi(\nabla\varphi, \nabla\varphi)}{|\nabla\varphi|^2}$ to shorten notations. Using that $|\nabla w|^2 = |\xi_1|^2 + |\xi_2|^2$, we see that (1.146) is equivalent to

$$2s\int_\Omega D^2\varphi(\nabla w, \nabla w)\,dx \geqslant 4s\int_\Omega G(\varphi)|\xi_2|^2\,dx + 2s\beta\int_\Omega |\xi_1|^2\,dx$$
$$- 2s\int_\Omega G(\varphi)|\nabla w|^2\,dx + 4s\int_\Omega D^2\varphi(\xi_1, \xi_2)\,dx. \tag{1.147}$$

Hence, equation (1.145) implies

$$(P_S w, P_A w)_{L^2(\Omega)} + s\int_{\Gamma_\varphi} \partial_n\varphi\,|\partial_n w|^2 ds(x)$$
$$\geqslant 2s^3\int_\Omega D^2\varphi(\nabla\varphi, \nabla\varphi)|w|^2\,dx - \frac{s}{2}\int_\Omega \Delta^2\varphi\,|w|^2\,dx + 4s\int_\Omega G(\varphi)|\xi_2|^2\,dx$$
$$+ 2s\beta\int_\Omega |\xi_1|^2\,dx - 2s\int_\Omega G(\varphi)|\nabla w|^2\,dx + 4s\int_\Omega D^2\varphi(\xi_1, \xi_2)\,dx,$$

from which we obtain, using Lemma 1.59,

$$(P_S w, P_A w)_{L^2(\Omega)} + s\int_{\Gamma_\varphi} \partial_n\varphi\,|\partial_n w|^2 ds(x)$$
$$\geqslant 2s\int_\Omega P_S w\,w\,G(\varphi)\,dx - s\int_\Omega \Delta G(\varphi)\,|w|^2\,dx - \frac{s}{2}\int_\Omega \Delta^2\varphi\,|w|^2\,dx$$
$$+ 4s\int_\Omega G(\varphi)|\xi_2|^2\,dx + 2s\beta\int_\Omega |\xi_1|^2\,dx + 4s\int_\Omega D^2\varphi(\xi_1, \xi_2)\,dx. \tag{1.148}$$

We have already seen in the proof of Proposition 1.56 that

$$\int_\Omega |\xi_1|^2\,dx = s^2\int_\Omega |\nabla\varphi|^2|w|^2\,dx - \int_\Omega |\xi_2|^2\,dx - \int_\Omega P_S w\,w\,dx,$$

hence (1.148) and $|\nabla\varphi| \geqslant \alpha$ gives

$$(P_S w, P_A w)_{L^2(\Omega)} + s\int_{\Gamma_\varphi} \partial_n\varphi\,|\partial_n w|^2 ds(x)$$
$$\geqslant 2s\int_\Omega P_S w\,w\,G(\varphi)\,dx - s\int_\Omega \Delta G(\varphi)\,|w|^2\,dx - \frac{s}{2}\int_\Omega \Delta^2\varphi\,|w|^2\,dx$$
$$+ 2s^3\beta\alpha^2\int_\Omega |w|^2\,dx - 2s\beta\int_\Omega |\xi_2|^2\,dx - 2s\beta\int_\Omega P_S w\,w\,dx$$
$$+ 4s\int_\Omega G(\varphi)|\xi_2|^2\,dx + 4s\int_\Omega D^2\varphi(\xi_1, \xi_2)\,dx. \tag{1.149}$$

As $\varphi \in \mathscr{C}^4(\overline{\Omega})$, there exists a constant C such that $|G(\varphi)| \leqslant C$ on $\overline{\Omega}$. Combined with estimate (1.133), we obtain that there exists a constant $C > 0$ such that

$$2s \left| \int_\Omega (2G(\varphi) - 1)|\xi_2|^2 \, dx \right| \leqslant \frac{C}{s} \|P_A w\|_{L^2(\Omega)}^2 + Cs \int_\Omega |w|^2 \, dx, \quad (1.150)$$

whereas a simple computation shows that

$$2s \left| \int_\Omega (G(\varphi) - 1) P_S w \, w \, dx \right| \leqslant Cs \int_\Omega |P_S w \, w| \, dw$$

$$\leqslant \frac{C}{2\sqrt{s}} \|P_S w\|_{L^2(\Omega)}^2 + \frac{Cs^{5/2}}{2} \int_\Omega |w|^2 \, dx. \quad (1.151)$$

Additionally, estimate (1.135) gives a constant $C > 0$ such that

$$4s \left| \int_\Omega D^2\varphi(\xi_1, \xi_2) \, dx \right| \leqslant Cs^{5/2} \int_\Omega |w|^2 + \frac{C}{\sqrt{s}} \left(\|P_S w\|_{L^2(\Omega)}^2 + \|P_A w\|_{L^2(\Omega)}^2 \right). \quad (1.152)$$

Inserting estimates (1.150), (1.151) and (1.152) in (1.149), and furthermore using that there exists a constant C such that

$$s \left| \int_\Omega \left(\Delta G(\varphi) + \frac{1}{2} \Delta^4 \varphi \right) |w|^2 \, dx \right| \leqslant Cs \int_\Omega |w|^2 \, dx,$$

we obtain that there exist constants $c_* > 0$ and $C > 0$ such that

$$(P_S w, P_A w)_{L^2(\Omega)} + s \int_{\Gamma_\varphi} \partial_n\varphi \, |\partial_n w|^2 ds(x)$$

$$\geqslant (c_* s^3 - Cs^{5/2} - Cs) \int_\Omega |w|^2 \, dx - \frac{C}{\sqrt{s}} \left(\|P_S w\|_{L^2(\Omega)}^2 + \|P_A w\|_{L^2(\Omega)}^2 \right). \quad (1.153)$$

We obtain the result using the existence of $C > 0$ such that $|\partial_n\varphi| \leqslant C$ on $\partial\Omega$ and choosing $s \geqslant s_0$, with $s_0 > 0$ large enough such that $c_* s^3/2 - Cs^{5/2} - Cs \geqslant 0$ for all $s \geqslant s_0$. $\qquad \square$

1.4.4. *Some comments on the Hörmander's strict pseudo-convexity condition*

In this section, we give some comments on the Hörmander's pseudo-convexity condition.

1.4.4.1. *Case of a weight function depending on $x \cdot e$ only*

We suppose in this section that $\varphi \in \mathscr{C}^4(\overline{\Omega})$ depends only on the coordinate $x \cdot e$, where e is a fixed direction, that is $e \in \mathbb{S}^{d-1}$. Let us denote

$$m = \min_{x \in \overline{\Omega}} x \cdot e, \quad M = \max_{x \in \overline{\Omega}} x \cdot e,$$

and suppose $\varphi(x) = f(x \cdot e)$, for some $f \in \mathscr{C}^4([m, M])$. Then $\nabla \varphi = f'(x \cdot e) e$, hence $|\nabla \varphi| > \alpha > 0$ directly reads $|f'| > \alpha$ on $[m, M]$, that is a strict monotonicity condition for f on $[m, M]$. Furthermore, for all $a, b \in \mathbb{R}^d$, it is straightforward to see that

$$D^2 \varphi(a, b) = (a \cdot e)(b \cdot e) f''(x \cdot e).$$

As furthermore all $\xi \in \mathbb{R}^d$ verifying $\xi \cdot \nabla \varphi = 0$ necessarily verifies $\xi \cdot e = 0$, the condition (1.129) simply rewrites

$$f''(x \cdot e) > \beta \text{ on } \overline{\Omega},$$

which reads as a strict convexity condition for f on $[m, M]$.

These are precisely the two conditions verified by the weight function we used in our study of the previous section, in the case of a strip.

1.4.4.2. *Construction of a weight function*

Let us now consider Ω an arbitrary smooth bounded domain of $\mathbb{R}^d$, and $\Gamma \subset \Omega$ nonempty and open. Suppose that we want to use Theorem 1.51 with an observation localized on Γ, that is such that

$$\Gamma_\varphi = \{x \in \partial \Omega, \ \partial_n \varphi(x) > 0\} \subset \Gamma,$$

where φ is the weight function in the Carleman estimate (1.130). The existence of such a weight function satisfying furthermore $|\nabla \varphi| \geqslant \alpha > 0$ on $\overline{\Omega}$ and Hörmander's strict pseudo-convexity condition (1.129) is not obvious. But it turns out it is always possible to construct a function φ verifying all the require properties. The key result to do so is the following proposition.

Proposition 1.61. *There exists a function ψ belonging to $\mathscr{C}^4(\overline{\Omega})$ such that*

$$\inf_{x \in \overline{\Omega}} \{|\nabla \psi(x)|\} > 0, \quad \text{and} \quad \forall x \in \partial \Omega \setminus \Gamma, \ \partial_n \psi(x) < 0. \qquad (1.154)$$

Proof. See for instance Appendix 3 of [Tucsnak and Weiss (2009)]. $\square$

The process to construct φ from ψ is often called convexification. It consists in defining φ as $e^{\lambda \psi}$, for a parameter $\lambda > 0$ chosen sufficiently large.

Proposition 1.62. *Let ψ be given by Proposition 1.61, and $\varphi = e^{\lambda \psi}$ with $\lambda > 0$. There exists $\lambda_0 > 0$ such that for all $\lambda \geqslant \lambda_0$, φ satisfies $\Gamma_\varphi \subset \Gamma$,*

$$\inf_{x \in \overline{\Omega}} |\nabla \varphi| > 0,$$

and Hörmander's strict pseudo-convexity condition (1.129).

Proof. First of all, as $\nabla \varphi = \lambda \varphi \nabla \psi$, $\lambda \varphi > 0$ on $\overline{\Omega}$ and $\Gamma_\psi \subset \Gamma$, we clearly have

$$\inf_{x \in \overline{\Omega}} |\nabla \varphi| = \alpha > 0 \text{ and } \Gamma_\varphi \subset \Gamma.$$

It then remains to consider Hörmander's pseudo convexity condition.

A simple computation shows that for all a, $b \in \mathbb{R}^d$, one has

$$D^2 \varphi(a, b) = \lambda^2 \varphi (a \cdot \nabla \psi)(b \cdot \nabla \psi) + \lambda \varphi D^2 \psi(a, b).$$

Therefore, for all $x \in \overline{\Omega}$ and all $\xi \in \mathbb{R}^d$ such that

$$|\xi| = |\nabla \varphi(x)| = \lambda \varphi(x) |\nabla \psi(x)|,$$

and $\xi \cdot \nabla \varphi = 0$, or equivalently $\xi \cdot \nabla \psi = 0$, we have

$$D^2 \varphi(\nabla \varphi(x), \nabla \varphi(x)) + D^2 \varphi(\xi, \xi)$$
$$= |\nabla \varphi(x)|^2 \varphi(x) \left[\lambda^2 |\nabla \psi|^2 + \lambda D^2 \psi \left(\frac{\nabla \psi}{|\nabla \psi|}, \frac{\nabla \psi}{|\nabla \psi|} \right) + \lambda D^2 \psi \left(\frac{\xi}{|\xi|}, \frac{\xi}{|\xi|} \right) \right],$$

hence

$$D^2 \varphi(\nabla \varphi(x), \nabla \varphi(x)) + D^2 \varphi(\xi, \xi) \geqslant |\nabla \varphi(x)|^2 e^{\lambda \sigma_\psi} \left[\lambda^2 \alpha_\psi - 2 \lambda \gamma_\psi \right],$$

where σ_ψ, α_ψ and γ_ψ stand for

$$\sigma_\psi = \inf_{x \in \overline{\Omega}} \psi(x), \quad \alpha_\psi = \inf_{x \in \overline{\Omega}} |\nabla \psi| > 0, \quad \gamma_\psi = \sup_{e \in \mathbb{S}^{d-1}} \left| D^2 \psi(e, e) \right|.$$

Choosing $\lambda_0 > 2 \dfrac{\gamma_\psi}{\alpha_\psi}$ gives the result. $\qquad \square$

Remark 1.63. Interestingly, choosing a weight function φ of the form $e^{\lambda \psi}$ gives a new parameter λ that can be chosen arbitrarily large. This extra parameter is of paramount importance in some applications, see e.g. [Le Rousseau (2015)].

Remark 1.64. In fact, once a function ψ is constructed satisfying the conditions of Proposition 1.61, it is interesting to look for φ under the form $\varphi = f(\psi)$ for some strictly monotonic function f. In particular, such process will not modify the level sets of the function ψ, so that the quantity

$D^2\varphi_x(\xi,\xi)$ for ξ satisfying $\xi \cdot \nabla\varphi(x) = 0$, which corresponds to curvatures of the level set of the function, is barely modified, while the convexity in the direction of $\nabla\varphi$ will be strongly modified. In fact, condition (1.130) then reads: for all $x \in \overline{\Omega}$, $\forall \xi \in \mathbb{R}^d$ with $|\xi| = 1$ and $\xi \cdot \nabla\psi(x) = 0$,

$$f''(\psi(x)) + f'(\psi(x)) \left(D^2\psi_x \left(\frac{\nabla_x\psi(x)}{|\nabla_x\psi(x)|}, \frac{\nabla_x\psi(x)}{|\nabla_x\psi(x)|} \right) + D^2\psi_x(\xi,\xi) \right) \geqslant \beta.$$

It is then clear that this could be achieved by taking f'' large enough compared to f', which is precisely what the choice $f(y) = \exp(\lambda y)$ does when λ is large enough.

This choice "convexifies" the function ψ. In particular, one needs f to be sufficiently convex in the direction of $\nabla\psi$ compared to the curvature of the level sets $\{\psi = c\}$. This process is called a "convexification" process.

1.4.4.3. *Easier set of assumptions*

Here, we consider again the setting of Theorem 1.51, and we mention that if we impose some additional conditions, then we can derive an easier proof than the one of Theorem 1.51.

Theorem 1.65. *Let $\varphi \in \mathscr{C}^4(\overline{\Omega})$ and assume that there exist $\alpha > 0$ and $\beta > 0$ for which we have*

$$\inf_{x\in\overline{\Omega}} |\nabla\varphi(x)| \geqslant \alpha, \tag{1.155}$$

and

$$\begin{cases} -|\nabla\varphi|^2\Delta\varphi + 2D^2\varphi(\nabla\varphi, \nabla\varphi) \geqslant \beta|\nabla\varphi|^2, \\ \inf_{\substack{x\in\overline{\Omega} \\ \xi\in\mathbb{R}^d, |\xi|=1}} \{\Delta\varphi|\xi|^2 + 2D^2\varphi(\xi,\xi)\} \geqslant \beta. \end{cases} \tag{1.156}$$

Then there exists a constant $C > 0$ such that for all $u \in L^2(\Omega)$ solution of (1.106) with source term $f \in L^2(\Omega)$, for all $s \geqslant 1$,

$$s^3 \|e^{s\varphi}u\|^2_{L^2(\Omega)} + s \|e^{s\varphi}u\|^2_{L^2(\Omega)} \leqslant C \left(\|e^{s\varphi}f\|^2_{L^2(\Omega)} + s \|e^{s\varphi}\partial_n u\|^2_{L^2(\Gamma_\varphi)} \right), \tag{1.157}$$

where

$$\Gamma_\varphi = \{x \in \partial\Omega, \, \partial_n\varphi(x) > 0\}.$$

Conditions (1.155)–(1.156) have to be compared with (1.128)–(1.129), which are more general.

Before going into the proof of Theorem 1.65, let us mention the following result, which indicates that the additional observation may be done on any arbitrary subset of the boundary:

Lemma 1.66. *Let Γ be a non-empty open subset of the boundary $\partial\Omega$. Then there exists a smooth function φ satisfying condition (1.155)–(1.156) and $\Gamma_\varphi \subset \Gamma$.*

Proof. Similarly as for the proof of Proposition 1.62, we start by taking ψ as in Proposition 1.61, then look for φ of the form $\varphi(x) = e^{\lambda\psi(x)}$ for some parameter $\lambda \geqslant 1$ large enough. Explicit computations yield

$$\nabla\varphi = \lambda\nabla\psi\varphi,$$
$$\Delta\varphi = \lambda^2|\nabla\psi|^2\varphi + \lambda\Delta\psi\varphi,$$
$$D^2\varphi = \lambda^2\nabla\psi^T\nabla\psi\varphi + \lambda D^2\psi\varphi.$$

First, according to condition (1.154), we easily check that $\Gamma_\varphi \subset \Gamma$.

We then compute

$$-|\nabla\varphi|^2\Delta\varphi + 2D^2\varphi(\nabla\varphi, \nabla\varphi)$$
$$= -\lambda^4|\nabla\psi|^4\varphi^3 - \lambda^3 D^2\psi(\nabla, \nabla\psi)\varphi^3 + 2\lambda^4|\nabla\psi|^4\varphi^3 + \lambda^3\Delta\psi|\nabla\psi|^2\varphi^3$$
$$= \lambda^4|\nabla\psi|^4\varphi^3 - \lambda^3 D^2\psi(\nabla, \nabla\psi)\varphi^3 + \lambda^3\Delta\psi|\nabla\psi|^2\varphi^3.$$

According to condition (1.154), condition $(1.156)_1$ is satisfied for $\lambda \geqslant \lambda_1$ for large enough $\lambda_1 > 0$.

Similarly, explicit computations yield

$$\Delta\varphi|\xi|^2 + 2D^2\varphi(\xi, \xi)$$
$$= \lambda^2|\nabla\psi|^2|\xi|^2\varphi + \lambda\Delta\psi|\xi|^2\varphi + 2\lambda^2(\nabla\psi \cdot \xi)^2\varphi + \lambda D^2\psi(\xi, \xi)\varphi.$$

Again, due to condition (1.154), we can make this term positive on $\overline{\Omega}$ by taking $\lambda \geqslant \lambda_1$ large enough.

Therefore, the choice $\varphi = e^{\lambda\psi}$ for λ sufficiently large satisfies all the assumptions of Theorem 1.65 with $\Gamma_\varphi \subset \Gamma$. □

Proof of Theorem 1.65. We set $U = ue^{s\varphi}$ and $F = fe^{s\varphi}$. Then

$$F = e^{s\varphi}\Delta u = e^{s\varphi}\Delta(e^{-s\varphi}U) = \Delta U - 2s\nabla\varphi \cdot \nabla U + s^2|\nabla\varphi|^2 U - s\Delta\varphi U,$$

so that U solves

$$\begin{cases} \Delta U - 2s\nabla\varphi \cdot \nabla U + s^2|\nabla\varphi|^2 U - s\Delta\varphi U = F, & \text{in } \Omega, \\ U = 0, & \text{on } \partial\Omega. \end{cases} \tag{1.158}$$

We thus write

$$P_1 U = \Delta U + s^2|\nabla\varphi|^2 U, \quad P_2 U = -2s\Delta\varphi U - 2s\nabla\varphi\cdot\nabla U, \quad RU = -s\Delta\varphi U,$$

so that

$$P_1 U + P_2 U = F + RU,$$

and P_1, P_2 roughly correspond to the self-adjoint and skew-adjoint parts of the operator in (1.158).

It follows

$$2 \int_\Omega P_1 U\, P_2 U dx \leqslant 2 \int_\Omega |F|^2 dx + 2 \int_\Omega |RU|^2 dx. \qquad (1.159)$$

We thus compute the product of $P_1 U$ by $P_2 U$. We will denote by $I_{i,j}$ the cross product between the i-th term of P_1 and the j-th term of P_2. Of course, our computations will strongly use the fact that u and thus U vanish on the boundary.

$$I_{11} = -2s \int_\Omega \Delta\varphi\, U\, \Delta U dx = 2s \int_\Omega \Delta\varphi |\nabla U|^2 dx - s \int_\Omega \Delta^2\varphi\, |U|^2 dx,$$

$$I_{12} = -2s \int_\Omega \Delta U\, \nabla\varphi \cdot \nabla U dx$$

$$= 2s \int_\Omega \nabla U \cdot \nabla(\nabla\varphi \cdot \nabla U) dx - 2s \int_{\partial\Omega} \partial_n\varphi |\partial_n U|^2 d\sigma$$

$$= 2s \int_\Omega D^2\varphi(\nabla U, \nabla U) dx - s \int_\Omega \Delta\varphi\, |\nabla U|^2 dx - s \int_{\partial\Omega} \partial_n\varphi |\partial_n U|^2 d\sigma,$$

$$I_{21} = -2s^3 \int_\Omega |\nabla\varphi|^2 \Delta\varphi |U|^2 dx,$$

$$I_{22} = -2s^3 \int_\Omega |\nabla\varphi|^2 \nabla\varphi \cdot \nabla U\, U dx = -s^3 \int_\Omega |\nabla\varphi|^2 \nabla\varphi \cdot \nabla(|U|^2) dx$$

$$= s^3 \int_\Omega \left(|\nabla\varphi|^2 \Delta\varphi + 2D^2\varphi(\nabla\varphi, \nabla\varphi) \right) |U|^2 dx.$$

Combining these estimates,

$$\int_\Omega P_1 U\, P_2 U dx$$

$$= s^3 \int_\Omega \left(-|\nabla\varphi|^2 \Delta\varphi + 2D^2\varphi(\nabla\varphi, \nabla\varphi) \right) |U|^2 dx - s \int_\Omega \Delta^2\varphi |U|^2 dx$$

$$+ s \int_\Omega \Delta\varphi |\nabla U|^2 dx + 2s \int_\Omega D^2\varphi(\nabla U, \nabla U) dx - s \int_{\partial\Omega} \partial_n\varphi |\partial_n U|^2 d\sigma. \qquad (1.160)$$

Using assumptions (1.155)–(1.156), we obtain, for $s \geqslant s_0$ large enough, that

$$s^3 \int_\Omega |U|^2 dx + s \int_\Omega |\nabla U|^2 dx \leqslant C \int_\Omega P_1 U\, P_2 U dx + Cs \int_{\Gamma_\varphi} |\partial_n U|^2 d\sigma.$$

Using (1.159), we obtain

$$s^3 \int_\Omega |U|^2 dx + s \int_\Omega |\nabla U|^2 dx \leqslant C \int_\Omega |F|^2 dx + Cs^2 \int_\Omega |U|^2 dx + Cs \int_{\Gamma_\varphi} |\partial_n U|^2 d\sigma.$$

Hence taking s_0 larger if necessary, for all $s \geqslant s_0$,

$$s^3 \int_\Omega |U|^2 dx + s \int_\Omega |\nabla U|^2 dx \leqslant C \int_\Omega |F|^2 dx + Cs \int_{\Gamma_\varphi} |\partial_n U|^2 d\sigma. \quad (1.161)$$

We now remark that $U = ue^{s\varphi}$ hence

$$|u|^2 e^{2s\varphi} \leqslant |U|^2 \quad \text{and} \quad |\nabla u|^2 e^{2s\varphi} \leqslant 2|\nabla U|^2 + 2s^2 |U|^2,$$

that yields the claimed result. □

Remark 1.67. Note that we could have incorporated $\|P_1 U\|_{L^2(\Omega)}^2 + \|P_2 U\|_{L^2(\Omega)}^2$ on the left hand side of (1.159). Remarking then that

$$\|P_2 U\|_{L^2(\Omega)}^2 \geqslant 2s^2 \int_\Omega |\nabla\varphi \cdot \nabla U|^2 \, dx - Cs^2 \int_\Omega |U|^2 dx,$$

following the above proof, we easily get that one could add the term

$$s^2 \int_\Omega |\nabla\varphi \cdot \nabla U|^2 \, dx$$

in the left hand side of (1.161). We thus recover that the gradient of U can be estimated with a power s^2, similarly to what we proved in Theorem 1.55.

1.4.5. *Application to unique continuation*

The main result of this section is the following one:

Theorem 1.68. *Let $q \in L^\infty(\Omega)$ and $W \in (L^\infty(\Omega))^d$, and let Γ be a non-empty open subset of the boundary $\partial\Omega$. Then any solution $u \in H_0^1(\Omega)$ of*

$$\begin{cases} \Delta u + qu + W \cdot \nabla u = 0, & \text{for } x \in \Omega, \\ u = 0, & \text{for } x \in \partial\Omega, \end{cases}$$

which further satisfies $\partial_n u = 0$ on Γ vanishes identically on Ω.

This property is the so-called unique continuation property through Γ.

Proof. Choose a weight function φ so that $\Gamma_\varphi \subset \Gamma$ according to Lemma 1.66, and apply the Carleman estimate of Theorem 1.65 with $f = -qu - W \cdot \nabla u$: for all $s \geqslant 1$,

$$s^3 \|e^{s\varphi} u\|_{L^2(\Omega)}^2 + s \|e^{s\varphi}\nabla u\|_{L^2(\Omega)}^2$$
$$\leqslant C \left(\|q\|_{L^\infty}^2 \|e^{s\varphi} u\|_{L^2(\Omega)}^2 + \|W\|_{L^\infty}^2 \|e^{s\varphi}\nabla u\|_{L^2(\Omega)}^2 \right).$$

Taking s large enough, we easily obtain $u = 0$ in Ω. □

Problem 1.69. *Arguing as in Corollary 1.25, check that if $u \in H_0^1(\Omega)$ solves*

$$\begin{cases} \Delta u + qu + W \cdot \nabla u = f, & \text{for } x \in \Omega, \\ u = 0, & \text{for } x \in \partial\Omega, \end{cases}$$

with a source term $f \in L^2(\Omega)$ that vanishes on a set of the form $\{\varphi > a\}$ for some φ satisfying (1.78)–(1.156) and $\partial_n u$ vanishing on Γ_φ and $q \in L^\infty(\Omega)$, $W \in (L^\infty(\Omega))^d$, then u vanishes on the whole set $\{\varphi > a\}$.

1.5. Conclusion

In this lecture, we have explained how Carleman estimates can be derived for the Laplace operator, using several different techniques and approaches.

We emphasize that Carleman estimates can be derived in much more general settings, regarding the PDE under consideration or the question at stake. The approaches we presented here are quite far from being general, and somehow shortcuts semi-classical arguments by a careful study of what happens in the case of a vertical strip and coefficients depending only on x_1.

For more about Carleman estimates, in particular regarding more PDE and presenting several applications, we refer to the textbooks [Lebeau *et al.* (2022); Choulli (2009); Fu *et al.* (2019); Lerner (2019); Klibanov and Timonov (2004); Isakov (2006)], and to the survey articles [Lebeau and Le Rousseau (2011); Fernández-Cara and Guerrero (2006)], among many others.

This lecture was focused on Carleman estimates considered as a tool, for which we give only few applications, namely to the Calderón problem and to the unique continuation for elliptic equations. We should nevertheless close this lecture by emphasizing that Carleman estimates appeared in many fields and have much more applications.

As we discussed, Carleman estimates can be applied to derive *unique continuation properties* in many settings, and yield several related properties. For instance, they can be use to derive *propagation of smallness* of solutions of PDE, in which the question is the following: if the solution u of some PDE is small in some set ω, can we prove that the solution is small in the whole domain? See an instance of such result in Theorem 1.45. As a matter of fact, this question is also related to the Cauchy problem for the Laplace equation that we saw in the beginning, in which this question can be recast into a stability problem. Carleman estimates can also be used to prove the so-called three sphere inequality, which is a classical property for harmonic functions, and can be generalized to solutions of elliptic PDE, see e.g. the survey article [Alessandrini *et al.* (2009)]: there exist $C > 0$ and $\theta \in (0,1)$, such that for $u \in L^2(B(4))$ solution of $\Delta u = 0$ in the ball $B(4)$ of radius 4,

$$\|u\|_{L^2(B(2))} \leqslant C \|u\|_{L^2(B(1))}^{\theta} \|u\|_{L^2(B(4))}^{1-\theta} .$$

Related to these unique continuation properties, we may quote the *Landis conjecture*: is it true that, if $u \in L^1_{loc}(\mathbb{R}^d; \mathbb{R})$ satisfies $|\Delta u| \leqslant |u|$ in $\mathbb{R}^d$ and for some $C > 0$ sufficiently large, $|u(x)| \exp(C|x|) \in L^\infty(\mathbb{R}^d)$, then u vanishes? This example should be thought of as a unique continuation problem at ∞, meaning that the solution is supposed to be appropriately small in a neighborhood of the infinity. It should be emphasized that the Landis conjecture is still a conjecture, except in the 1d case, but decisive results have been made using Carleman estimates, see e.g. [Bourgain and Kenig (2005)] for results in general dimensions, used in the proof of Anderson localization for the Bernoulli models, and the recent work [Logunov *et al.* (2020)] for a proof of a slightly weaker result in the 2d case.

As we have explained for the Calderón problem, Carleman estimates can also be used to derive stability results for many *inverse problems*, and we refer to the textbooks [Isakov (2006); Klibanov and Timonov (2004); Choulli (2009); Bellassoued and Yamamoto (2017)] for many applications in the context of inverse problems. Of course, this is a very rich field and there is no time to present it in more details.

Another broad application of Carleman estimates is to derive *observability estimates* for PDE, which basically corresponds to quantification of the unique continuation for PDE. Such observability properties are often used to derive *controllability results*, which are equivalent one to another through classical duality arguments (see for instance [Tucsnak and Weiss (2009)]). Therefore, Carleman estimates are among the main tools used to derive controllability results for PDE, in particular since the pioneering works [Fursikov and Imanuvilov (1996); Lebeau and Robbiano (1995)] which proved null-controllability of the heat equation in arbitrary geometric settings.

We will not list more the various applications which use Carleman estimates, but we would like to point out that this list is not exhaustive (by far!), and they can be encountered as well to establish the logarithmic decay rate of waves when no geometric control condition is satisfied ([Lebeau and Robbiano (1997)]), or to prove that a weak solution of the Navier-Stokes equations which belongs to $L^\infty(0, T; L^3(\mathbb{R}^d))$ is smooth up to the time T ([Iskauriaza *et al.* (2003); Tao (2020)]).

Acknowledgements

Both authors have been supported by the CIMI Labex, Toulouse, France, under grant ANR-11-LABX-0040-CIMI, the MATH AmSud program ACIPDE, and the ANR Project TRECOS, grant ANR-20-CE40-0009.

References

Alessandrini, G., Rondi, L., Rosset, E. and Vessella, S. (2009). The stability for the Cauchy problem for elliptic equations, *Inverse Problems* **25**, 12, pp. 123004, 47, doi:10.1088/0266-5611/25/12/123004, URL http://dx.doi.org/10.1088/0266-5611/25/12/123004.

Astala, K. and Päivärinta, L. (2006). Calderón's inverse conductivity problem in the plane, *Ann. of Math. (2)* **163**, 1, pp. 265–299, doi:10.4007/annals.2006.163.265, URL https://doi.org/10.4007/annals.2006.163.265.

Bellassoued, M. and Yamamoto, M. (2017). *Carleman estimates and applications to inverse problems for hyperbolic systems*, Springer Monographs in Mathematics, Springer, Tokyo, ISBN 978-4-431-56598-7; 978-4-431-56600-7, doi:10.1007/978-4-431-56600-7, URL https://doi.org/10.1007/978-4-431-56600-7.

Bourgain, J. and Kenig, C. E. (2005). On localization in the continuous Anderson-Bernoulli model in higher dimension, *Invent. Math.* **161**, 2, pp. 389–426, doi:10.1007/s00222-004-0435-7, URL http://dx.doi.org/10.1007/s00222-004-0435-7.

Brezis, H. (1983). *Analyse fonctionnelle*, Collection Mathématiques Appliquées pour la Maitrise. [Collection of Applied Mathematics for the Master's Degree], Masson, Paris, ISBN 2-225-77198-7, théorie et applications. [Theory and applications].

Calderón, A.-P. (1958). Uniqueness in the Cauchy problem for partial differential equations, *Amer. J. Math.* **80**, pp. 16–36, doi:10.2307/2372819, URL https://doi.org/10.2307/2372819.

Choulli, M. (2009). *Une introduction aux problèmes inverses elliptiques et paraboliques, Mathématiques & Applications (Berlin)*, Vol. 65, Springer-Verlag, Berlin, ISBN 978-3-642-02459-7, doi:10.1007/978-3-642-02460-3, URL http://dx.doi.org/10.1007/978-3-642-02460-3.

Dos Santos Ferreira, D. (2005). Sharp L^p Carleman estimates and unique continuation, *Duke Math. J.* **129**, 3, pp. 503–550, doi:10.1215/S0012-7094-05-12933-7, URL http://dx.doi.org/10.1215/S0012-7094-05-12933-7.

Dos Santos Ferreira, D., Kenig, C. E., Salo, M. and Uhlmann, G. (2009). Limiting Carleman weights and anisotropic inverse problems, *Invent. Math.* **178**,

1, pp. 119–171, doi:10.1007/s00222-009-0196-4, URL http://dx.doi.org/10.1007/s00222-009-0196-4.

Evans, L. C. (1998). *Partial differential equations*, Graduate Studies in Mathematics, Vol. 19, American Mathematical Society, Providence, RI, ISBN 0-8218-0772-2.

Fernández-Cara, E. and Guerrero, S. (2006). Global Carleman inequalities for parabolic systems and applications to controllability, *SIAM J. Control Optim.* **45**, 4, pp. 1399–1446 (electronic), doi:10.1137/S0363012904439696, URL http://dx.doi.org/10.1137/S0363012904439696.

Fu, X., Lü, Q. and Zhang, X. (2019). *Carleman estimates for second order partial differential operators and applications*, SpringerBriefs in Mathematics, Springer, Cham, ISBN 978-3-030-29529-5; 978-3-030-29530-1, doi:10.1007/978-3-030-29530-1, URL https://doi.org/10.1007/978-3-030-29530-1, a unified approach, BCAM SpringerBriefs.

Fursikov, A. V. and Imanuvilov, O. Y. (1996). *Controllability of evolution equations, Lecture Notes Series*, Vol. 34, Seoul National University Research Institute of Mathematics Global Analysis Research Center, Seoul.

Gel'fand, I. M. and Shilov, G. E. (1964). *Generalized functions. Vol. I: Properties and operations*, Academic Press, New York-London, translated by Eugene Saletan.

Imanuvilov, O. Y. and Puel, J.-P. (2003). Global Carleman estimates for weak solutions of elliptic nonhomogeneous Dirichlet problems, *Int. Math. Res. Not.* **16**, pp. 883–913.

Isakov, V. (2006). *Inverse problems for partial differential equations*, Applied Mathematical Sciences, Vol. 127, 2nd edn., Springer, New York, ISBN 978-0387-25364-0; 0-387-25364-5.

Iskauriaza, L., Serëgin, G. A. and Shverak, V. (2003). $L_{3,\infty}$-solutions of Navier-Stokes equations and backward uniqueness, *Uspekhi Mat. Nauk* **58**, 2(350), pp. 3–44, doi:10.1070/RM2003v058n02ABEH000609, URL http://dx.doi.org/10.1070/RM2003v058n02ABEH000609.

Kenig, C. E., Ruiz, A. and Sogge, C. D. (1987). Uniform Sobolev inequalities and unique continuation for second order constant coefficient differential operators, *Duke Math. J.* **55**, 2, pp. 329–347, doi:10.1215/S0012-7094-87-05518-9, URL https://doi.org/10.1215/S0012-7094-87-05518-9.

Kenig, C. E., Sjöstrand, J. and Uhlmann, G. (2007). The Calderón problem with partial data, *Ann. of Math. (2)* **165**, 2, pp. 567–591, doi:10.4007/annals.2007.165.567, URL http://dx.doi.org/10.4007/annals.2007.165.567.

Klibanov, M. V. and Timonov, A. (2004). *Carleman estimates for coefficient inverse problems and numerical applications*, Inverse and Ill-posed Problems Series, VSP, Utrecht, ISBN 90-6764-405-6.

Koch, H. and Tataru, D. (2001). Carleman estimates and unique continuation for second-order elliptic equations with nonsmooth coefficients, *Comm. Pure Appl. Math.* **54**, 3, pp. 339–360.

Le Rousseau, J. (2015). On Carleman estimates with two large parameters, *Indiana Univ. Math. J.* **64**, 1, pp. 55–113, doi:10.1512/iumj.2015.64.5397, URL https://doi.org/10.1512/iumj.2015.64.5397.

Le Rousseau, J. and Lerner, N. (2013). Carleman estimates for anisotropic elliptic operators with jumps at an interface, *Anal. PDE* **6**, 7, pp. 1601–1648, doi:10.2140/apde.2013.6.1601, URL http://dx.doi.org/10.2140/apde.2013.6.1601.

Le Rousseau, J. and Robbiano, L. (2010). Carleman estimate for elliptic operators with coefficients with jumps at an interface in arbitrary dimension and application to the null controllability of linear parabolic equations, *Arch. Ration. Mech. Anal.* **195**, 3, pp. 953–990, doi:10.1007/s00205-009-0242-9, URL http://dx.doi.org/10.1007/s00205-009-0242-9.

Lebeau, G. and Le Rousseau, J. (2011). On Carleman estimates for elliptic and parabolic operators. Applications to unique continuation and control of parabolic equations, *ESAIM Control Optim. Calc. Var.* **18**, 3, pp. 712–747, doi:10.1051/cocv/2011168, URL http://doi.org/10.1051/cocv/2011168.

Lebeau, G., Le Rousseau, J. and Robbiano, L. (2022). *Elliptic Carleman Estimates and Applications to Stabilization and Controllability, Volume I: Dirichlet Boundary Conditions on Euclidean Space*, PNLDE Subseries in Control, Birkhäuser, ISBN 978-3-030-88669-1.

Lebeau, G., Le Rousseau, J. and Robbiano, L. (2022). *Elliptic Carleman Estimates and Applications to Stabilization and Controllability, Volume II: General Boundary Conditions on Riemannian Manifold*, PNLDE Subseries in Control, Birkhäuser, ISBN 978-3-030-88673-8.

Lebeau, G. and Robbiano, L. (1995). Contrôle exact de l'équation de la chaleur, *Comm. Partial Differential Equations* **20**, 1-2, pp. 335–356.

Lebeau, G. and Robbiano, L. (1997). Stabilisation de l'équation des ondes par le bord, *Duke Math. J.* **86**, 3, pp. 465–491, doi:10.1215/S0012-7094-97-08614-2, URL http://dx.doi.org/10.1215/S0012-7094-97-08614-2.

Lerner, N. (2019). *Carleman inequalities, Grundlehren der Mathematischen Wissenschaften*, Fundamental Principles of Mathematical Sciences, Vol. 353, Springer, Cham, ISBN 978-3-030-15992-4; 978-3-030-15993-1, doi:10.1007/978-3-030-15993-1, URL https://doi.org/10.1007/978-3-030-15993-1, an introduction and more.

Lions, J.-L. and Magenes, E. (1968). *Problèmes aux limites non homogènes et applications. Vol. 1*, Travaux et Recherches Mathématiques, No. 17, Dunod, Paris.

Logunov, A., Malinnikova, E., Nadirashvili, N. and Nazarov, F. (2020). The Landis conjecture on exponential decay, doi:10.48550/arXiv.2007.07034, URL https://doi.org/10.48550/arXiv.2007.07034.

Salo, M. (2008). http://www.rni.helsinki.fi/~msa/lecturenotes/calderon_lectures.pdf.

Sogge, C. D. (1993). *Fourier integrals in classical analysis*, Cambridge Tracts in Mathematics, Vol. 105, Cambridge University Press, Cambridge, ISBN 0-521-43464-5, doi:10.1017/CBO9780511530029, URL https://doi.org/10.1017/CBO9780511530029.

Tao, T. (2020). Quantitative bounds for critically bounded solutions to the Navier-Stokes equations, doi:10.48550/arXiv.1908.04958, URL https://doi.org/10.48550/arXiv.1908.04958.

Tucsnak, M. and Weiss, G. (2009). *Observation and control for operator semi-groups*, Birkhäuser Advanced Texts: Basler Lehrbücher. [Birkhäuser Advanced Texts: Basel Textbooks], Birkhäuser Verlag, Basel, ISBN 978-3-7643-8993-2, doi:10.1007/978-3-7643-8994-9, URL http://dx.doi.org/10.1007/978-3-7643-8994-9.

Uhlmann, G. (1999). Developments in inverse problems since Calderón's foundational paper, in *Harmonic analysis and partial differential equations (Chicago, IL, 1996)*, pp. 295–345, Chicago Lectures in Math., Univ. Chicago Press, Chicago, IL.

Uhlmann, G. (2009). Visibility and invisibility, in *ICIAM 07—6th International Congress on Industrial and Applied Mathematics*, pp. 381–408, Eur. Math. Soc., Zürich, doi:10.4171/056-1/18, URL http://dx.doi.org/10.4171/056-1/18.

Saturated Boundary Stabilization of Partial Differential Equations Using Control-Lyapunov Functions

Hugo Lhachemi

Université Paris-Saclay, CNRS, CentraleSupélec, Laboratoire des signaux et systèmes, 91190, Gif-sur-Yvette, France
hugo.lhachemi@centralesupelec.fr

Christophe Prieur*

Université Grenoble Alpes, CNRS, Grenoble-INP, GIPSA-lab, F-38000, Grenoble, France
christophe.prieur@gipsa-lab.fr

Abstract. This chapter reviews some recent results on the boundary stabilization of different classes of partial differential equations. In order to provide a self-content chapter with consistent control objectives and notation, we first review the finite-dimensional case. Controllability and observability conditions for linear ordinary differential equations are recalled together with some basic Lyapunov theory for the stability analysis and the design of saturated controllers. Then we address the boundary control problem for the stabilization of a reaction-diffusion equation by means of numerically tractable design methods while considering different norms and possible constraints on the amplitude of the inputs. Finally similar control design problems will be studied for the stabilization of the Korteweg-de Vries equation and the wave equation.

1. Introduction

The goal of this chapter is to review some recent results on boundary stabilization of distributed parameter systems as those modeled by parabolic partial differential equations or hyperbolic partial differential equations. No

---

*This work has been partially supported by MIAI@Grenoble Alpes (ANR-19-P3IA-0003).

prerequisite on control theory will be necessary, only basic knowledge on control objectives. However, background in nonlinear dynamical systems and essentials on partial differential equations (PDEs) would be helpful, even if some references will be given throughout the text.

The topics covered in this chapter embrace different potential applications such as control and stability theory of reaction-diffusion phenomenon as those modeled by parabolic PDEs. Some control techniques presented in this chapter will be useful for stability theory of physical dynamics described by balance laws and modeled by hyperbolic partial differential equations. Different control objectives will be studied and solved such as the design of stabilizing control laws ensuring that all the trajectories of the closed-loop systems converge to a given equilibrium. Different control schemes are considered, covering in-domain control (the control input appears directly in the main part of the PDE) and boundary control (the control input applies at the boundary of the domain as it appears through the boundary conditions). Moreover, when possible, the described control laws will be designed based on the only knowledge of a prescribed and limited part of the state, the so-called output.

For each of the different numerical illustrations reported in this chapter, the Python code of the numerical simulations is provided, allowing the readers to easily modify the control objectives and further experience the control theory of the considered dynamical systems.

The outline of this chapter is as follows. First finite-dimensional control systems will be considered and some basic definitions will be given on stability, attractivity, etc., providing a sharp introduction to basics of control systems theory. Then in Section 3, parabolic PDEs are considered for the design of finite-dimensional output-feedback controllers towards saturated control schemes. Section 4 is devoted to the wave and Korteweg-de Vries equation, and the use of finite-dimensional controllers to solve the stabilization problems. In these both sections, linear feedback laws and also cone-bounded controllers are designed. Section 5 contains a concluding discussion on current research activities and presents some possible research directions emanating from this chapter.

This chapter has been written following an online course given in LIASFMA school by the second author in April 2021. We would like to thank the Organizing Committee of this school that was composed of Jean-Michel Coron (Sorbonne Université), Tatsien Li (Fudan University), and Zhiqiang Wang (Fudan University). The help of Xinyue Feng has been very much appreciated.

Notation used in this chapter

Spaces $\mathbb{R}^n$ are endowed with the Euclidean norm denoted by $\|\cdot\|$. The associated induced norms of matrices are also denoted by $\|\cdot\|$. Given two vectors X and Y, $\mathrm{col}(X,Y)$ denotes the vector $[X^\top, Y^\top]^\top$. $L^2(0,1)$ stands for the space of square integrable functions on $(0,1)$ and is endowed with the inner product $\langle f, g \rangle = \int_0^1 f(x)g(x)\,\mathrm{d}x$ with associated norm denoted by $\|\cdot\|_{L^2(0,1)}$ or simply $\|\cdot\|_{L^2}$ when there is no ambiguity. For an integer $m \geq 1$, the m-order Sobolev space is denoted by $H^m(0,1)$ and is endowed with its usual norm denoted by $\|\cdot\|_{H^m}$. For a symmetric matrix $P \in \mathbb{R}^{n \times n}$, $P \succeq 0$ (resp. $P \succ 0$) means that P is positive semi-definite (resp. positive definite) while $\lambda_M(P)$ (resp. $\lambda_m(P)$) denotes its maximal (resp. minimal) eigenvalue. For a symmetric matrix, $\star$ stands for the symmetric term. For instance, $\begin{bmatrix} A & B \\ \star & C \end{bmatrix}$ stands for $\begin{bmatrix} A & B \\ B^\top & C \end{bmatrix}$.

For any Hilbert basis $\{\phi_n, n \geq 1\}$ of $L^2(0,1)$ and any integers $1 \leq N < M$, we define the operators of projection $\pi_N : L^2(0,1) \to \mathbb{R}^N$ and $\pi_{N,M} : L^2(0,1) \to \mathbb{R}^{M-N}$ by setting $\pi_N f = \left[\langle f, \phi_1 \rangle \ \ldots \ \langle f, \phi_N \rangle \right]^\top$ and $\pi_{N,M} f = \left[\langle f, \phi_{N+1} \rangle \ \ldots \ \langle f, \phi_M \rangle \right]^\top$. We also define $\mathcal{R}_N : L^2(0,1) \to L^2(0,1)$ by $\mathcal{R}_N f = f - \sum_{n=1}^N \langle f, \phi_n \rangle \phi_n = \sum_{n \geq N+1} \langle f, \phi_n \rangle \phi_n$.

2. Finite-dimensional systems

2.1. Stability notions of nonlinear finite-dimensional systems

This section is devoted to the introduction of control theory for finite-dimensional systems, as those described by nonlinear dynamics. To be more specific, let us consider the following dynamical system:

$$\dot{z}(t) = f(z(t)) \tag{1}$$

where the state $z(t)$ is a vector from a finite-dimensional state-space $\mathbb{R}^n$ and f is a nonlinear function from $\mathbb{R}^n$ to $\mathbb{R}^n$. Under suitable regularity assumptions, such as locally Lipschitz continuity of f with respect to z, for any given initial condition $z_0 \in \mathbb{R}^n$ there exists a unique solution $x : [0,T) \to \mathbb{R}^n$ to the Cauchy problem:

$$\begin{aligned} \dot{z}(t) &= f(z(t)), \quad t > 0, \\ z(0) &= z_0 \end{aligned} \tag{2}$$

defined on a maximal interval of existence $[0,T)$ for some $T > 0$ (which depends on z_0). See e.g. [30, Theorem 3.1] for such a existence and uniqueness result. The value z_0 is called the initial condition and, at any time $t \in [0, T)$, the value $z(t)$ is called the state at time t.

Assume further that $f(0) = 0$. This implies that the constant trajectory $z(t) = 0$, for all $t \geq 0$, is a particular solution to (1) associated with the initial condition $z_0 = 0$. The point $0 \in \mathbb{R}^n$, sometimes referred to as the origin, is called an equilibrium for (1). In control theory, the nature of an equilibrium is characterized by certain "stability" properties. Some basic definitions related to the concept of "stability" are introduced in the following definition.

Definition 1. Assume that $f(0) = 0$. Then the equilibrium 0 of (1) is said to be

- stable if for any $\varepsilon > 0$, there exists $\delta > 0$ such that
$$|z(0)| \leq \delta \Rightarrow |z(t)| \leq \varepsilon, \ \forall t \geq 0;$$
- attractive if there exists $\delta > 0$ such that
$$|z(0)| \leq \delta \Rightarrow z(t) \to_{t \to +\infty} 0;$$
- asymptotically stable if it is both stable and attractive.

In the previous definition, it is implicitly required that the solutions exist, are unique, and are well defined for all $t \geq 0$. Even implicit, these requirements are of primary importance. Some of them can be difficult to check in practice depending upon the nature of the studied system.

Assuming that 0 is an attractive equilibrium of (1), an important concept is the notion of *basin of attraction*. This is defined as the set of all initial conditions $z_0 \in \mathbb{R}^n$ such that the solution to (2) tends to 0 as $t \to \infty$. In addition, we say that the equilibrium is *globally attractive* if it is attractive and the basin of attraction coincides with the whole state-space $\mathbb{R}^n$. When 0 is not globally attractive, we often write that 0 is *locally asymptotically stable* (LAS) to emphasize the "local" nature of the property. Finally, we say that 0 is *globally asymptotically stable* (GAS) if it is asymptotically stable and globally attractive. It is worth being noted that the notions of attractivity and stability are disconnected. More specifically, there exist systems for which 0 is stable but not attractive (the most simple example being $\dot{z} = 0$) while there are also systems such that 0 is attractive but not stable (see for instance the example of [24, Paragraph 40]).

Instead of (1), let us now consider the case where the dynamics depends on an external signal, called the *control* or the *input*. More specifically, consider the dynamics described by

$$\dot{z}(t) = f(z(t), u(t)) \tag{3}$$

where $u(t)$ is a vector of $\mathbb{R}^m$. The input u is seen as a way to influence the dynamics of the system, which can significantly vary depending on the choice of the control. As an example, consider the following control system described by

$$\dot{z}(t) = u(t)z(t) \tag{4}$$

with $u(t) \in \mathbb{R}$. If $u(t) = u \in \mathbb{R}$ is constant control, the trajectories of the system stating at time $t = 0$ from the initial condition $z_0 \in \mathbb{R}^n$ can be expressed as $z(t) = e^{ut}z_0$ for all $t \geq 0$. For $u = -1$ (more generally for any constant control $u < 0$), the equilibrium 0 is globally asymptotically stable. For $u = 0$, any point of $\mathbb{R}^n$ is an equilibrium (they are stable but not attractive). For $u = 1$ (more generally for any constant control $u > 0$), all solutions to (4) with non zero initial condition $z_0 \neq 0$ diverge to infinity (the equilibrium 0 is neither stable nor attractive).

In the more general setting of a time-varying control, i.e., $u = u(t)$ for a suitable function u of the time, (3) is a time-varying system. This implies that the solution starting from an initial condition z_0 at time t_0 differs, in general, from the trajectory starting from the same initial condition z_0 but at a different time $t_1 \neq t_0$. The behavior of these different solutions can be very different.

Assuming that f in (1) is linear, the system dynamics reduces to

$$\dot{z}(t) = Az(t) \tag{5}$$

where A is a matrix in $\mathbb{R}^{n \times n}$. In this case, the stability of the origin is intimately related to the position of the eigenvalues of the matrix A in the complex plane (see [26, Theorem 6.1]). More specifically, it can be proven that the origin of (5) is stable if and only if (i) all eigenvalues of A have a non-positive real part and (ii) for all eigenvalues with a zero real part, their algebraic multiplicity (exponent associated with the eigenvalue when computing the characteristic polynomial) coincides with their geometric multiplicity[a] (dimension of the eigenspace associated with the eigenvalue). Moreover, the origin of (5) is asymptotically stable if and only if all eigenvalues of A have a negative real part. In that case we say that the matrix A is Hurwitz. Finally, for such linear systems, the attractivity of the origin of (5) implies that the origin is stable and also asymptotically stable.

In this lecture notes, we will first study finite-dimensional control systems, and then dynamical control systems described by linear partial dif-

---

[a]Condition (ii) is crucial as it can be seen by considering the case $A = \begin{bmatrix} 0 & 1 \\ 0 & 0 \end{bmatrix}$.

ferential equations (PDEs) for which some nonlinear control problems will be solved.

2.2. *Control systems: a basic tour*

We focus in the the first part of this section on systems described by

$$\dot{z} = Az + Bu \qquad (6)$$

where $z(t) \in \mathbb{R}^n$ is the state, $u(t) \in \mathbb{R}^m$ is the control, A, B are two matrices of appropriate dimensions. One natural question is the design of a so-called stabilizing state-feedback law. That is, can we compute state-feedback law $z \mapsto u(z)$ so that the resulting closed-loop system

$$\dot{z} = Az + Bu(z) \qquad (7)$$

is asymptotically stable? In this context, due to the linearity of the system, it is natural to try to determine a state-feedback law $z \mapsto u(z)$ that is also linear, i.e., which takes the form $u = Kz$ where K is a matrix that is referred to as the feedback gain. In this setting, the closed-loop system dynamics reads

$$\dot{z} = (A + BK)z. \qquad (8)$$

Consequently the stability properties of the closed-loop system are fully characterized by the spectrum of the closed-loop matrix $A + BK$. The question is: can we compute a matrix K in order to impose the spectrum of $A + BK$ to ensure stability properties for the closed-loop system?

For linear finite-dimensional systems, the control theory is complete and the design of stabilizing state-feeback laws is fully solved.[1,26] More specifically, assuming the following *Kalman rank condition* (for controllability)

$$\mathrm{rank}\begin{bmatrix} B, & AB, \ldots, & A^{n-1}B \end{bmatrix} = n,$$

there exists a matrix K so that $u(z) = Kz$ makes the system (7) asymptotically stable. Furthermore, the matrix gain K can be selected to impose any arbitrary spectrum assignment for the closed-loop matrix $A + BK$. This result is not only an existence result, but it is also a practical design method. Indeed, it is the base of efficient numerical algorithms to compute the control matrix K. This is the so-called pole-shifting theorem (see [73] for an existence result and [71] for a constructive algorithm), which is stated in the next result.

Theorem 1. *Under the Kalman rank condition assumption, for any polynomial Π of degree n and with unit dominant coefficient, there exists a matrix K such that the characteristic polynomial of $A + BK$ is Π.*

With the previous result, computing a matrix K so that the linear state-feedback law $z \mapsto Kz$ renders the origin of the closed-loop system (8) asymptotically stable is numerically tractable.

Example 1. Let us see how to solve this control problem in practice using the programming language Python. In the next lines, with dimension $n = 3$, first a randomly chosen control system is selected (lines 7–8), the controllability condition is checked and a pole-placement controller is computed using the *Python Control Systems Library* (lines 10–18). Then the differential equation is integrated numerically and the phase-portrait of the solution is plotted (lines 27–33). This gives Fig. 1 where it can be checked that a solution converges to the equilibrium 0 in $\mathbb{R}^3$.

```
 1  import numpy as np
 2  import control
 3  from scipy.integrate import odeint
 4  import matplotlib as mpl
 5  import matplotlib.pyplot as plt
 6
 7  n= 3 # dimension of the state space
 8  A= np.random.random([n,n])
 9  B= np.random.random([n,1])
10  CtrbMatrix= control.ctrb(A,B) # compute the controlability matrix
11
12  if np.linalg.matrix_rank(CtrbMatrix)== n:
13      print('controllable system')
14  else:
15      print('uncontrollable system')
16
17  p= np.linspace(-n,-1,n) # choice of the eigenvalues of the closed-
        loop system
18  K=-control.place(A,B,p)
19
20  def ode(z,t):
21      return np.dot((A+np.dot(B,K)),z)
22
23  z0=np.random.random([n,1]); z0=z0.reshape(n,)
24  t=np.linspace(0,10,1000)
25  sol=odeint(ode,z0,t)
26
27  if n==3: # plot3D
28      from mpl_toolkits.mplot3d import Axes3D
29      mpl.rcParams['legend.fontsize'] = 10
30      fig = plt.figure(); ax = fig.gca(projection='3d')
31      x, y, z =sol.T
32      ax.plot(x, y, z, label='solution'); ax.legend()
33      plt.savefig('solution.png',bbox_inches='tight')
```

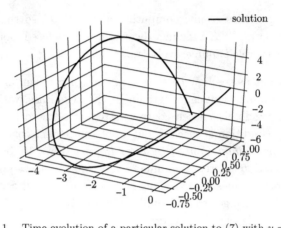

Fig. 1.   Time-evolution of a particular solution to (7) with $u = Kz$

We repeat the same procedure for 10 randomly chosen initial conditions. See the lines 35–41 of the code and the corresponding Fig. 2.

```
35  fig = plt.figure(); ax = fig.gca(projection='3d')
36  for i in range(10):
37      z0=np.random.random([n,1]); z0=z0.reshape(n,)
38      sol=odeint(ode,z0,t)
39      x, y, z =sol.T
40      ax.plot(x, y, z, label='solution'+str(i));
41  ax.legend()
42
43  plt.savefig('solutions.png',bbox_inches='tight')
```

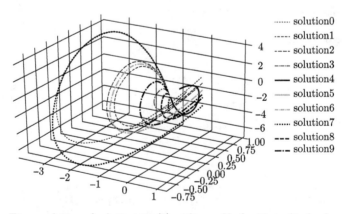

Fig. 2.   Time-evolution of solutions to (7) with $u = Kz$ for 10 randomly chosen initial conditions

2.3. *Lyapunov direct method*

The previous part of this section was devoted to linear systems for which the situation is relatively simple as the stability of the origin is fully characterized by the spectrum of the matrix A. When considering general nonlinear systems such as (1), the situation becomes much more complex. Here we need tools that allow studying the stability properties of an equilibrium condition without being able to write down the system trajectories in closed form (in general, very few nonlinear systems can be analytically integrated to obtain the closed form of the trajectories). In this context, an important tool to prove the attractivity of the equilibrium is the so-called *Direct Lyapunov method* which relies on the concept of *Lyapunov functions*. To explain this method, let us come back to the nonlinear system described by (1). The so-called Lyapunov stability theorem can be stated as follows (see [30, Theorem 4.1] for a proof).

Theorem 2. *Assume that $f(0) = 0$ and let D be an open and connected subset of $\mathbb{R}^n$ containing 0. Assume that $V : D \to \mathbb{R}$ is a $\mathcal{C}^1$ function such that*

$$V(0) = 0 \text{ and } V(z) > 0, \ \forall z \in D \setminus \{0\},$$

$$\frac{\partial V}{\partial z}(z) \cdot f(z) \leq 0, \ \forall z \in D.$$

Then $z = 0$ is stable. Moreover, if we have

$$\frac{\partial V}{\partial z}(z) \cdot f(z) < 0, \ \forall z \in D \setminus \{0\},$$

then $z = 0$ is locally asymptotically stable.

We often denote

$$\dot{V} = \frac{\partial V}{\partial z}(z) \cdot f(z)$$

since $\frac{\partial V}{\partial z}(z(t)) \cdot f(z(t))$ is the time-derivative of $V(z(t))$ along the solutions to (1).

If the Lyapunov theorem applies with the domain D specified as

$$D = \{z, \ V(z) < r\}$$

for some given $r > 0$, then the level set $\{z, \ V(z) < r\}$ is contained in the basin of attraction. Hence V can be used in order to estimate the basin of attraction while trying to maximize the value of $r > 0$ such that Theorem 2 applies with $D = \{z, \ V(z) < r\}$.

It is worth noting that, for finite-dimensional systems as the ones that are considered in this section, all norms are equivalent and, somehow, V is "equivalent" to any norm (say, e.g., the Euclidian norm). Thus establishing a stability by considering a particular norm is actually the same as establishing a stability by considering any other norm. Such an equivalence fails in infinite-dimension, which will be the topic of the next sections.

As we saw, Lyapunov functions are very convenient to prove asymptotic stability since all we need is to consider a suitable Lyapunov function candidate $V : D \to \mathbb{R}$, that is a $\mathcal{C}^1$ function such that

$$V(0) = 0 \text{ and } V(z) > 0, \ \forall z \in D \setminus \{0\}$$

and then compute the following vectors in $\mathbb{R}^n$:

$$\frac{\partial V}{\partial z}(z) \cdot f(z), \ \forall z \in D \setminus \{0\},$$

and evaluate its sign. Obviously, finding such functions V highly depends on the nature of the studied nonlinear system and can be very complex in practice. Some basic techniques for finding such functions will be reviewed in this notes, as well as associated numerically tractable methods.

In the context of linear systems as described by (5), the Lyapunov theorem is rewritten as follows. Using the Lyapunov function candidate $V(z) = z^\top P z$ for some symmetric positive definite matrix P, and computing its time derivative along the system trajectories, the origin of (5) is asymptotically stable if and only if there exists a symmetric positive definite matrix P such that

$$A^\top P + P A^\top \preceq -I.$$

Let us emphasize the "if and only if" condition from the previous statement, as well as the class of quadratic function $V(z) = z^\top P z$ as sufficient Lyapunov function candidates. In other words, for linear systems, there is no need to consider other classes of Lyapunov function candidates. This result is one of the so-called *converse Lyapunov theorems*. Such converse results of the direct method also exist for nonlinear systems under certain regularity conditions on the function f (see e.g. [2, Theorem 2.4]). Note however that converse Lyapunov theorems can hardly be applied to actually find Lyapunov function candidates since these converse results are generally not constructive (even if some design methods exist as reviewed in particular in the references [14, 63]).

Example 2 (Example 1 continued). In this extension of Example 1, we compute the eigenvalues of the previous closed-loop system (see line 45) and we compute a Lyapunov matrix P.

```
45 AA=A+np.dot(B,K); e, v= np.linalg.eig(AA) # eigen-values, -vectors
46 m=max(e.real)
47 print('Largest real part for the closed-loop system:',"{:.2f}".
       format(m))
48
49 P=control.lyap(AA.T,np.eye(n))
```

2.4. *Separation principle for linear systems*

Up to now we only considered the control problem of dynamical systems such as the ones described by (3). In this context, we made the implicit assumption that the full state $z(t)$ is known in real time at any time $t \geq 0$ so that we can use this information to implement the control law $z \mapsto u(z)$. We say that this control strategy takes the form of a state-feedback. However in many applications the full state is not available in real-time. Only partial information is available under the form of sensor measurements $y(t) \in \mathbb{R}^p$ which are somehow related to the state $x(t) \in \mathbb{R}^n$ of the system. For control linear system described by (6), the relation between the output y and the state x generally takes the form:

$$y = Cz \tag{9}$$

where C is a matrix of appropriate dimensions. We say that y is the output of the system. This output represents the measurements that are assumed to be available at each time instant. In this context, a natural question is whether the knowledge of the system (i.e., the matrices A, B, and C), of the control $u(t)$, and of the measurements $y(t)$, is sufficient to asymptotically estimate the state $z(t)$. This problem is a so-called observation problem. For linear systems, this problem is also fully solved and is strongly connected to the so-called *Kalman rank condition for observability*, which is written as

$$\text{rank} \begin{bmatrix} C \\ \vdots \\ CA \\ CA^{n-1} \end{bmatrix} = n.$$

Note that this assumption is equivalent to the controllability of the pair $(A^\top, C^\top)$. This is why observability and controllability properties are said to be dual properties.

Consider now the dynamics described by

$$\dot{\hat{z}} = A\hat{z} + Bu + L(C\hat{z} - y) \tag{10}$$

where L is a matrix with suitable dimensions. Equation (10) mimics the dynamics of the system (6) while adding an extra term used to correct the dynamics of the observation in function of the error between the actual measurement $y(t)$ and its estimation $\hat{y}(t) = C\hat{x}(t)$ obtained from (10). Introducing the error of observation $e = z - \hat{z}$, this error satisfies the dynamics described by

$$\dot{e} = (A + LC)e. \tag{11}$$

Under the above mentioned observability assumption, there exists a matrix L so that $A + LC$ is Hurwitz. Selecting this way the observer gain L, the origin of (11) is asymptotically stable, meaning that the observation error $e(t) = z(t) - \hat{z}(t)$ asymptotically converges to zero. In other words, the state of the observer $\hat{z}(t)$ "asymptotically observes" the actual (unmeasured) stated of the system $z(t)$. We say that (10) is an observer for (6).

So far, we detailed (i) how an state-feedback $u = Kz$ can be designed to stabilize the linear system (6) and (ii) how an observer of the form (10) can be designed in order to compute $\hat{z}$ an estimate of the state z of the system (6) from its outputs y given by (9). A natural question is whether we can reunite these two approaches to obtain a stabilizing output-feedback. In other words, under the controllability assumption of (A, B) and the observability assumption of (A, C), can we separately design a feedback gain K and an observer gain L so that the origin of the system (6) in closed-loop with $u = K\hat{z}$ where the dynamics of $\hat{z}$ is given by (10) is asymptotically stable? The answer to this question is positive and is referred to as the *separatation principle* for linear finite-dimensional systems.

Theorem 3. *Let us consider the dynamics:*

$$\begin{aligned} \dot{z} &= Az + Bu, \\ y &= Cz, \end{aligned} \tag{12}$$

where $z \in \mathbb{R}^n$, $y \in \mathbb{R}^p$ and A, B, C are matrices with suitable dimensions. Assume that the pair (A, B) is controllable and the pair (A, C) is observable. Then for any matrices K and L such that $A + BK$ and $A + LC$ are Hurwitz, the equilibrium $(0, 0)$ of

$$\begin{aligned} \dot{z} &= Az + BK\hat{z}, \\ \dot{\hat{z}} &= (A + BK)\hat{z} + L(C\hat{z} - y) \end{aligned} \tag{13}$$

is asymptotically stable.

This theorem provides a design method for a stabilizing dynamic output-feedback controller whose architecture is described by

$$\dot{\hat{z}} = A\hat{z} + Bu + L(C\hat{z} - y),$$
$$u = K\hat{z}.$$

Proof of Theorem 3. For proving Theorem 3, it is convenient not to study the asymptotic stability of the origin of (13) in the coordinates $(z, \hat{z})$, but rather in the coordinates $(\hat{z}, e)$ which give

$$\dot{\hat{z}} = (A + BK)\hat{z} + L(y - C\hat{z}),$$
$$\dot{e} = (A + LC)e. \tag{14}$$

Since $A + LC$ is Hurwitz, there exists a symmetric positive definite matrix Q such that

$$(A + LC)^{\top}Q + Q(A + LC)^{\top} \preceq -I \tag{15}$$

and so $W(e) = e^{\top}Qe$ satisfies

$$\dot{W} \leq -e^{\top}e$$

along the trajectories of $\dot{e} = (A + LC)e$. Hence the e-component of (14) converges to 0 as time goes to $+\infty$. Now pick a symmetric positive definite matrix P such that

$$(A + BK)^{\top}P + P(A + BK)^{\top} \preceq -I. \tag{16}$$

Letting $V(\hat{z}) = \hat{z}^{\top}P\hat{z}$ we have

$$\dot{W} \leq -\hat{z}^{\top}\hat{z} + 2\hat{z}LCe$$

along the trajectories of $\dot{\hat{z}} = A\hat{z} + BK\hat{z} + L(y - C\hat{z})$. Invoking now Young's inequality and the fact that $e(t) \to 0$ gives that $\hat{z}(t)$ goes to 0 as well when time goes to $+\infty$. Therefore, the origin of the linear system (14) is attractive and thus asymptotically stable.

Note that another proof of the asymptotic stability of the origin of (14) is based on proving that $V + 4\|PLC\|^2 W$ is actually a Lyapunov function. To do that denote $\mathcal{V}(\hat{z}, e) = V(\hat{z}) + 4\|PLC\|^2 W(e)$ and compute the time derivative of $\mathcal{V}$ along the solutions to (14):

$$\dot{\mathcal{V}} = \hat{z}^{\top}((A + BK)^{\top}P + P(A + BK))\hat{z} + 2\hat{z}^{\top}PLCe$$
$$+ 4\|PLC\|^2 e^{\top}((A + LC)^{\top}Q + (A + LC)Q)e$$
$$\leq -\frac{1}{2}\|\hat{z}\|^2 + 2\|PLCe\|^2 - 4\|PLC\|^2\|e\|^2,$$

where Young's inequality, (15) and (16) have been used for the previous inequality. Therefore $\dot{\mathcal{V}} \preceq -\dfrac{1}{2}I$, and $\mathcal{V}$ is a Lyapunov function for (14). □

The computation done in the proof of Theorem 3 will be generalized for PDEs in the next sections.

Example 3 (Example 1 continued). In this part of the example, we first select a randomly chosen matrix, and we check the Kalman rank condition for observability (lines 54–57). Then we compute a matrix L by placing the eigenvalues of the matrix $A + LC$ (line 60), and finally we plot solutions of (13) for 10 randomly chosen initial conditions (lines 72–79).

```
51  C= np.random.random([1,n])
52  ObsvMatrix= control.obsv(A,C) # compute the observability matrix
53
54  if np.linalg.matrix_rank(ObsvMatrix)== n:
55      print('observable system')
56  else:
57      print('unobservable system')
58
59  q= np.linspace(-n-1,-2,n) # choice of the eigenvalues of the closed
        -loop system
60  L=-control.place(A.T,C.T,q).T
61
62  def ode2(ztot,t):
63      z=ztot[:n]; zhat=ztot[n:]
64      u= np.dot(np.dot(B,K),zhat)
65      return np.concatenate((np.dot(A,z)+u ,np.dot(A,zhat)+u
            -np.dot(L,np.dot(C,z)-np.dot(C,zhat))))
66
67  # set up a figure twice as wide as it is tall
68  fig = plt.figure(figsize=plt.figaspect(0.5))
69  ax0 = fig.add_subplot(1, 2, 1, projection='3d')
70  ax1 = fig.add_subplot(1, 2, 2, projection='3d')
71
72  for i in range(10):
73      z0=np.random.random([n,1]); z0=z0.reshape(n,)
74      zhat0=np.random.random([n,1]); zhat0=zhat0.reshape(n,)
75      ztot0=np.concatenate((z0,zhat0))
76      sol=odeint(ode2,ztot0,t)
77      ztot =sol.T; z=ztot[:n]; zhat=ztot[n:];
78      ax0.plot(z[0], z[1], z[2]);
79      ax1.plot(zhat[0], zhat[1], zhat[2]);
80
81  ax0.set_title('z'); ax1.set_title('$\hat z$')
82  plt.savefig('solutions2.png',bbox_inches='tight')
```

Fig. 3 presents several solutions to (14) for randomly chosen initial conditions $(z(0), \hat{z}(0))$, and confirms the attractivity of the origin for this system.

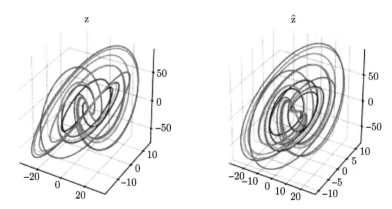

Fig. 3.   Time-evolution of solutions to (13) for 10 randomly chosen initial conditions

The Lyapunov function that is considered at the end of the proof of Theorem 3 is computed on lines 49, 85 and 86. It is checked on Fig. 4 that this function decreases and converges to 0 along the solutions to (14) for the initial conditions used for Fig. 3.

```
84  AE=A+np.dot(L,C);
85  Q=control.lyap(AE.T,np.eye(n))
86  M= 4*np.linalg.norm(np.dot(P,np.dot(L,C)))**2
87
88  fig , ax= plt.subplots()
89  ax.set_title('Lyapunov function')
90
91  for i in range(10):
92      z0=np.random.random([n,1]); z0=z0.reshape(n,)
93      zhat0=np.random.random([n,1]); zhat0=zhat0.reshape(n,)
94      ztot0=np.concatenate((z0,zhat0))
95      sol=odeint(ode2,ztot0,t)
96      ztot=sol.T; z=ztot[:n]; zhat=ztot[n:]; e=z-zhat; lyapu=[]
97      for tt in range(len(t)):
98          lyapu.append(np.dot(np.dot(zhat[:,tt].T,P),zhat[:,tt])+M*np
            .dot(np.dot(e[:,tt].T,Q),e[:,tt]))
99      ax.plot(t,lyapu)
100 plt.savefig('lyapu2.png',bbox_inches='tight')
```

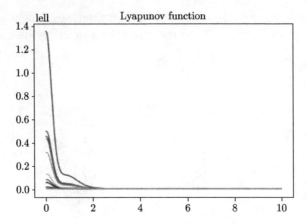

Fig. 4. Time-evolution of the designed Lyapunov function along several solutions to (14)

2.5. *Saturated control*

For many applications of control problems, the input values are limited in amplitude. Instead of applying $u = Kz$, only

$$u = \mathtt{sat}(Kz)$$

can actually be applied, where $\mathtt{sat}\colon \mathbb{R}^m \to \mathbb{R}^m$ is the saturation map defined componentwise by, for all $i = 1, \ldots, m$,

$$\mathtt{sat}_i(\sigma_i) = \begin{cases} \sigma_i & \text{if } |\sigma_i| < s_i, \\ \mathtt{sign}(\sigma_i)s_i, & \text{else} \end{cases} \qquad (17)$$

for a fixed vector s in $\mathbb{R}^m$ with positive components $s_i > 0$. Such function is a decentralized nonlinear map that makes the closed-loop system as follows:

$$\dot{z} = Az + B\mathtt{sat}(Kz). \qquad (18)$$

In the presence of a saturation, system (18) can exhibit various behaviors. Even if the matrix $A + BK$ is Hurwitz, there may exist several equilibrium points, some limit cycle may appear, and there may exist diverging trajectories. See [70, 74] for introductory references on stability of such dynamical systems.

Example 4. As an example, consider

$$\dot{z} = Az + B\mathtt{sat}(Kz) \qquad (19)$$

with $A = \begin{pmatrix} 0 & 1 \\ 1 & 0 \end{pmatrix}$, $B = \begin{pmatrix} 0 \\ -1 \end{pmatrix}$, $K = (13\ 7)$, and $s = 5$ as saturation level. The matrix A is unstable (eigenvalues located at -1 and $+1$), and the matrix $A + BK$ is Hurwitz (eigenvalues located at -1 and -13). As noted in [70, Example 1.1], the nonlinear system (19) exhibits several equilibriums and presents different behaviors depending on the initial condition. These behaviors are illustrated on Fig. 5 based on different initial conditions. The first trajectory converges to 0 in $\mathbb{R}^2$, the second trajectory converges to the non zero equilibrium point $\begin{pmatrix} -5 \\ 0 \end{pmatrix}$, and the last trajectory diverges as the time increases.

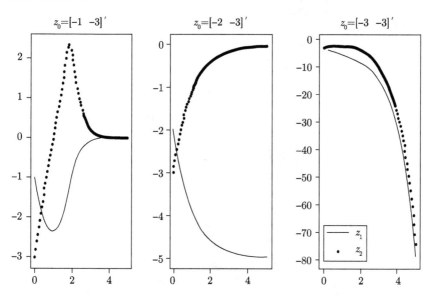

Fig. 5.   Time-evolutions of three solutions to (19) for three different initial conditions

The simulation code is given below.

```
import numpy as np
import control
from scipy.integrate import odeint
import matplotlib as mpl
import matplotlib.pyplot as plt

n= 3 # dimension of the state space
A= np.random.random([n,n])
B= np.random.random([n,1])
CtrbMatrix= control.ctrb(A,B) # compute the controllability matrix
```

```
11
12  if np.linalg.matrix_rank(CtrbMatrix)== n:
13      print('controllable system')
14  else:
15      print('uncontrollable system')
16
17  p= np.linspace(-n,-1,n) # choice of the eigenvalues of the closed-
        loop system
18  K=-control.place(A,B,p)
19
20  def ode(z,t):
21      return np.dot((A+np.dot(B,K)),z)
22
23  z0=np.random.random([n,1]); z0=z0.reshape(n,)
24  t=np.linspace(0,10,1000)
25  sol=odeint(ode,z0,t)
26
27  if n==3: # plot3D
28      from mpl_toolkits.mplot3d import Axes3D
29      mpl.rcParams['legend.fontsize'] = 10
30      fig = plt.figure(); ax = fig.gca(projection='3d')
31      x, y, z =sol.T
32      ax.plot(x, y, z, label='solution'); ax.legend()
33      plt.savefig('solution.png',bbox_inches='tight')
34  # for 10 randomnly chose initial conditions
35  fig = plt.figure(); ax = fig.gca(projection='3d')
36  for i in range(10):
37      z0=np.random.random([n,1]); z0=z0.reshape(n,)
38      sol=odeint(ode,z0,t)
39      x, y, z =sol.T
40      ax.plot(x, y, z, label='solution'+str(i));
41  ax.legend()
42
43  plt.savefig('solutions.png',bbox_inches='tight')
```

To analyze the stability of the equilibrium 0 of (18), let us consider the following Lyapunov function candidate $V : z \mapsto z^\top P z$, where $P \in \mathbb{R}^{n \times n}$ is a symmetric definite positive matrix. The computation of its time derivative along the solutions of (18) gives

$$\dot{V} = z^\top (A^\top P + PA)z + 2z^\top PB\mathrm{sat}(Kz).$$

To ease the comparison of the dynamics of $\dot{z} = Az + Bkz$ and of (18), we introduce the deadzone function ϕ defined by

$$\phi(\sigma) = \mathrm{sat}(\sigma) - \sigma, \ \forall \sigma \in \mathbb{R}^m. \tag{20}$$

Using this notation we get

$$\dot{V} = z^{\top}((A+BK)^{\top}P + P(A+BK))z + 2z^{\top}PB\phi(Kz)$$

$$= \begin{pmatrix} z \\ \phi(Kz) \end{pmatrix}^{\top} \begin{pmatrix} (A+BK)^{\top}P + P(A+BK) & PB \\ \star & 0 \end{pmatrix} \begin{pmatrix} z \\ \phi(Kz) \end{pmatrix}.$$

Note that the matrix $\begin{pmatrix} (A+BK)^{\top}P + P(A+BK) & PB \\ \star & 0 \end{pmatrix}$ can not be in general negative semidefinite (except, e.g., for the trivial case $B = 0$). Consequently, in order to use the Lyapunov function candidate V to analyze the stability of the origin of (18), we need to find a relation between z and $\mathsf{sat}(Kz)$. This can be done by using the geometric conditions of the saturation map, as described by the so-called local and global sector conditions.

As introduced in [23], for any given $G \in \mathbb{R}^{m \times n}$ and any given diagonal positive definite matrix $T \in \mathbb{R}^m$, the following *local sector condition* holds:

$$(\mathsf{sat}(Kz) - Kz)T(\mathsf{sat}(Kz) - (K-G)z) \le 0, \ \forall z \text{ such that } |((K-G)z)_i| \le s_i, \tag{21}$$

for all i, where $(K-G)_{(i)}$ denotes the ith row of $K - G$.

Letting in particular $G = K$ in (21), the following *global sector condition* holds for any diagonal positive definite matrix T,

$$(\mathsf{sat}(Kz) - Kz)T\mathsf{sat}(Kz) \le 0, \ \forall z \in \mathbb{R}^m. \tag{22}$$

From the local sector condition, we obtain that for any $G \in \mathbb{R}^{n \times m}$ and any diagonal positive definite matrix T, as long as $|((K-G)z)_i| \le s_i$, for all i

$$\dot{V} \le z^{\top}((A+BK)^{\top}P + P(A+BK))z + 2z^{\top}PB\phi(Kz)$$

$$- 2\phi(Kz)^{\top}T(\phi(Kz) + Gz)$$

$$\le \begin{pmatrix} z \\ \phi(Kz) \end{pmatrix}^{\top} \begin{pmatrix} (A+BK)^{\top}P + P(A+BK) & PB - G^{\top}T \\ \star & -2T \end{pmatrix} \begin{pmatrix} z \\ \phi(Kz) \end{pmatrix}.$$

Considering the special case where $G = K$, we obtain the following theorem.

Theorem 4. *If there exist a symmetric definite matrix P in $\mathbb{R}^{n \times n}$ and a diagonal positive definite matrix T in $\mathbb{R}^m$ such that*

$$\begin{pmatrix} (A+BK)^{\top}P + P(A+BK) & PB - K^{\top}T \\ \star & -2T \end{pmatrix} \prec 0,$$

then the origin of (19) is globally asymptotically stable.

Remark 1. Some observations are in order.

Checking the existence of such matrices P and T is numerically tractable. This is a convex problem that could be solved using different solvers and method as interior-point method,[22] or a primal/dual method.[21] See also [6].

As discussed in [67], the existence of a globally stabilizing saturating control is subject to a number of strong conditions such as: (i) A has no eigenvalues with positive real part, and (ii) the pair (A, B) is stabilizable in the ordinary sense, that there exists a matrix K such that $A + BK$ is asymptotically stable.

Setting $G = K$ is generally restrictive since global asymptotic stability is generally a too strong property for saturated systems. To derive a sufficient condition for the weaker property of local asymptotic stability, we use the local sector condition (21). Given i in $\{1, \cdots, n\}$ to ensure the condition $|((K - G)z)_i| \le s_i$, we note that $\{z, \ z^\top P z \le 1\} \subset \{z, \ |((K - G)z)_i| \le s_i\}$ provided the LMI condition

$$\begin{pmatrix} P & (K - G)^\top_{(i)} \\ \star & s_i^2 \end{pmatrix} \succeq 0$$

holds. This result is a direct consequence of the Schur complement (see [6, Page 7]). Returning now to the linear matrix inequality

$$\begin{pmatrix} (A + BK)^\top P + P(A + BK) & PB - G^\top T \\ \star & -2T \end{pmatrix} \prec 0,$$

we note that there is a product $G^\top T$ of unknown variables, making the problem nonlinear. Nevertheless, the problem can be made linear by introducing a simple change of variable. Indeed, using the change of variables $S = T^{-1}$, $W = P^{-1}$, and $H = GP^{-1}$, we obtain the equivalent condition

$$\begin{pmatrix} W(A + BK)^\top + (A + BK)W & BS - H^\top \\ \star & -2S \end{pmatrix} \prec 0.$$

We have thus proven the following sufficient condition for local asymptotic stability of (19).

Theorem 5. *If there exist* $W = W^\top > 0$, S *diagonal definite positive and* G *such that*

$$\begin{pmatrix} W & WK_{(i)}^\top - H_{(i)}^\top \\ \star & s_i^2 \end{pmatrix} \succeq 0, \tag{23}$$

$$\begin{pmatrix} W(A + BK)^\top + (A + BK)W & BS - H^\top \\ \star & -2S \end{pmatrix} \prec 0, \tag{24}$$

then the origin of (19) *is locally asymptotically stable with a basin of attraction containing* $\{z, \ z^\top W^{-1} z \le 1\}$.

Remark 2. Checking the condition of Theorem 5 reduces to solving a convex problem. Different optimization criterion can be considered in order to maximize the estimation of the basin of attraction, as e.g., maximizing the trace of the matrix W. This idea is illustrated in Example 5 below.

Note that this sufficient condition for local asymptotic stability of the closed-loop system can also be used in order to compute the matrix of feedback gain K. See [70, Chapter 3] and in the next sections for infinite-dimensional dynamics.

Example 5 (Example 4 continued). Solving the matrix inequalities of Theorem 5 is done with the code below, where the *Python cvxpy Library* has been used to write the matrix conditions in lines 52–64 with the unknown variables introduced in lines 46–48. The optimization problem

$$\max_{W,S,H} \ \text{trace}(W) \text{ such that (23) and (24) hold}$$

is solved in line 56, using the default solver.

```
44 n=len(A); m= 1
45
46 W=cp.Variable((n,n),PSD=True)
47 S=cp.Variable((m,m),diag=True)
48 H=cp.Variable((m,n))
49 B=B.reshape(2,1)
50 K=K.reshape(1,2)
51
52 M11=W @ (np.transpose(A+np.dot(B,K)))
53 M11=M11+M11.T
54 M12=B @ S - H.T
55 matrixConstr1 = cp.bmat([[ M11 , M12],
56                          [M12.T, -2 *S]])
57 M22= W @ K.T  - H.T
58 matrixConstr2 = cp.bmat([[W,       M22    ],
59                          [M22.T ,s0 ** 2*np.array([[1]])]])
60
61 constr = [S >> 0]
62 constr += [matrixConstr1<<0] + [matrixConstr2>>0]
63 prob = cp.Problem(cp.Maximize(cp.trace(W)),constr)
64 prob.solve()
65
66 P=np.linalg.inv(W.value)
67 z0=z0tot[0]
68 print("z0^T P z0 is "+str(np.dot(np.dot(z0.T,P),z0)))
```

It gives

$$P = \begin{pmatrix} 0.19727007 & 0.11506782 \\ \star & 0.08307019 \end{pmatrix}$$

for which, due to Theorem 5, $\{z, \ z^\top P z \leq 1\}$ is included in the basin of attraction.

For the first initial condition of Fig. 5, that is with $z_0 = [-1 - 3]^\top$, we have

$$z_0^\top P z_0 = 0.99 < 1.$$

Thus z_0 is indeed in the basin of attraction, as confirmed by the time-evolution of the first plot in Fig. 5.

2.6. *Section conclusion*

This section was devoted to finite-dimensional control systems by recalling some basic definitions and techniques for the stability analysis of equilibrium of such dynamical systems. In particular we reviewed the direct Lyapunov method for the asymptotic stability analysis. The control systems with saturated inputs have been also considered, and some sufficient conditions for the local (and global) asymptotic stability of the origin have been recalled. The next section will develop these techniques for the boundary stabilization of parabolic and hyperbolic systems.

3. Parabolic equations

3.1. *Introduction*

This section considers parabolic partial differential equations modeling reaction-diffusion phenomenon. This class of dynamical systems may be unstable in open loop. We focus on 1D parabolic equations for which spectral decomposition can be easily handled since the eigenvalues are simple and the eigenfunctions form a Hilbert basis of the state-space. For further studies on abstract parabolic PDEs in several dimensional spaces, see [7] in particular for controllability properties of such systems.

Based on the basic tools presented in the previous section, we present design methods for the design of output-feedback laws rendering the equilibrium asymptotically stable. The approach is based on modal approximation methods that have been shown to be efficient for other control problems related to parabolic PDEs; see [61] as well as more recent references including [15, 16, 37, 43, 50, 57]. The rationale behind the design method presented in this section is split into several steps. First a finite-dimensional state-feedback is computed only with a finite number of selected modes of the model. Then a finite-dimensional observer is designed in a separate fashion in order to estimate a finite number of modes that include in particular the modes used for the state-feedback design. Such a control design approach roots back to the pioneer papers [3, 18, 25, 62] which essentially rely on small gain arguments. Taking advantage of the controller architecture reported in [62], the possibility to recast this control design problem into a LMI framework was shown in [29] for a particular set of input/output maps and specific norms for the asymptotic stability estimates. This procedure was enhanced and generalized in a systematic manner in [40, 41] for general reaction-diffusion PDEs with Dirichlet/Neumann/Robin boundary control and Dirichlet/Neumann measurement while performing the control design directly with the control input instead of its time derivative (see [17] for an introduction to boundary control systems). This generalized and systematic approach has been shown to be key and very efficient for the predictor-based compensation of arbitrarily long input and output delays,[38,42] the domination of state-delays,[44] the local output-feedback stabilization of linear reaction-diffusion PDEs in the presence of a saturation,[34] the global stabilization of linear-reaction-diffusion PDEs in the presence of a Lipschitz continuous sector nonlinearity in the application of the boundary control,[40] as well as the global stabilization of semilinear reaction-diffusion PDE with globally Lipschitz nonlinearity.[39]

In this framework, the proof of stability of the closed-loop system (composed of the PDE, the finite-dimensional observer, and the state-feedback) is assessed using the Lyapunov direct method presented in the previous section, but adapted to the distributed nature of the state. This approach can be seen as an alternative output-feedback design method for reaction parabolic PDEs to other very efficient tools, such as backstepping transformations for PDEs (see the introductory textbook [31]) for which a form of separation principle between controller and observer designs generally exists. Nevertheless, the infinite-dimensional nature of the observer obtained using backstepping methods implies the necessity to resort to late lumping approximations in order to obtain a finite-dimensional control strategy that is suitable for practical implementation, inducing in general the loss of the stability performance guarantees originally obtained during the synthesis phase. The benefit of the approach reported in this section is that the observer obtained during the synthesis phase is directly finite-dimensional.

The material presented in this section of the lecture notes is widely inspired from [41] in the linear case and from [34] for the saturated input scenario.

The rest of this section is organized as follows. After introducing a number of notations and properties, the case of Dirichlet boundary control with a bounded observation operator is considered in Section 3.2. The control design procedure is then extended to the cases of a boundary Dirichlet observation in Section 3.3. The case of in-domain control in the presence of an input saturation is discussed in Section 3.4.

Reminders on Sturm Liouville theory

Let us conclude this introduction with some reminders on Sturm-Liouville theory for parabolic operators in one space dimension. See [51] for a reference on the mathematical properties that will be extensively used in this section.

Let $\theta_1, \theta_2 \in [0, \pi/2]$, $p \in \mathcal{C}^1([0,1])$, and $q \in \mathcal{C}^0([0,1])$ with $p > 0$ and $q \geq 0$. Consider the Sturm-Liouville operator $\mathcal{A} : D(\mathcal{A}) \subset L^2(0,1) \to L^2(0,1)$ defined by

$$\mathcal{A}f = -(pf')' + qf$$

on the domain

$$D(\mathcal{A}) = \{f \in H^2(0,1) : \cos(\theta_1)f(0) - \sin(\theta_1)f'(0) = 0,$$
$$\cos(\theta_2)f(1) + \sin(\theta_2)f'(1) = 0\}.$$

The eigenvalues $(\lambda_n)_{n \geq 1}$ of $\mathcal{A}$ are simple, non negative, and form an increasing sequence with $\lambda_n \to +\infty$ as $n \to +\infty$. The associated unit eigenvectors $\Phi_n \in L^2(0,1)$ form a Hilbert basis. The operator $\mathcal{A}$ and its domain can be characterized by these eigenstructures in the sense that

$$\mathcal{A}f = \sum_{n \geq 1} \lambda_n \langle f, \Phi_n \rangle, \quad \forall f \in D(\mathcal{A})$$

and

$$D(\mathcal{A}) = \{f \in L^2(0,1) : \sum_{n \geq 1} |\lambda_n|^2| \langle f, \Phi_n \rangle|^2 < +\infty\}$$

where $\langle f, g \rangle = \int_0^1 f(x)g(x)\mathrm{d}x$, for any $f, g \in L^2(0,1)$, stands for the inner product of $L^2(0,1)$. Hence, using an integration by parts, it can be seen that, for any $f \in D(\mathcal{A})$,

$$\sum_{n \geq 1} \lambda_n \langle f, \Phi_n \rangle^2 = \langle \mathcal{A}f, f \rangle$$

$$= p(0)f(0)f'(0) - p(1)f(1)f'(1) + \int_0^1 \left(p(x)f'(x)^2 + q(x)f(x)^2 \right) dx.$$

Using the boundary conditions involved in the definition of $D(\mathcal{A})$, we infer the existence of a constant $C_2 > 0$ such that

$$\sum_{n \geq 1} \lambda_n \langle f, \Phi_n \rangle^2 = \langle \mathcal{A}f, f \rangle \leq C_2 \|f\|_{H^1}.$$

Moreover, if either (i) $\theta_1, \theta_2 \in \{0, \pi/2\}$ with $\theta_i = 0$ for at least one $i \in \{0, 1\}$; or (ii) $q > 0$, this implies the existence of a constant $C_1 > 0$ such that

$$C_1 \|f\|_{H^1} \leq \sum_{n \geq 1} \lambda_n \langle f, \Phi_n \rangle^2 = \langle \mathcal{A}f, f \rangle \leq C_2 \|f\|_{H^1}. \tag{25}$$

Hence, for any $f \in D(\mathcal{A})$, the series expansion $f = \sum_{n \geq 1} \langle f, \Phi_n \rangle \Phi_n$ holds in $H^1(0,1)$ norm. Then, using the definition of $\mathcal{A}$ and the fact that it is a Riesz-spectral operator, we obtain that the latter series expansion holds in $H^2(0,1)$ norm. Due to the continuous embedding $H^1(0,1) \subset L^\infty(0,1)$, we obtain that

$$f(0) = \sum_{n \geq 1} \langle f, \Phi_n \rangle \Phi_n(0), \quad f'(0) = \sum_{n \geq 1} \langle f, \Phi_n \rangle \Phi'_n(0).$$

Let $p_*, p^*, q^* \in \mathbb{R}$ be such that $0 < p_* \leq p(x) \leq p^*$ and $0 \leq q(x) \leq q^*$ for all $x \in [0,1]$. Then we have:[51]

$$0 \leq \pi^2(n-1)^2 p_* \leq \lambda_n \leq \pi^2 n^2 p^* + q^* \tag{26}$$

for all $n \geq 1$. If we further assume that $p \in \mathcal{C}^2([0,1])$, we have (see again [51]) that

$$\Phi_n(0) = O_{n \to +\infty}(1), \quad \Phi'_n(0) = O_{n \to +\infty}(\sqrt{\lambda_n}). \tag{27}$$

3.2. *Bounded observation operator*

We first consider the reaction-diffusion system described by

$$z_t(t,x) = (p(x)z_x(t,x))_x + (q_c - q(x))z(t,x), \tag{28a}$$

$$z_x(t,0) = 0, \quad z(t,1) = u(t), \tag{28b}$$

$$z(0,x) = z_0(x), \tag{28c}$$

$$y(t) = \int_0^1 c(x)z(t,x)\, dx \tag{28d}$$

for $t > 0$ and $x \in (0,1)$. Here $q_c \in \mathbb{R}$ is a constant, $u(t) \in \mathbb{R}$ is the control input, $y(t) \in \mathbb{R}$ with $c \in L^2(0,1)$ is the measurement, $z_0 \in L^2(0,1)$ is the initial condition, and $z(t, \cdot) \in L^2(0,1)$ is the state.

3.2.1. *Spectral reduction*

In (28), the control input u appears in the right boundary condition. Let us transfer the control input from the boundary into the PDE by invoking the change of variable:

$$w(t,x) = z(t,x) - x^2 u(t). \tag{29}$$

It has been specifically selected in order to ensure that we still have the left boundary condition $w_x(t,0) = 0$ while enforcing $w(t,0) = 0$. Hence, we have

$$w_t(t,x) = (p(x)w_x(t,x))_x + (q_c - q(x))w(t,x) + a(x)u(t) + b(x)\dot{u}(t), \tag{30a}$$

$$w_x(t,0) = 0, \quad w(t,1) = 0, \tag{30b}$$

$$w(0,x) = w_0(x), \tag{30c}$$

$$\tilde{y}(t) = \int_0^1 c(x)w(t,x)\,dx. \tag{30d}$$

Here $a, b \in L^2(0,1)$ are defined by $a(x) = 2p(x) + 2xp'(x) + (q_c - q(x))x^2$ and $b(x) = -x^2$, respectively, while $\tilde{y}(t) = y(t) - \left(\int_0^1 x^2 c(x)\,dx\right)u(t)$ and $w_0(x) = z_0(x) - x^2 u(0)$.

The parabolic equation (30) presents homogeneous boundary conditions (30b) that are much easier to deal with. However, the price of this transfer is the occurrence of the time derivative $\dot{u}$ of the control input u in the PDE (30a). This is why we introduce the auxiliary control input $v(t) = \dot{u}(t)$, that will be used as the control input for control design. In other words, v will be used as the control input for the design of the control strategy. However, for final implementation of the control strategy, u remains the actual control input of the plant. In this context, the dynamics of the system reads

$$\dot{u}(t) = v(t), \tag{31a}$$

$$\frac{dw}{dt}(t, \cdot) = -\mathcal{A}w(t, \cdot) + q_c w(t, \cdot) + au(t) + bv(t) \tag{31b}$$

with $D(\mathcal{A}) = \{f \in H^2(0,1) : f'(0) = f(1) = 0\}$. Introducing the coefficients of projection $w_n(t) = \langle w(t, \cdot), \Phi_n \rangle$, $a_n = \langle a, \Phi_n \rangle$, $b_n = \langle b, \Phi_n \rangle$, and

$c_n = \langle c, \Phi_n \rangle$, the projection of the PDE solutions into the Hilbert basis of eigenfunctions $(\Phi_n)_{n \geq 1}$ gives

$$\dot{u}(t) = v(t), \tag{32a}$$

$$\dot{w}_n(t) = (-\lambda_n + q_c)w_n(t) + a_n u(t) + b_n v(t), \quad n \geq 1, \tag{32b}$$

$$\tilde{y}(t) = \sum_{i \geq 1} c_i w_i(t). \tag{32c}$$

Note that (32) has been obtained from (31) by 1) multiplying (31) by Φ_n; 2) integrating on the space domain; and 3) performing two integration by parts while using the boundary conditions coming from the definition of $D(\mathcal{A})$.

3.2.2. *Control design*

We start by fixing an integer $N_0 \geq 1$ and positive real number $\delta > 0$ such that $-\lambda_n + q_c < -\delta < 0$ for all $n \geq N_0+1$. Let $N \geq N_0+1$ be arbitrary. The general idea, borrowed to [62], is to compute a stabilizing output-feedback controller in three steps. Firstly an observer to estimate the N first modes of the plant is designed. Secondly the state-feedback is only performed on the N_0 first estimated modes of the plant. Finally a dedicated stability analysis is performed to prove that the origin of the closed-loop is asymptotically stable. In this context, inspired by the controller architecture first reported in [62], the adopted control strategy takes the form:

$$\dot{\hat{w}}_n(t) = (-\lambda_n + q_c)\hat{w}_n(t) + a_n u(t) + b_n v(t) + l_n \left(\hat{y}(t) - \tilde{y}(t) \right), \quad 1 \leq n \leq N_0, \tag{33a}$$

$$\dot{\hat{w}}_n(t) = (-\lambda_n + q_c)\hat{w}_n(t) + a_n u(t) + b_n v(t), \quad N_0 + 1 \leq n \leq N, \tag{33b}$$

$$\hat{y}(t) = \int_0^1 c(x) \sum_{i=1}^N \hat{w}_i(t)\Phi_i(x)\,\mathrm{d}x = \sum_{i=1}^N c_i \hat{w}_i(t), \tag{33c}$$

$$v(t) = \dot{u}(t) = \sum_{i=1}^{N_0} k_i \hat{w}_i(t) + k_u u(t), \tag{33d}$$

where $k_i, k_u \in \mathbb{R}$ are the feedback gains while $l_n \in \mathbb{R}$ are the observer gains. Signals $\hat{w}_n$ stand for the estimations of the modes w_n for $1 \leq n \leq N$. These estimations are used for the computation of $\hat{y}$ that represents the estimation of the actual system measurement $\tilde{y}$. Note that the feedback law (33d) is computed only based on the observations $\hat{w}_n$ for $1 \leq n \leq N_0$. The remaining observations, namely $\hat{w}_n$ for $N_0 + 1 \leq n \leq N$, are only used in (33c) to improve the estimation $\hat{y}$ of the actual system output $\tilde{y}$. This estimation $\hat{y}$ is used to introduce a correction term in the observation dynamics (33a) related to the mismatch between the estimation $\hat{y}$ and the measurement $\tilde{y}$. Note that no such correction is applied in (33b) for the observed modes associated with $N_0 + 1 \leq n \leq N$.

In order to study the validity of the control strategy (33), we need to introduce a number of definitions. Introducing

$$W^{N_0}(t) = \begin{bmatrix} w_1(t) \\ \vdots \\ w_{N_0}(t) \end{bmatrix}, \quad B_{0,a} = \begin{bmatrix} a_1 \\ \vdots \\ a_{N_0} \end{bmatrix}, \quad B_{0,b} = \begin{bmatrix} b_1 \\ \vdots \\ b_{N_0} \end{bmatrix},$$

$$A_0 = \mathrm{diag}(-\lambda_1 + q_c, \ldots, -\lambda_{N_0} + q_c),$$

we have from (32b) that

$$\dot{W}^{N_0}(t) = A_0 W^{N_0}(t) + B_{0,a} u(t) + B_{0,b} v(t). \tag{34}$$

Hence, defining

$$W_a^{N_0}(t) = \begin{bmatrix} u(t) \\ W^{N_0}(t) \end{bmatrix}, \quad A_1 = \begin{bmatrix} 0 & 0 \\ B_{0,a} & A_0 \end{bmatrix}, \quad B_1 = \begin{bmatrix} 1 \\ B_{0,b} \end{bmatrix},$$

we obtain that

$$\dot{W}_a^{N_0}(t) = A_1 W_a^{N_0}(t) + B_1 v(t).$$

We now define for $1 \leq n \leq N$ the observation error as

$$e_n(t) = w_n(t) - \hat{w}_n(t). \tag{35}$$

With $\zeta(t) = \sum_{i \geq N+1} c_i w_i(t)$, we infer from (33a) that

$$\dot{\hat{w}}_n(t) = (-\lambda_n + q_c)\hat{w}_n(t) + a_n u(t) + b_n v(t) - l_n \sum_{i=1}^{N} c_i e_i(t) - l_n \zeta(t) \tag{36}$$

for $1 \leq n \leq N_0$. Inspired by Section 2, we write the dynamics in coordinates of the observer state and of the error variable. To do so, we introduce

$$\hat{W}^{N_0}(t) = \begin{bmatrix} \hat{w}_1(t) \\ \vdots \\ \hat{w}_{N_0}(t) \end{bmatrix}, \quad E^{N_0}(t) = \begin{bmatrix} e_1(t) \\ \vdots \\ e_{N_0}(t) \end{bmatrix}, \quad E^{N-N_0}(t) = \begin{bmatrix} e_{N_0+1}(t) \\ \vdots \\ e_N(t) \end{bmatrix},$$

$$C_0 = \begin{bmatrix} c_1 & c_2 & \ldots & c_{N_0} \end{bmatrix}, \quad C_1 = \begin{bmatrix} c_{N_0+1} & \ldots & c_N \end{bmatrix}, \quad L = \begin{bmatrix} l_1 \\ \vdots \\ l_{N_0} \end{bmatrix}.$$

Hence we have

$$\dot{\hat{W}}^{N_0}(t) = A_0 \hat{W}^{N_0}(t) + B_{0,a}u(t) + B_{0,b}v(t) \tag{37}$$
$$- LC_0 E^{N_0}(t) - LC_1 E^{N-N_0}(t) - L\zeta(t).$$

With

$$\hat{W}_a^{N_0}(t) = \begin{bmatrix} u(t) \\ \hat{W}^{N_0}(t) \end{bmatrix}, \quad \tilde{L} = \begin{bmatrix} 0 \\ L \end{bmatrix}, \tag{38}$$

we deduce that

$$\dot{\hat{W}}_a^{N_0}(t) = A_1 \hat{W}_a^{N_0}(t) + B_1 v(t) - \tilde{L}C_0 E^{N_0}(t) - \tilde{L}C_1 E^{N-N_0}(t) - \tilde{L}\zeta(t). \tag{39}$$

In view of (33d) we deduce that

$$v(t) = K \hat{W}_a^{N_0}(t), \tag{40}$$

where $K \in \mathbb{R}^{1 \times (N_0+1)}$. Hence we obtain that

$$\dot{\hat{W}}_a^{N_0}(t) = (A_1 + B_1 K) \hat{W}_a^{N_0}(t) - \tilde{L}C_0 E^{N_0}(t) - \tilde{L}C_1 E^{N-N_0}(t) - \tilde{L}\zeta(t) \tag{41}$$

and, using (34) and (37),

$$\dot{E}^{N_0}(t) = (A_0 + LC_0) E^{N_0}(t) + LC_1 E^{N-N_0}(t) + L\zeta(t). \tag{42}$$

Claim 1. *The pair* (A_1, B_1) *is controllable.*

Proof of Claim 1. Let us compute the Kalman matrix $\mathcal{C}$ for controllability of (A_1, B_1) as introduced in Section 1. Denoting by $\mu_n = -\lambda_n + q_c$, we get

$$\mathcal{C} = \begin{bmatrix} 1 & 0 & \cdots & 0 \\ b_1 & a_1 + \mu_1 b_1 & \cdots & (a_1 + \mu_1 b_1)\mu_1^{N_0-1} \\ b_2 & a_2 + \mu_2 b_2 & \cdots & (a_2 + \mu_2 b_2)\mu_2^{N_0-1} \\ \vdots & \vdots & & \vdots \\ b_{N_0} & a_{N_0} + \mu_{N_0} b_{N_0} & \cdots & (a_{N_0} + \mu_{N_0} b_{N_0})\mu_{N_0}^{N_0-1} \end{bmatrix}$$

whose determinant is

$$\det(\mathcal{C}) = \Pi_{n=1}^{N_0}(a_n + \mu_n b_n) \begin{vmatrix} 1 & \mu_1 & \cdots & \mu_1^{N_0-1} \\ 1 & \mu_2 & \cdots & \mu_2^{N_0-1} \\ \vdots & \vdots & & \vdots \\ 1 & \mu_{N_0} & \cdots & \mu_{N_0}^{N_0-1} \end{vmatrix}.$$

The determinant appearing in the latter equation is known as the Vandermonde determinant. Since the μ_n are distinct, the Vandermonde determinant is non zero hence the pair (A_1, B_1) is controllable if and only if $\Pi_{n=1}^{N_0}(a_n + \mu_n b_n) \neq 0$. To check this latter condition, let us compute, for each $n = 1, \ldots, N_0$, the quantity $a_n + \mu_n b_n$. Recalling $\mu_n = -\lambda_n + q_c$ and from the definitions of the function a and b, we obtain that

$$
\begin{aligned}
a_n + \mu_n b_n &= \int_0^1 [2p(x) + 2xp'(x) + (q_c - q(x))x^2]\Phi_n(x)\mathrm{d}x \\
&\quad + (-\lambda_n + q_c)\int_0^1 -x^2\Phi_n(x)\mathrm{d}x \\
&= \int_0^1 [(2p(x) + 2xp'(x))\Phi_n(x) - x^2q(x)\Phi_n(x)]\mathrm{d}x \\
&\quad + \int_0^1 [-x^2(p(x)\Phi'_n(x))' + x^2q(x)\Phi_n(x)]\mathrm{d}x \\
&= \int_0^1 (2p(x) + 2xp'(x))\Phi_n(x)\mathrm{d}x - \int_0^1 x^2(p(x)\Phi'_n(x))'\mathrm{d}x \\
&= -p(1)\Phi'_n(1).
\end{aligned}
$$

Recalling that Φ_n is a non-trivial solution to a second order ODE with $\Phi_n(1) = 0$, we must have $\Phi'_n(1) \neq 0$. Therefore $a_n + \mu_n b_n \neq 0$ hence the pair (A_1, B_1) is controllable.   □

Claim 2. *Assuming $c_n \neq 0$ for all $1 \leq n \leq N_0$, the pair (A_0, C_0) is observable.*

Proof of Claim 2. Let us compute the Kalman matrix for observation of the pair (A_0, C_0):

$$
\begin{bmatrix}
c_1 & \cdots & c_{N_0} \\
\mu_1 c_1 & \cdots & \mu_{N_0} c_{N_0} \\
\vdots & & \vdots \\
\mu_1^{N_0-1} c_1 & \cdots & \mu_{N_0}^{N_0-1} c_{N_0}
\end{bmatrix}.
$$

Since $n \neq m$ implies $\mu_n \neq \mu_m$, this matrix is full rank if and only if $c_n \neq 0$ for all $n = 1, \ldots, N_0$.   □

We now define the vectors and matrices:

$$
\hat{W}^{N-N_0}(t) = \begin{bmatrix} \hat{w}_{N_0+1}(t) \\ \vdots \\ \hat{w}_N(t) \end{bmatrix}, \quad
B_{2,a} = \begin{bmatrix} a_{N_0+1} \\ \vdots \\ a_N \end{bmatrix}, \quad
B_{2,b} = \begin{bmatrix} b_{N_0+1} \\ \vdots \\ b_N \end{bmatrix},
$$

$$
A_2 = \mathrm{diag}(-\lambda_{N_0+1} + q_c, \ldots, -\lambda_N + q_c).
$$

From (33b) and (40) we obtain that

$$\dot{\hat{W}}^{N-N_0}(t) = A_2 \hat{W}^{N-N_0}(t) + B_{2,a}u(t) + B_{2,b}v(t)$$
$$= A_2 \hat{W}^{N-N_0}(t) + \left(B_{2,b}K + [B_{2,a} \ 0] \right) \hat{W}_a^{N_0}(t) \qquad (43)$$

and, using in addition (32b) and (35),

$$\dot{E}^{N-N_0}(t) = A_2 E^{N-N_0}(t). \qquad (44)$$

Putting together (41)–(44), we obtain with

$$X(t) = \mathrm{col}(\hat{W}_a^{N_0}(t), E^{N_0}(t), \hat{W}^{N-N_0}(t), E^{N-N_0}(t)) \qquad (45)$$

that

$$\dot{X}(t) = FX(t) + \mathcal{L}\zeta(t) \qquad (46)$$

where

$$F = \begin{bmatrix} A_1 + B_1 K & -\tilde{L}C_0 & 0 & -\tilde{L}C_1 \\ 0 & A_0 + LC_0 & 0 & LC_1 \\ B_{2,b}K + [B_{2,a} \ 0] & 0 & A_2 & 0 \\ 0 & 0 & 0 & A_2 \end{bmatrix}, \quad \mathcal{L} = \begin{bmatrix} -\tilde{L} \\ L \\ 0 \\ 0 \end{bmatrix}. \qquad (47)$$

Defining $E = \begin{bmatrix} 1 & 0 & \dots & 0 \end{bmatrix}$ and $\tilde{K} = \begin{bmatrix} K & 0 & 0 & 0 \end{bmatrix}$, we obtain from (38), (40), and (45) that

$$u(t) = EX(t), \quad v(t) = \tilde{K}X(t). \qquad (48)$$

Finally, defining $g = \|a\|_{L^2}^2 + \|b\|_{L^2}^2 \|K\|^2$, we can introduce

$$G = \|a\|_{L^2}^2 E^\top E + \|b\|_{L^2}^2 \tilde{K}^\top \tilde{K} \preceq gI. \qquad (49)$$

3.2.3. *Stability analysis*

Theorem 6. *Let $p \in \mathcal{C}^1([0,1])$ with $p > 0$, $q \in \mathcal{C}^0([0,1])$ with $q \geq 0$, $q_c \in \mathbb{R}$, and $c \in L^2(0,1)$. Consider the reaction-diffusion system described by (28). Let $N_0 \geq 1$ and $\delta > 0$ be given such that $-\lambda_n + q_c < -\delta < 0$ for all $n \geq N_0 + 1$. Assume that $c_n \neq 0$ for all $1 \leq n \leq N_0$. Let $K \in \mathbb{R}^{1 \times (N_0+1)}$ and $L \in \mathbb{R}^{N_0}$ be such that $A_1 + B_1 K$ and $A_0 + LC_0$ are Hurwitz with eigenvalues that have a real part strictly less than $-\delta < 0$. Assume that there exist $N \geq N_0 + 1$, $P \succ 0$, $\alpha > 1$, and $\beta, \gamma > 0$ such that*

$$\Theta = \begin{bmatrix} F^\top P + PF + 2\delta P + \alpha\gamma G & PL \\ \star & -\beta \end{bmatrix} \preceq 0, \qquad (50a)$$

$$\Gamma_{1,N+1} = -\lambda_{N+1} + q_c + \delta + \frac{1}{\alpha} + \frac{\beta\|c\|_{L^2}^2}{2\gamma} \leq 0, \qquad (50b)$$

$$\Gamma_{2,N+1} = -\left(1 - \frac{1}{\alpha}\right)\lambda_{N+1} + q_c + \delta + \frac{\beta\|c\|_{L^2}^2}{2\gamma\lambda_{N+1}} \leq 0, \qquad (50c)$$

for all $n \geq N + 1$. Then, for the closed-loop system composed of the plant (28) and the controller (33):

(1) *the origin is asymptotically stable in L^2-norm, that is there exists $M > 0$ such that, for any $\hat{w}_n(0) \in \mathbb{R}$, any $z_0 \in L^2(0,1)$ and any $u(0) \in \mathbb{R}$, the mild solution of the closed-loop system satisfies*

$$u(t)^2 + \sum_{n=1}^{N} \hat{w}_n(t)^2 + \|z(t,\cdot)\|_{L^2}^2 \leq M e^{-2\delta t} \left(u(0)^2 + \sum_{n=1}^{N} \hat{w}_n(0)^2 + \|z_0\|_{L^2}^2 \right)$$

for all $t \geq 0$.

(2) *the origin is asymptotically stable in H^1-norm, that is there exists $M > 0$ such that, for any $\hat{w}_n(0) \in \mathbb{R}$, any $z_0 \in H^2(0,1)$ and any $u(0) \in \mathbb{R}$ such that $z_0'(0) = 0$ and $z_0(1) = u(0)$, the classical solution of the closed-loop system satisfies*

$$u(t)^2 + \sum_{n=1}^{N} \hat{w}_n(t)^2 + \|z(t,\cdot)\|_{H^1}^2 \leq M e^{-2\delta t} \left(u(0)^2 + \sum_{n=1}^{N} \hat{w}_n(0)^2 + \|z_0\|_{H^1}^2 \right)$$

for all $t \geq 0$.

Moreover, the above constraints are always feasible for N large enough.

Remark 3. The feasibility problem of Theorem 6 is not linear due to the presence of some terms such as $\alpha\gamma$ and $\frac{1}{\alpha}$ involving the decision variables. However the use of Schur complement allows to rewrite (50b) as follows:

$$\begin{bmatrix} -\lambda_{N+1} + q_c + \delta + \dfrac{\beta\|c\|_{L^2}^2}{2\gamma} & 1 \\ \star & -\alpha \end{bmatrix} \leq 0,$$

and similarly for (50c). Therefore, as soon as γ is fixed, checking the conditions of Theorem 6 reduces to check linear matrix inequalities (LMIs). Thus, given a desired exponential decay rate $\delta > 0$ and a number of modes $N \geq N_0 + 1$ for the observer, the sufficient conditions of the previous theorem can be recasted as an efficient optimization problem to solve LMIs.

Proof of Theorem 6. Consider the Lyapunov function candidate

$$V(X, w) = X^\top P X + \gamma \sum_{n \geq N+1} \langle w, \Phi_n \rangle^2$$

for $X \in \mathbb{R}^{2N+1}$ and $w \in L^2(0,1)$. The first term accounts for the dynamics of the N first modes of the PDE and the dynamics of the observer, while the series accounts for the dynamics of the modes corresponding to $n \geq N + 1$.

The computation of the time derivative of V along the system solutions (32b) and (46) gives

$$\dot{V} + 2\delta V = X^\top \left(F^\top P + PF + 2\delta P \right) X + 2X^\top P\mathcal{L}\zeta$$
$$+ 2\gamma \sum_{n \geq N+1} (-\lambda_n + q_c + \delta)w_n^2 + 2\gamma \sum_{n \geq N+1} (a_n u + b_n v)w_n.$$

The use of Young's inequality gives

$$2 \sum_{n \geq N+1} a_n w_n u \leq \frac{1}{\alpha} \sum_{n \geq N+1} w_n^2 + \alpha \|a\|_{L^2}^2 u^2,$$

$$2 \sum_{n \geq N+1} b_n w_n(t)v(t) \leq \frac{1}{\alpha} \sum_{n \geq N+1} w_n^2 + \alpha \|b\|_{L^2}^2 v^2,$$

for any $\alpha > 0$. From (48)–(49), we infer that

$$\dot{V} + 2\delta V \leq \begin{bmatrix} X \\ \zeta \end{bmatrix}^\top \begin{bmatrix} F^\top P + PF + 2\delta P + \alpha\gamma G & P\mathcal{L} \\ \star & 0 \end{bmatrix} \begin{bmatrix} X \\ \zeta \end{bmatrix}$$
$$+ 2\gamma \sum_{n \geq N+1} \left(-\lambda_n + q_c + \delta + \frac{1}{\alpha} \right) w_n^2.$$

Recalling the definition $\zeta(t) = \sum_{n \geq N+1} c_n w_n(t)$, we obtain from Cauchy-Schwarz inequality that $\zeta(t)^2 \leq \|c\|_{L^2}^2 \sum_{n \geq N+1} w_n(t)^2$. Hence, for any $\beta > 0$,

$$\beta\|c\|_{L^2}^2 \sum_{n \geq N+1} w_n^2 - \beta\zeta^2 \geq 0. \tag{51}$$

Combining the two latter inequalities, we obtain that

$$\dot{V} + 2\delta V \leq \begin{bmatrix} X \\ \zeta \end{bmatrix}^\top \Theta \begin{bmatrix} X \\ \zeta \end{bmatrix} + 2\gamma \sum_{n \geq N+1} \Gamma_{1,n} w_n^2 \leq 0$$

where $\Gamma_{1,n} = -\lambda_n + q_c + \delta + \frac{1}{\alpha} + \frac{\beta\|c\|_{L^2}^2}{2\gamma} \leq \Gamma_{1,N+1}$ for all $n \geq N+1$. The assumptions (50) imply that $V(t) \leq e^{-2\delta t}V(0)$ for all $t \geq 0$, giving the claimed stability estimate for PDE trajectories evaluated in L^2-norm.

We now address the stability assessment of the system trajectories when evaluated in H^1-norm. To do so, in view of (25), we introduce the Lyapunov functional candidate:

$$V(X, w) = X^\top PX + \gamma \sum_{n \geq N+1} \lambda_n \langle w, \Phi_n \rangle^2 \tag{52}$$

with $X \in \mathbb{R}^{2N+1}$ and $w \in D(\mathcal{A})$. The computation of the time derivative of V along the system solutions (32b) and (46) gives

$$\dot{V} + 2\delta V = X^\top \left(F^\top P + PF + 2\delta P \right) X + 2X^\top P \mathcal{L} \zeta \tag{53}$$

$$+ 2\gamma \sum_{n \geq N+1} \lambda_n (-\lambda_n + q_c + \delta) w_n^2 + 2\gamma \sum_{n \geq N+1} \lambda_n (a_n u + b_n v) w_n(t).$$

Using again Young's inequality, we obtain

$$2 \sum_{n \geq N+1} \lambda_n a_n w_n u \leq \frac{1}{\alpha} \sum_{n \geq N+1} \lambda_n^2 w_n^2 + \alpha \|a\|_{L^2}^2 u^2, \tag{54a}$$

$$2 \sum_{n \geq N+1} \lambda_n b_n w_n v \leq \frac{1}{\alpha} \sum_{n \geq N+1} \lambda_n^2 w_n^2 + \alpha \|b\|_{L^2}^2 v^2, \tag{54b}$$

for any $\alpha > 0$. Hence, owing to (48)–(49) and (51), we deduce that

$$\dot{V} + 2\delta V \leq \begin{bmatrix} X \\ \zeta \end{bmatrix}^\top \Theta \begin{bmatrix} X \\ \zeta \end{bmatrix} + 2\gamma \sum_{n \geq N+1} \lambda_n \Gamma_{2,n} w_n^2 \leq 0$$

with $\Gamma_{2,n} = -\lambda_n + q_c + \delta + \frac{\lambda_n}{\alpha} + \frac{\beta \|c\|_{L^2}^2}{2\gamma \lambda_n} \leq \Gamma_{2,N+1}$ for all $n \geq N+1$ where it has been used that $\alpha > 1$. Thus (50) implies that $V(t) \leq e^{-2\delta t} V(0)$ for all $t \geq 0$. The claimed stability estimate in H^1-norm is now obtained from (25), (29), and (52).

We conclude the proof by showing that one can always select the order of the observer $N \geq N_0 + 1$ large enough and find $P \succ 0$, $\alpha > 1$, and $\beta, \gamma > 0$ such that $\Theta \preceq 0$, $\Gamma_{1,N+1} \leq 0$, and $\Gamma_{2,N+1} \leq 0$. Owing to the Schur complement, we have $\Theta \preceq 0$ if and only if $F^\top P + PF + 2\delta P + \alpha \gamma G + \frac{1}{\beta} P \mathcal{L} \mathcal{L}^\top P^\top \preceq 0$. We now note that $A_1 + B_1 K + \delta I$ and $A_0 - LC_0 + \delta I$ are Hurwitz and $\|e^{(A_2 + \delta I)t}\| \leq e^{-\kappa_0 t}$ with $\kappa_0 = \lambda_{N_0+1} - q_c - \delta > 0$. Moreover, $\|\tilde{L} C_1\| \leq \|L\| \|c\|_{L^2}$, $\|L C_1\| \leq \|L\| \|c\|_{L^2}$, and $\|B_{2,b} K + \begin{bmatrix} B_{2,a} & 0 \end{bmatrix}\| \leq \|b\|_{L^2} \|K\| + \|a\|_{L^2}$. The right-hand sides of all the previous inequalities are independent of the order of the observer $N \geq N_0 + 1$. Hence, Lemma 1, which is reported immediately after this proof, applied to $F + \delta I$ shows for any $N \geq N_0 + 1$ the existence of $P \succ 0$ such that $F^\top P + PF + 2\delta P = -I$ with $\|P\| = O(1)$ as $N \to +\infty$. Finally, we have (49) and $\|\mathcal{L}\| = \sqrt{2} \|L\|$ with g and L that are independent of N. Hence, with $\alpha = N^{1/4}$, $\beta = N$, and $\gamma = N^{-1/2}$, we infer from (26) the existence of a sufficiently large integer $N \geq N_0 + 1$, independent of the initial conditions, such that $\Theta \preceq 0$, $\Gamma_{1,N+1} \leq 0$, and $\Gamma_{2,N+1} \leq 0$.   □

A technical lemma

The following lemma generalizes the statement of a result presented in [29] while the proof, reported below, remains essentially identical.

Lemma 1. *Let* $n, m, N \geq 1$, $M_{11} \in \mathbb{R}^{n \times n}$ *and* $M_{22} \in \mathbb{R}^{m \times m}$ *Hurwitz,* $M_{12} \in \mathbb{R}^{n \times m}$, $M_{14}^N \in \mathbb{R}^{n \times N}$, $M_{24}^N \in \mathbb{R}^{m \times N}$, $M_{31}^N \in \mathbb{R}^{N \times n}$, $M_{33}^N, M_{44}^N \in \mathbb{R}^{N \times N}$, *and*

$$
F^N = \begin{bmatrix}
M_{11} & M_{12} & 0 & M_{14}^N \\
0 & M_{22} & 0 & M_{24}^N \\
M_{31}^N & 0 & M_{33}^N & 0 \\
0 & 0 & 0 & M_{44}^N
\end{bmatrix}.
$$

We assume that there exist constants $C_0, \kappa_0 > 0$ *such that* $\|e^{M_{33}^N t}\| \leq C_0 e^{-\kappa_0 t}$ *and* $\|e^{M_{44}^N t}\| \leq C_0 e^{-\kappa_0 t}$ *for all* $t \geq 0$ *and all* $N \geq 1$. *Moreover, we assume that there exists a constant* $C_1 > 0$ *such that* $\|M_{14}^N\| \leq C_1$, $\|M_{24}^N\| \leq C_1$, *and* $\|M_{31}^N\| \leq C_1$ *for all* $N \geq 1$. *Then there exists a constant* $C_2 > 0$ *such that, for any* $N \geq 1$, *there exists a symmetric matrix* $P^N \in \mathbb{R}^{n+m+2N}$ *with* $P^N \succ 0$ *such that* $(F^N)^\top P^N + P^N F^N = -I$ *and* $\|P^N\| \leq C_2$.

Proof of Lemma 1. It is sufficient to show the existence of constants $\tilde{C}_0, \eta > 0$ such that $\|e^{F^N t}\| \leq \tilde{C}_0 e^{-\eta t}$ for all $t \geq 0$ and all $N \geq 1$. Indeed, in that case, $P^N = \int_0^\infty e^{(F^N)^\top t} e^{F^N t} \, dt$ is well defined and satisfies the claimed properties. We introduce $F^N = F_1^N + F_2^N$ with

$$
F_1^N = \begin{bmatrix}
M_{11} & M_{12} & 0 & 0 \\
0 & M_{22} & 0 & 0 \\
0 & 0 & M_{33}^N & 0 \\
0 & 0 & 0 & M_{44}^N
\end{bmatrix}, \quad
F_2^N = \begin{bmatrix}
0 & 0 & 0 & M_{14}^N \\
0 & 0 & 0 & M_{24}^N \\
M_{31}^N & 0 & 0 & 0 \\
0 & 0 & 0 & 0
\end{bmatrix}.
$$

Then there exist constants $\kappa, \tilde{C}_1, \tilde{C}_2 > 0$ such that $\|e^{F_1^N t}\| \leq \tilde{C}_1 e^{-\kappa t}$ and $\|F_2^N\| \leq \tilde{C}_2$ for all $t \geq 0$ and all $N \geq 1$. One can check that $(F_2^N)^3 = 0$ and

$$
(F_1^N)^{n_i} = \begin{bmatrix}
\bullet & \bullet & 0 & 0 \\
0 & \bullet & 0 & 0 \\
0 & 0 & \bullet & 0 \\
0 & 0 & 0 & \bullet
\end{bmatrix}
$$

for any $n_i \geq 0$ and where "$\bullet$" denotes a possibly non zero element, that is not needed in this proof. Hence

$$
(F_1^N)^{n_i} F_2^N = \begin{bmatrix}
0 & 0 & 0 & \bullet \\
0 & 0 & 0 & \bullet \\
\bullet & 0 & 0 & 0 \\
0 & 0 & 0 & 0
\end{bmatrix}
$$

for any $n_i \geq 0$. We deduce that

$$\prod_{i=1}^{3}(F_1^N)^{n_i} F_2^N = \begin{bmatrix} 0 & 0 & 0 & \bullet \\ 0 & 0 & 0 & \bullet \\ \bullet & 0 & 0 & 0 \\ 0 & 0 & 0 & 0 \end{bmatrix}^3 = 0$$

for any $n_i \geq 0$. Therefore,

$$\prod_{i=1}^{3} e^{F_1^N t_i} F_2^N = \sum_{k_1 \geq 0} \sum_{k_2 \geq 0} \sum_{k_3 \geq 0} \frac{t_1^{k_1} t_2^{k_2} t_3^{k_3}}{k_1! k_2! k_3!} \prod_{i=1}^{3} (F_1^N)^{k_i} F_2^N = 0 \qquad (55)$$

for all $t_1, t_2, t_3 \geq 0$. Now we note that[b], for any square matrices A, B, $e^{(A+B)t} = e^{At} + \int_0^t e^{A(t-\tau)} B e^{(A+B)\tau} \, d\tau$. Hence we have, using the last identity three times consecutively,

$$e^{F^N t} = e^{F_1^N t} + \int_0^t e^{F_1^N(t-t_1)} F_2^N e^{F^N t_1} \, dt_1$$

$$= e^{F_1^N t} + \int_0^t e^{F_1^N(t-t_1)} F_2^N e^{F_1^N t_1} \, dt_1$$

$$+ \int_0^t \int_0^{t_1} e^{F_1^N(t-t_1)} F_2^N e^{F_1^N(t_1-t_2)} F_2^N e^{F^N t_2} \, dt_2 dt_1$$

$$= e^{F_1^N t} + \int_0^t e^{F_1^N(t-t_1)} F_2^N e^{F_1^N t_1} \, dt_1$$

$$+ \int_0^t \int_0^{t_1} e^{F_1^N(t-t_1)} F_2^N e^{F_1^N(t_1-t_2)} F_2^N e^{F_1^N t_2} \, dt_2 dt_1$$

where the last identity has been obtained by using (55). Recalling that $\|e^{F_1^N t}\| \leq \tilde{C}_1 e^{-\kappa t}$ and $\|F_2^N\| \leq \tilde{C}_2$ for all $t \geq 0$ and all $N \geq 1$, the claimed conclusion holds. $\qquad \square$

3.3. Dirichlet boundary measurement

We extend the result of the previous section to the case of a reaction-diffusion PDE with Dirichlet boundary observation described by

$$z_t(t,x) = (p(x) z_x(t,x))_x + (q_c - q(x)) z(t,x), \qquad (56a)$$

$$z_x(t,0) = 0, \quad z(t,1) = u(t), \qquad (56b)$$

$$z(0,x) = z_0(x), \qquad (56c)$$

$$y(t) = z(t,0), \qquad (56d)$$

---

[b] $x(t) = e^{(A+B)t} x_0$ is such that $\dot{x}(t) = Ax(t) + u(t)$ with $u(t) = Bx(t)$. The claimed formula follows from $x(t) = e^{At} x_0 + \int_0^t e^{A(t-\tau)} u(\tau) \, d\tau$.

for $t > 0$ and $x \in (0,1)$. We make throughout this subsection the assumption that $p \in C^2([0,1])$ in order to use the asymptotic behavior (27).

3.3.1. *Spectral reduction*

The only change compared to the previous subsection is the modification of the nature of the observation operator. Hence, the spectral reduction is conducted identically but the observation (30d) is replaced by $\tilde{y}(t) = w(t,0) = y(t)$. Considering classical solutions for the PDE, we have $w(t,\cdot) \in D(\mathcal{A})$ for all $t \geq 0$. Hence, (32c) is simply replaced by $\tilde{y}(t) = \sum_{i \geq 1} \Phi_i(0) w_i(t)$.

3.3.2. *Control design*

Let $N_0 \geq 1$ and $\delta > 0$ be fixed so that $-\lambda_n + q_c < -\delta < 0$ for all $n \geq N_0+1$. Let $N \geq N_0 + 1$ be arbitrary and to be determined later. We proceed as in the previous subsection: we design an observer to estimate the N first modes of the plant while the state-feedback is performed on the N_0 first modes of the plant. Hence, the controller dynamics is described by

$$\dot{\hat{w}}_n(t) = (-\lambda_n + q_c)\hat{w}_n(t) + a_n u(t) + b_n v(t) + l_n\left(\hat{y}(t) - \tilde{y}(t)\right), \quad 1 \leq n \leq N_0,$$

(57a)

$$\dot{\hat{w}}_n(t) = (-\lambda_n + q_c)\hat{w}_n(t) + a_n u(t) + b_n v(t), \quad N_0 + 1 \leq n \leq N,$$ (57b)

$$\hat{y}(t) = \sum_{i=1}^{N} \Phi_i(0)\hat{w}_i(t),$$ (57c)

$$v(t) = \dot{u}(t) = \sum_{i=1}^{N_0} k_i \hat{w}_i(t) + k_u u(t),$$ (57d)

which is the same as the one described by (33) but with measurement, originally given by (33a), replaced by (57a). In this context, (36) is replaced by the following, defined for $1 \leq n \leq N_0$,

$$\dot{\hat{w}}_n(t) = (-\lambda_n + q_c)\hat{w}_n(t) + a_n u(t) + b_n v(t)$$
$$- l_n \sum_{i=1}^{N_0} \Phi_i(0)e_i(t) + l_n \sum_{i=N_0+1}^{N} \frac{\Phi_i(0)}{\sqrt{\lambda_i}}\tilde{e}_i(t) + l_n\zeta(t).$$

Here $\zeta(t)$ is defined by $\zeta(t) = \sum_{i \geq N+1} \Phi_i(0)w_i(t)$ while, following [41], we introduced the scaled error of observation $\tilde{e}_n(t) = \sqrt{\lambda_n}e_n(t)$ with e_n given by (35). The definitions of C_0 and C_1 are replaced by

$$C_0 = \begin{bmatrix} \Phi_1(0) \dots \Phi_{N_0}(0) \end{bmatrix}, \quad C_1 = \begin{bmatrix} \dfrac{\Phi_{N_0+1}(0)}{\sqrt{\lambda_{N_0+1}}} & \cdots & \dfrac{\Phi_N(0)}{\sqrt{\lambda_N}} \end{bmatrix} \quad (58)$$

and defining

$$\tilde{E}^{N-N_0}(t) = \begin{bmatrix} \tilde{e}_{N_0+1}(t) \dots \tilde{e}_N(t) \end{bmatrix}^\top,$$

we obtain in replacement of (37) and (39) that

$$\dot{\hat{W}}^{N_0}(t) = A_0 \hat{W}^{N_0}(t) + B_{0,a} u(t) + B_{0,b} v(t)$$
$$- LC_0 E^{N_0}(t) - LC_1 \tilde{E}^{N-N_0}(t) - L\zeta(t) \quad (59)$$

and

$$\dot{\hat{W}}_a^{N_0}(t) = A_1 \hat{W}_a^{N_0}(t) + B_1 v(t) - \tilde{L}C_0 E^{N_0}(t) - \tilde{L}C_1 \tilde{E}^{N-N_0}(t) - \tilde{L}\zeta(t), \quad (60)$$

respectively. In this framework, the command input is still given by (40). Using now (34) and (59), the error dynamics originally given by (42) is now replaced by

$$\dot{E}^{N_0}(t) = (A_0 + LC_0)E^{N_0}(t) + LC_1 \tilde{E}^{N-N_0}(t) + L\zeta(t). \quad (61)$$

Moreover, since $\dot{e}_n(t) = (-\lambda_n + q_c)e_n(t)$, we have $\dot{\tilde{e}}_n(t) = (-\lambda_n + q_c)\tilde{e}_n(t)$ for all $N_0 + 1 \le n \le N$. Then (44) is replaced by

$$\dot{\tilde{E}}^{N-N_0}(t) = A_2 \tilde{E}^{N-N_0}(t). \quad (62)$$

Putting together (40), (43), and (60)–(62), the introduction of

$$X(t) = \text{col}(\hat{W}_a^{N_0}(t), E^{N_0}(t), \hat{W}^{N-N_0}(t), \tilde{E}^{N-N_0}(t))$$

shows that (46) holds with the different matrices defined by (47).

Remark 4. Based on the arguments of Claim 1 and Claim 2, we have that (A_1, B_1) is controllable and (A_0, C_0) is observable.

3.3.3. *Stability analysis*

We introduce the constant $M_{1,\Phi} = \sum_{n \ge 2} \frac{\Phi_n(0)^2}{\lambda_n}$. Note that this constant is well defined (i.e., finite) when $p \in \mathcal{C}^2([0,1])$ due to (26)–(27).

Theorem 7. *Let* $p \in \mathcal{C}^2([0,1])$ *with* $p > 0$, $q \in \mathcal{C}^0([0,1])$ *with* $q \ge 0$, *and* $q_c \in \mathbb{R}$. *Consider the reaction-diffusion system described by* (56). *Let* $N_0 \ge 1$ *and* $\delta > 0$ *be given such that* $-\lambda_n + q_c < -\delta < 0$ *for all* $n \ge N_0 + 1$. *Let* $K \in \mathbb{R}^{1 \times (N_0+1)}$ *and* $L \in \mathbb{R}^{N_0}$ *be such that* $A_1 + B_1 K$ *and* $A_0 - LC_0$ *are*

Hurwitz with eigenvalues that have a real part strictly less than $-\delta < 0$. *Assume that there exist* $N \geq N_0 + 1$, $P \succ 0$, $\alpha > 1$, *and* $\beta, \gamma > 0$ *such that* $\Theta \preceq 0$, *where* Θ *is defined by* (50a), *and*

$$\Gamma_{3,N+1} = -\left(1 - \frac{1}{\alpha}\right)\lambda_{N+1} + q_c + \delta + \frac{\beta M_{1,\Phi}}{2\gamma} \leq 0. \qquad (63)$$

Then the origin of the closed-loop system composed of the plant (56) *and the controller* (57) *is exponentially stable in* H^1*-norm in the sense that there exists* $M > 0$ *such that, for any* $\hat{w}_n(0) \in \mathbb{R}$, *for any* $z_0 \in H^2(0,1)$ *and any* $u(0) \in \mathbb{R}$ *such that* $z'_0(0) = 0$ *and* $z_0(1) = u(0)$, *the classical solution of the closed-loop system satisfies*

$$u(t)^2 + \sum_{n=1}^{N} \hat{w}_n(t)^2 + \|z(t,\cdot)\|_{H^1}^2 \leq M e^{-2\delta t}\left(u(0)^2 + \sum_{n=1}^{N} \hat{w}_n(0)^2 + \|z_0\|_{H^1}^2\right)$$

for all $t \geq 0$. *Moreover, the above constraints are always feasible for* N *large enough.*

Remark 5. The previous result deals with the exponential stability of the closed-loop system in H^1-norm. This type of approach can be extended to a number of control design problems such as:

- L^2 stability using the same control strategy;[40]
- Robin boundary conditions;[34,40]
- Neumann boundary observations;[41]
- input/output delayed boundary control;[38,42]
- nonlinearities;[34,39,40]
- regulation problems.[37]

Proof. Consider again the Lyapunov function candidate defined by (52). The computation of its time derivative along the system solutions (32b) and (46) gives (53). Since $\zeta(t) = \sum_{n \geq N+1} \Phi_n(0)w_n(t)$, we have by Cauchy-Schwarz inequality that $\zeta(t)^2 \leq M_{1,\Phi} \sum_{n \geq N+1} \lambda_n w_n(t)^2$ hence $\beta M_{1,\Phi} \sum_{n \geq N+1} \lambda_n w_n(t)^2 - \beta\zeta(t)^2 \geq 0$ for any $\beta > 0$. Using this latter estimate into (53) and invoking Young's inequality as in (54) along with (48–49), we deduce that

$$\dot{V} + 2\delta V \leq \begin{bmatrix} X \\ \zeta \end{bmatrix}^\top \Theta \begin{bmatrix} X \\ \zeta \end{bmatrix} + 2\gamma \sum_{n \geq N+1} \lambda_n \Gamma_{3,n} w_n(t)^2 \leq 0$$

where $\Gamma_{3,n} = -\left(1 - \frac{1}{\alpha}\right)\lambda_n + q_c + \delta + \frac{\beta M_{1,\Phi}}{2\gamma} \leq \Gamma_{3,N+1}$ for all $n \geq N+1$. Hence the assumptions give $V(t) \leq e^{-2\delta t}V(0)$ for all $t \geq 0$. Proceeding as in the previous proof, we obtain the claimed estimate.

To complete the proof, it remains to show that one can always select $N \geq N_0 + 1$ large enough, $P \succ 0$, $\alpha > 1$, and $\beta, \gamma > 0$, such that $\Theta \preceq 0$ and $\Gamma_{3,N+1} \leq 0$. Owing to the Schur complement, $\Theta \preceq 0$ is equivalent to $F^{\top}P + PF + 2\delta P + \alpha\gamma G + \frac{1}{\beta}P\mathcal{L}\mathcal{L}^{\top}P^{\top} \preceq 0$. Applying Lemma 1 to[c] $F + \delta I$, we have for any $N \geq N_0 + 1$ the existence of $P \succ 0$ such that $F^{\top}P + PF + 2\delta P = -I$ with $\|P\| = O(1)$ as $N \to +\infty$. Moreover, we have (49) and $\|\mathcal{L}\| = \sqrt{2}\|L\|$ with g and L that are independent of N. Hence, setting $\alpha = \beta = \sqrt{N}$ and $\gamma = N^{-1}$, we obtain from (26) the existence of a sufficiently large integer $N \geq N_0 + 1$ such that $\Theta \preceq 0$ and $\Gamma_{3,N+1} \leq 0$. $\square$

Example 6. Consider the Dirichlet boundary measurement setting described by (56). Let $p = 1$, $q = 0$, and $q_c = 3$, giving an unstable open-loop system. To obtain the closed loop exponential decay rate $\delta = 0.5$, we set $N_0 = 1$. Then we run the following Python code. On lines 14–18, we compute the eigenvalues and eigenvectors of the problem. On lines 26–29, we check whether N_0 is selected adequately.

```
1  import numpy as np
2  import control
3  import scipy.integrate as integrate
4  import cvxpy as cp
5  import matplotlib as mpl
6  import matplotlib.pyplot as plt
7  from mpl_toolkits.mplot3d import Axes3D
8
9  # Parameters of the PDE
10 p = 1
11 q = 3 # this is q_c (q is zero)
12 delta = 0.5
13
14 # Eigenstructures
15 def lam(n):
16     return (n-1/2)**2*np.pi**2*p
17 def phi(n,x):
18     return np.sqrt(2)*np.cos((n-1/2)*np.pi*x)
19
20 # Equivalent bounded input operators
21 def input_a(x):
22     return 2*p + q*x**2 # case of p constant and q=0
23 def input_b(x):
24     return -x**2
25
26 # Number of modes to be stabilized
27 N0 = 0
```

---

[c]This is possible because, owing to the definition (58) of the matrix C_1, it ensures that $\|C_1\| = O(1)$ as $N \to +\infty$. This remark is key to allow the application of Lemma 1.

```
28  while ( -lam(N0+1) + q) >= -delta:
29      N0 = N0 + 1;
```

On line 36, we set to define the number N of modes for the observer. Then we start building the matrices necessary to check the conditions of Theorem 7 after line 49.

```
30  if N0 == 0:
31      print('All the modes of the open-loop system are < -delta')
32  else:
33      print('The number of modes to be stabilized is N_0='+str(N0))
34
35  # Select the number of modes for the observer
36  N = N0+2
37
38  # Matrices of the truncated model
39  tmp=[]
40  for i in range(1,N+1):
41      #print(i)
42      tmp.append(-lam(i)+q)
43
44  A0  = np.diag(tmp[0:N0])
45  A2  = np.diag(tmp[N0:N+1])
46
47  B0a = []; B0b = []; B2a = []; B2b = []; C0  = []; C1  = []
48
49  for k in range(1,N0+1):
50      def fun(x):
51          return input_a(x)*phi(k,x)
52      y,err= integrate.quad(fun,0,1)
53      B0a.append(y)
54      def fun(x):
55          return input_b(x)*phi(k,x)
56      y,err=integrate.quad(fun,0,1)
57      B0b.append(y)
58      C0.append(phi(k,0))
59
60  for k in range(1,N-N0+1):
61      def fun(x):
62          return input_a(x)*phi(N0+k,x)
63      y,err=integrate.quad(fun,0,1)
64      B2a.append(y)
65      def fun(x):
66          return input_b(x)*phi(N0+k,x)
67      y,err=integrate.quad(fun,0,1)
68      B2b.append(y)
69
70  for k in range(N0+1,N+1):
71      C1.append(phi(k,0)/np.sqrt(lam(k)))
72
```

```
73 B0a = np.array(B0a).reshape((N0,1))
74 B0b = np.array(B0b).reshape((N0,1))
75 B2a = np.array(B2a).reshape((N-N0,1))
76 B2b = np.array(B2b).reshape((N-N0,1))
77 C0  = np.array(C0).reshape((1,N0))
78 C1  = np.array(C1).reshape((1,N-N0))
79
80 A1 = np.vstack((np.zeros((1,N0+1)),np.hstack((B0a,A0))))
81 B1 = np.vstack((np.ones((1,1)),B0b))
```

The control matrix K and the observation matrix L are chosen separately on lines 84 and 88. The feedback gain is $K = \begin{bmatrix} -5.0058 & -2.7748 \end{bmatrix}$, and the observer gain is $L = 1.4373$. The matrix inequalities in Theorem 7 are built after line 104. The Schur complement is used to rewrite (63) into a linear matrix inequality in the unknown variables as described in Remark 3.

```
82 # Pole placement for the state feedback
83 Pdes = np.linspace(-N0-1-delta,-1-delta,N0+1)
84 K    = -control.place(A1,B1,Pdes);
85
86 # Pole placement for the observer
87 Qdes = np.linspace(-N0-delta,-1-delta,N0)
88 L0   = -control.place(A0.T,C0.T,Qdes).T;
89
90 tL0 =  np.vstack((np.zeros((1,1)),L0))
91
92 def hcont(A,B,C,D): # help to build F by concatenate horizontally
93     return np.hstack((np.hstack((np.hstack((A,B)),C)),D))
94
95 F1=hcont(A1+np.dot(B1,K),np.dot(tL0,C0),np.zeros((N0+1,N-N0)),np.
       dot(tL0,C1))
96 F2=hcont(np.zeros((N0,N0+1)),A0+np.dot(L0,C0),np.zeros((N0,N-N0)),
       np.dot(L0,C1))
97 F3=hcont(np.dot(B2b,K)+np.hstack((B2a,np.zeros((N-N0,N0)))),np.
       zeros((N-N0,N0)),A2,np.zeros((N-N0,N-N0)))
98 F4=hcont(np.zeros((N-N0,N0+1)),np.zeros((N-N0,N0)),np.zeros((N-N0,
       N-N0)),A2)
99
100 F = np.vstack((np.vstack((np.vstack((F1,F2)),F3)),F4))
101
102 cL = np.vstack((np.vstack((np.vstack((tL0,L0)),np.zeros((N-N0,1))))
       ,np.zeros((N-N0,1))))
103
104 # LMI conditions
105 gamma = 0.00155 #Fix the decision variable gamma > 0
106 M_phi = 12/(np.pi**2*p)
107 E   = np.hstack((np.ones((1,1)),np.zeros((1,2*N))))
108 tK  = np.hstack((K,np.zeros((1,2*N-N0))))
```

```
109
110 def fun(x):
111     return input_a(x)**2
112 y,err=integrate.quad(fun,0,1)
113 norm_a = np.sqrt(y)
114
115 def fun(x):
116     return input_b(x)**2
117 y,err=integrate.quad(fun,0,1)
118 norm_b = np.sqrt(y)
119
120 # check Matrix inequalities
121 alpha = cp.Variable()
122 beta  = cp.Variable()
123 P     = cp.Variable((2*N+1,2*N+1),PSD=True)
124
125 # build the constraints
126 constr = [alpha >= 1]
127 constr += [beta >= 0]
128 M11= (-lam(N+1)+ q +delta+beta*M_phi/(2*gamma))*np.ones((1,1))
129 M12 =np.sqrt(np.abs(-lam(N+1)))*np.ones((1,1))
130 Gamma=cp.bmat([[ M11, M12],
131                [M12.T,-alpha*np.ones((1,1))] ])
132 constr += [Gamma <<0]
133 # build the last constraint
134 G = norm_a**2*(np.dot(E.T,E))+norm_b**2*(np.dot(tK.T,tK))
135 M11=F.T @ P +P @ F+ 2* delta * P + alpha*gamma*G
136 M12=P@ cL
137 matConstr = cp.bmat([[ M11 , M12],
138                      [M12.T, -beta*np.ones((1,1))]])
139 constr +=  [matConstr << 0]
```

The feasability of the convex problem is checked on line 142 using the solver in CVXOPT with cvxpy package. The conditions of Theorem 7 are thus feasible for $N = 3$. Then the values of the unknown variables are stored, and matrix constraints are checked before line 166.

```
140 prob = cp.Problem(cp.Maximize(1),constr)
141 prob.solve(solver='CVXOPT')
142 print(prob.status)
143
144 P=P.value; alpha=alpha.value; beta=beta.value
145 w,v=np.linalg.eig(P)
146 # w should be positive as all constraints
147 m=np.min([alpha-1,np.min([beta,np.min(w)])])
148 # first matrix inequality
149 M11= (-lam(N+1)+ q +delta+beta*M_phi/(2*gamma))
150 M12 =np.sqrt(np.abs(-lam(N+1)))
151 M1=np.hstack((M11,M12))
152 M2=np.hstack((M12,-alpha))
```

```
153 M=np.vstack((M1,M2))
154 w1,v=np.linalg.eig(M) #max w1 should be negative
155 # second matrix inequality
156 M11=np.dot(F.T,P)+np.dot(P,F)+ 2*delta*P +alpha*gamma*G
157 M12=np.dot(P,cL)
158 M1=np.hstack((M11,M12))
159 M2=np.hstack((M12.T,-beta*np.ones((1,1))))
160 M=np.vstack((M1,M2))
161 w2,v=np.linalg.eig(M) #max w2 should be negative
162 mm=np.min([m,-np.max(w2),-np.max(w2)])
163 if mm<0:
164     print('Matrix inequalities not satisfied')
165 else:
166     print('Matrix inequalities satisfied')
```

The code to numerically compute the behavior of the closed-loop system associated with the initial condition $z_0(x) = 1 + x^2$, and with zero initial condition for the observer, obtained based on the 50 dominant modes of the plant is given after line 167.

```
167 # Simulation
168 # Number of modes for simulation
169 Nsim = 50;
170
171 if Nsim < 2*N:
172     print('Number of modes for simulation strictly less than 2*N')
173
174 tmp=[]
175 for i in range(1,Nsim+1):
176     tmp.append(-lam(i)+q)
177
178 # Matrices of the truncated model
179 Ass    = np.diag(tmp[0:Nsim]) # System used for simulations based on
                Nsim dominant modes
180 Aobs   = np.diag(tmp[0:N])
181
182 Bssa = []; Bssb  = []; Bobsa = []; Bobsb = []; Css = []; Cobs = []
183 for k in range(1,Nsim+1):
184     Css.append(phi(k,0))
185     if k<N+1:
186         Cobs.append(phi(k,0))
187
188 Lobs = np.vstack((L0,np.zeros((N-NO,1))))
189 Lobs=Lobs.reshape(N,)
190
191 for k in range(1,Nsim+1):
192     def fun(x):
193         return input_a(x)*phi(k,x)
194     y,err= integrate.quad(fun,0,1)
195     Bssa.append(y)
```

```
196    if k<N+1:
197        Bobsa.append(y)
198    def fun(x):
199        return input_b(x)*phi(k,x)
200    y,err=integrate.quad(fun,0,1)
201    Bssb.append(y)
202    if k<N+1:
203        Bobsb.append(y)
204
205 Bssa  = np.array(Bssa).reshape((Nsim,))
206 Bssb  = np.array(Bssb).reshape((Nsim,))
207 Bobsa = np.array(Bobsa).reshape((N,))
208 Bobsb = np.array(Bobsb).reshape((N,))
209
210 # Initial condition (IC)
211 def z0(x):
212    return 1+x**2 # IC of the PDE
213
214 u0 = z0(1) # IC of the control
215
216 def w0(x):
217    return z0(x) - x**2*u0 #IC of the homogeneous Dirichlet system
218
219 zsim0=[]
220 for k in range(1,Nsim+1):
221    def fun(x):
222        return w0(x)*phi(k,x)
223    y,err=integrate.quad(fun,0,1)
224    zsim0.append(y) # Coefficients of projection of the IC w0
225
226 zsim0 = np.array(zsim0).reshape((Nsim,1))
227
228 # Check the validity of the projection (graph)
229 space = np.linspace(0,1,100)
230 check = np.zeros((len(space),1))
231
232 for kx in range(len(space)):
233    for k in range(Nsim):
234        check[kx,0] = check[kx,0] + zsim0[k,0]*phi(k,space[kx])
235
236 fig , ax= plt.subplots()
237 ax.set_title('To check')
238
239 ax.plot(space, w0(space),'g-', label='w0')
240 ax.plot(space, check,'r.', label='approx')
241 ax.legend()
242 # plt.savefig('to_check.png',bbox_inches='tight')
243
244 # time discretization
245 Tsim = 6;
```

```
246
247  def ode(z,t):
248      u=float(z[0])
249      zsim=z[1:Nsim+1]
250      zhat=z[Nsim+1:]
251      whata=np.vstack((u,zhat[0:N0]))
252      v=float(np.dot(K,whata))
253      udot=v
254      zsimdot=np.dot(Ass,zsim)+np.dot(Bssa,u)+np.dot(Bssb,v)
255      zhatdot=np.dot(Aobs,zhat)+np.dot(Bobsa,u)+np.dot(Bobsb,v)
256      zhatdot+=-Lobs*float(np.dot(Css,zsim)-np.dot(Cobs,zhat))
257      zdot=np.hstack((np.hstack((udot*np.ones(1),zsimdot)),zhatdot))
258      return zdot
259
260  t=np.linspace(0,Tsim,60)
261
262  # Initial condition of the full state
263  zhat0=np.zeros((N,1))
264  z0tot=np.vstack((np.vstack((u0,zsim0)),zhat0))
265  z0tot=z0tot.reshape(len(z0tot),)
266  sol=integrate.odeint(ode,z0tot,t)
267
268  Mstate_z  = np.zeros((len(space),len(t))) # PDE in original
         coordinates z
269  Mstate_w  = np.zeros((len(space),len(t))) # PDE in homogeneous
         coordinates w
270  MstateObs = np.zeros((len(space),len(t))) # State of the observer
271
272  for k_time in range(len(t)):
273      for k_space in range(len(space)):
274          for k in range(Nsim):
275              Mstate_w[k_space,k_time] +=  sol[k_time,k+1]*phi(k,
         space[k_space])
276          Mstate_z[k_space,k_time] = Mstate_w[k_space,k_time] + space
         [k_space]**2*sol[k_time,0]
277          for k in range(N):
278              MstateObs[k_space,k_time] += MstateObs[k_space,k_time]
         + sol[k_time,k+1+Nsim]*phi(k,space[k_space])
279
280  mpl.rcParams['legend.fontsize'] = 10
281  fig = plt.figure(); ax = fig.add_subplot(111, projection='3d')
282  SX, ST = np.meshgrid(space, t)
283  ax.plot_surface(SX, ST, Mstate_z.T, cmap='jet')
284  ax.set_xlabel('x')
285  ax.set_ylabel('t')
286  ax.set_zlabel('z(t,x)')
287  ax.view_init(elev=15, azim=20) # adjust view so it is easy to see
288  plt.savefig('pde-3d.png')
289
290  fig = plt.figure(); ax = fig.add_subplot(111, projection='3d')
```

```
291 ax.plot_surface(SX, ST, Mstate_w.T-MstateObs.T, cmap='jet')
292 ax.set_xlabel('x')
293 ax.set_ylabel('t')
294 ax.set_zlabel('$e(t,x)$')
295 ax.view_init(elev=15, azim=20) # adjust view so it is easy to see
296 plt.savefig('pde-error.png')
```

Fig. 6 depicts the corresponding solution. The convergence of the state towards 0 can be observed, confirming the predictions of Theorem 7. The observation error is given on the same figure.

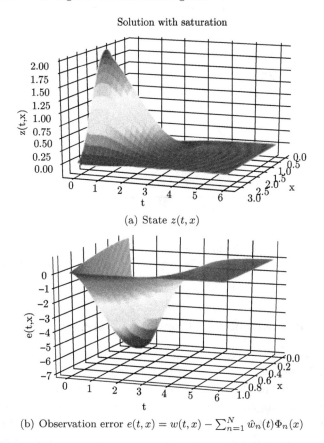

(a) State $z(t, x)$

(b) Observation error $e(t, x) = w(t, x) - \sum_{n=1}^{N} \hat{w}_n(t) \Phi_n(x)$

Fig. 6. State z and observation error e in closed-loop with Dirichlet boundary measurement feedback control for the reaction-diffusion system (56)

The time-evolutions of the control and output variables are given in Fig. 7.

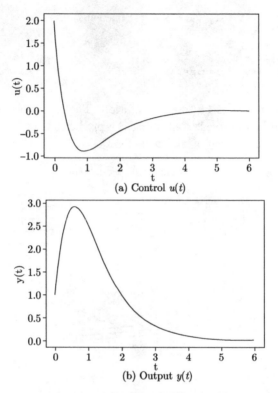

Fig. 7.   Control $u(t) = z(t, 1)$ and output $y(t, x) = z(t, 0)$ for the closed-loop system with Dirichlet boundary measurement feedback control for the reaction-diffusion system (56)

3.4. *Saturated control with internal measurement*

In this section we consider the stability analysis of parabolic PDEs when controlled in the presence of input saturations. In this setting, the control inputs apply in the domain by means of a bounded operator while the observation can take the form of either a bounded or an unbounded measurement operator. As in the previous section, the adopted approach relies on spectral-reduction methods. The presence of the input saturation is handled in the stability analysis by invoking the generalized sector condition reported in Section 2. This type of control design problem was reported in [48] in the case of a state-feedback. We consider here the case of an output-feedback by combining the Lyapunov-based analysis procedure discussed in the previous sections and the previously generalized sector condition. This allows the derivation of a set of sufficient conditions ensuring the local exponential stability of the origin of the closed-loop system. A subset of the domain of attraction is characterized by the decision variables of the abovementioned sufficient constraints.

Problem description

Let the reaction-diffusion equation with Robin boundary conditions be described by

$$z_t(t,x) = (p(x)z_x(t,x))_x - (q(x) - q_c)z(t,x) + \sum_{k=1}^{m} b_k(x)u_{\text{sat},k}(t), \quad (64\text{a})$$

$$\cos(\theta_1)z(t,0) - \sin(\theta_1)z_x(t,0) = 0, \quad (64\text{b})$$

$$\cos(\theta_2)z(t,1) + \sin(\theta_2)z_x(t,1) = 0, \quad (64\text{c})$$

$$z(0,x) = z_0(x) \quad (64\text{d})$$

with measurement equation

$$y(t) = \int_0^1 c(x)z(t,x)\,\mathrm{d}x. \quad (65)$$

Here we have $\theta_1, \theta_2 \in [0, \pi/2]$, $p \in C^1([0,1])$ with $p > 0$, $q \in C^0([0,1])$ with $q \geq 0$, $q_c \in \mathbb{R}$, and $b_k \in L^2(0,1)$. The scalar control inputs $u_{\text{sat},k}(t) \in \mathbb{R}$ act on the system. Hence, (64) can be written as

$$z_t(t,\cdot) = -\mathcal{A}z(t,\cdot) + q_c z(t,\cdot) + \sum_{k=1}^{m} b_k u_{\text{sat},k}(t), \quad (66\text{a})$$

$$z(0,\cdot) = z_0, \quad (66\text{b})$$

where $\mathcal{A}$ is the Sturm-Liouville operator defined at the beginning of this section.

The control input is assumed to be subject to saturations; for a given vector $s = \begin{bmatrix} s_1 & s_2 & \ldots & s_m \end{bmatrix}^\top \in (\mathbb{R}_+^*)^m$, we define sat : $\mathbb{R}^m \to \mathbb{R}^m$ by (17). Hence, the input $u_{\text{sat},k}(t)$ that is applied to the plant is expressed in function of the actual control inputs $u_k(t)$ as

$$u_{\text{sat}}(t) = \text{sat}(u(t))$$

with $u_{\text{sat}}(t) = \begin{bmatrix} u_{\text{sat},1}(t) & u_{\text{sat},2}(t) & \ldots & u_{\text{sat},m}(t) \end{bmatrix}^\top$ and $u(t) = \begin{bmatrix} u_1(t) & u_2(t) & \ldots & u_m(t) \end{bmatrix}^\top$.

In this context and similarly to [48] in the case of a state-feedback, the objective is to study the local stabilization of (64) with measurement (65) for the controller architecture studied in the first part of this section but in the presence of the saturating control inputs while estimating the associated domain of attraction.

3.4.1. *Spectral analysis*

Consider again the coefficients of projection $z_n(t) = \langle z(t, \cdot), \Phi_n \rangle$, $b_{n,k} = \langle b_k, \Phi_n \rangle$, and $c_n = \langle c, \Phi_n \rangle$. As done for (31) without saturation, the projection of the system solutions (66) and the output equation (65) into the Hilbert basis $\{\Phi_n : n \geq 1\}$ gives the following representation:

$$\dot{z}_n(t) = (-\lambda_n + q_c) z_n(t) + \sum_{k=1}^{m} b_{n,k} u_{\mathrm{sat},k}(t), \qquad (67a)$$

$$y(t) = \sum_{n \geq 1} c_n z_n(t). \qquad (67b)$$

Proceeding as in the previous subsection, we consider the feedback law taking the form of a finite-dimensional state-feedback coupled with a finite-dimensional observer. More precisely, let $\delta > 0$ and $N_0 \geq 1$ be such that $-\lambda_n + q_c < -\delta$ for all $n \geq N_0 + 1$. For a given integer $N \geq N_0 + 1$ to be selected later, the controller architecture takes the form:

$$\dot{\hat{z}}_n(t) = (-\lambda_n + q_c)\hat{z}_n(t) + \sum_{k=1}^{m} b_{n,k} u_{\mathrm{sat},k}(t) \qquad (68a)$$

$$+ L_n \left\{ \sum_{k=1}^{N} c_k \hat{z}_k(t) - y(t) \right\}, \quad 1 \leq n \leq N,$$

$$u_k(t) = \sum_{l=1}^{N_0} K_{k,l} \hat{z}_l(t), \quad 1 \leq k \leq m \qquad (68b)$$

with $L_n, K_{k,l} \in \mathbb{R}$ where $L_n = 0$ for $N_0 + 1 \leq n \leq N$.

We define the errors of estimation $e_n(t) = z_n(t) - \hat{z}_n(t)$. As in the previous subsection, we introduce the vectors and matrices defined by $\hat{Z}^{N_0} = \begin{bmatrix} \hat{z}_1 \ldots \hat{z}_{N_0} \end{bmatrix}^\top$, $\hat{Z}^{N-N_0} = \begin{bmatrix} \hat{z}_{N_0+1} \ldots \hat{z}_N \end{bmatrix}^\top$, $E^{N_0} = \begin{bmatrix} e_1 \ldots e_{N_0} \end{bmatrix}^\top$, $E^{N-N_0} = \begin{bmatrix} e_{N_0+1} \ldots e_N \end{bmatrix}^\top$, $A_0 = \mathrm{diag}(-\lambda_1 + q_c, \ldots, -\lambda_{N_0} + q_c)$, $A_1 = \mathrm{diag}(-\lambda_{N_0+1} + q_c, \ldots, -\lambda_N + q_c)$, $B_0 = (b_{n,k})_{1 \leq n \leq N_0, 1 \leq k \leq m}$, $B_1 = (b_{n,k})_{N_0+1 \leq n \leq N, 1 \leq k \leq m}$, $C_0 = \begin{bmatrix} c_1 \ldots c_{N_0} \end{bmatrix}$, $C_1 = \begin{bmatrix} c_{N_0+1} \ldots c_N \end{bmatrix}$, $L = \begin{bmatrix} L_1 \ldots L_{N_0} \end{bmatrix}^\top$, and $K = (K_{k,l})_{1 \leq k \leq m, 1 \leq l \leq N_0}$. This leads to

$$\dot{\hat{Z}}^{N_0} = A_0 \hat{Z}^{N_0} + B_0 u_{\mathrm{sat}} - LC_0 E^{N_0} - LC_1 E^{N-N_0} - L\zeta,$$

$$\dot{E}^{N_0} = (A_0 + LC_0)E^{N_0} + LC_1 E^{N-N_0} + L\zeta,$$

$$\dot{\hat{Z}}^{N-N_0} = A_1 \hat{Z}^{N-N_0} + B_1 u_{\mathrm{sat}},$$

$$\dot{E}^{N-N_0} = A_1 E^{N-N_0},$$

$$u = K\hat{Z}^{N_0},$$

where $\zeta(t) = \sum_{n \geq N+1} c_n z_n(t)$ is the residue of measurement. Owing to the definition of the deadzone nonlinearity (20), we infer that

$$\dot{\hat{Z}}^{N_0} = (A_0 + B_0 K)\hat{Z}^{N_0} - LC_0 E^{N_0} - LC_1 E^{N-N_0} - L\zeta + B_0 \phi(K\hat{Z}^{N_0}),$$
$$\dot{E}^{N_0} = (A_0 + LC_0)E^{N_0} + LC_1 E^{N-N_0} + L\zeta,$$
$$\dot{\hat{Z}}^{N-N_0} = A_1 \hat{Z}^{N-N_0} + B_1 K\hat{Z}^{N_0} + B_1 \phi(K\hat{Z}^{N_0}),$$
$$\dot{E}^{N-N_0} = A_1 E^{N-N_0}.$$

Introducing the state-vector

$$X = \text{col}(\hat{Z}^{N_0}, E^{N_0}, \hat{Z}^{N-N_0}, E^{N-N_0})$$

and the matrices

$$F = \begin{bmatrix} A_0 + B_0 K & -LC_0 & 0 & -LC_1 \\ 0 & A_0 + LC_0 & 0 & LC_1 \\ B_1 K & 0 & A_1 & 0 \\ 0 & 0 & 0 & A_1 \end{bmatrix}, \quad \mathcal{L} = \begin{bmatrix} -L \\ L \\ 0 \\ 0 \end{bmatrix}, \quad \mathcal{L}_\phi = \begin{bmatrix} B_0 \\ 0 \\ B_1 \\ 0 \end{bmatrix},$$

we deduce that

$$\dot{X} = FX + \mathcal{L}\zeta + \mathcal{L}_\phi \phi(K\hat{Z}^{N_0}). \tag{69}$$

We finally define $E = \begin{bmatrix} I & 0 & 0 & 0 \end{bmatrix}$ and $\tilde{K} = \begin{bmatrix} K & 0 & 0 & 0 \end{bmatrix}$, which are such that

$$\hat{Z}^{N_0} = EX, \qquad u = \tilde{K}X.$$

3.4.2. *Stability results*

For $z \in L^2(0,1)$ and $\hat{z} \in \mathbb{R}^N$, we define

$$\Pi(z, \hat{z}) = \begin{bmatrix} \Pi_1(z, \hat{z}) \\ \Pi_2(z, \hat{z}) \\ \Pi_3(z, \hat{z}) \\ \Pi_4(z, \hat{z}) \end{bmatrix}$$

with

$$\Pi_1(z, \hat{z}) = \begin{bmatrix} \hat{z}_1 \\ \vdots \\ \hat{z}_{N_0} \end{bmatrix}, \quad \Pi_2(z, \hat{z}) = \begin{bmatrix} \langle z, \Phi_1 \rangle - \hat{z}_1 \\ \vdots \\ \langle z, \Phi_{N_0} \rangle - \hat{z}_{N_0} \end{bmatrix},$$

and

$$\Pi_3(z, \hat{z}) = \begin{bmatrix} \hat{z}_{N_0+1} \\ \vdots \\ \hat{z}_N \end{bmatrix}, \quad \Pi_4(z, \hat{z}) = \begin{bmatrix} \langle z, \Phi_{N_0+1} \rangle - \hat{z}_{N_0+1} \\ \vdots \\ \langle z, \Phi_N \rangle - \hat{z}_N \end{bmatrix}.$$

Stabilization in L^2 norm. Let us now state and prove a result providing a stabilization for (64) in L^2-norm. This result is extracted from [34].

Theorem 8. *Let $\theta_1, \theta_2 \in [0, \pi/2]$, $p \in \mathcal{C}^1([0,1])$ with $p > 0$, $q \in \mathcal{C}^0([0,1])$ with $q \geq 0$, $q_c \in \mathbb{R}$, and $s \in (\mathbb{R}_+^*)^m$. Let $c \in L^2(0,1)$ and $b_k \in L^2(0,1)$ for $1 \leq k \leq m$. Consider the reaction-diffusion system described by (64) with measured output (65). Let $N_0 \geq 1$ and $\delta > 0$ be given such that $-\lambda_n + q_c < -\delta < 0$ for all $n \geq N_0 + 1$. Assume that 1) for any $1 \leq n \leq N_0$, there exists $1 \leq k = k(n) \leq m$ such that $b_{n,k} \neq 0$; 2) $c_n \neq 0$ for all $1 \leq n \leq N_0$. Let $K \in \mathbb{R}^{m \times N_0}$ and $L \in \mathbb{R}^{N_0}$ be such that $A_1 + B_1 K$ and $A_0 + L C_0$ are Hurwitz with eigenvalues that have a real part strictly less than $-\delta < 0$. Assume that there exist $N \geq N_0 + 1$, a symmetric positive definite $P \in \mathbb{R}^{2N \times 2N}$, $\alpha, \beta, \gamma, \mu, \kappa > 0$, a diagonal positive definite $T \in \mathbb{R}^{m \times m}$, and $G \in \mathbb{R}^{m \times N_0}$ such that*

$$\Theta_1(\kappa) \preceq 0, \quad \Theta_2 \succeq 0, \quad \Theta_3(\kappa) \leq 0, \tag{70}$$

where

$$\Theta_1(\kappa) = \begin{bmatrix} \Theta_{1,1,1}(\kappa) & P\mathcal{L} & -E^\top G^\top T + P\mathcal{L}_\phi \\ \star & -\beta & 0 \\ \star & \star & \alpha\gamma \sum_{k=1}^m \|\mathcal{R}_N b_k\|_{L^2}^2 I - 2T \end{bmatrix},$$

$$\Theta_2 = \begin{bmatrix} P & E^\top (K-G)^\top \\ \star & \mu \operatorname{diag}(s)^2 \end{bmatrix},$$

$$\Theta_3(\kappa) = 2\gamma \left\{ -\lambda_{N+1} + q_c + \kappa + \frac{1}{\alpha} \right\} + \beta \|\mathcal{R}_N c\|_{L^2}^2$$

with $\Theta_{1,1,1}(\kappa) = F^\top P + PF + 2\kappa P + \alpha\gamma \sum_{k=1}^m \|\mathcal{R}_N b_k\|_{L^2}^2 \tilde{K}^\top \tilde{K}$. Define the ellipsoid

$$\mathcal{E}_1 = \left\{ (z, \hat{z}) \in L^2(0,1) \times \mathbb{R}^N : \Pi_N(z, \hat{z})^\top P\Pi_N(z, \hat{z}) + \gamma \|\mathcal{R}_N z\|_{L^2}^2 \leq \frac{1}{\mu} \right\}.$$

Then, the origin of the closed-loop system composed of the plant (64) with measured output (65) and the control law (68) is locally exponentially stable in L^2-norm with exponential decay rate κ and with a basin of attraction including $\mathcal{E}_1$. More precisely, there exists $M > 0$ such that for any initial condition $(z_0, \hat{z}(0)) \in \mathcal{E}_1$, the solution satisfies

$$\|z(t, \cdot)\|_{L^2}^2 + \sum_{n=1}^N \hat{z}_n(t)^2 \leq Me^{-2\kappa t} \left(\|z_0\|_{L^2}^2 + \sum_{n=1}^N \hat{z}_n(0)^2 \right) \tag{71}$$

for all $t \geq 0$. Moreover, for any fixed $\kappa \in (0, \delta]$, the constraints (70) are always feasible for N large enough.

Proof of Theorem 8. Let the Lyapunov function candidate be defined by $V(X, z) = X^\top P X + \gamma \sum_{n \geq N+1} \langle z, \Phi_n \rangle^2$ for $X \in \mathbb{R}^{2N}$ and $z \in L^2(0,1)$. The computation of the time derivative of V along the system solutions to (67) and (69) gives

$$\dot{V} + 2\kappa V = X^\top \left(F^\top P + PF + 2\kappa P \right) X + 2X^\top P \mathcal{L} \zeta$$
$$+ 2X^\top P \mathcal{L}_\phi (K\hat{Z}^{N_0}) + 2\gamma \sum_{n \geq N+1} (-\lambda_n + q_c + \kappa) z_n^2$$
$$+ 2\gamma \sum_{n \geq N+1} z_n L_n^b \tilde{K} X + 2\gamma \sum_{n \geq N+1} z_n L_n^b \phi(K\hat{Z}^{N_0}),$$

where $L_n^b = [b_{n,1} \ldots b_{n,m}]$. From Young's inequality, we obtain for any $\alpha > 0$ and any $w \in \mathbb{R}^m$ that $2\sum_{n \geq N+1} z_n L_n^b w \leq \dfrac{1}{\alpha} \sum_{n \geq N+1} z_n^2 + \alpha \sum_{k=1}^m \|\mathcal{R}_N b_k\|_{L^2}^2 \|w\|^2$. Hence, introducing $\tilde{X} = \mathrm{col}(X, \zeta, \phi(K\hat{Z}^{N_0}))$, we deduce that

$$\dot{V} + 2\kappa V \leq \tilde{X}^\top \begin{bmatrix} \Theta_{1,1,1} & P\mathcal{L} & P\mathcal{L}_\phi \\ \star & 0 & 0 \\ \star & \star & \alpha\gamma \sum_{k=1}^m \|\mathcal{R}_N b_k\|_{L^2}^2 I \end{bmatrix} \tilde{X}$$
$$+ 2\gamma \sum_{n \geq N+1} \left(-\lambda_n + q_c + \kappa + \frac{1}{\alpha} \right) z_n^2.$$

Since, by definition, $\zeta = \sum_{n \geq N+1} c_n z_n$, we obtain that $\zeta^2 \leq \|\mathcal{R}_N c\|_{L^2}^2 \sum_{n \geq N+1} z_n^2$. Moreover, if $\hat{Z}^{N_0} \in \mathbb{R}^{N_0}$ satisfies $|(K - G)\hat{Z}^{N_0}| \leq s$, we deduce from (21) that $\phi(K\hat{Z}^{N_0})^\top T(\phi(K\hat{Z}^{N_0}) + G\hat{Z}^{N_0}) \leq 0$. Combining the latter estimates, we obtain for all $X \in \mathbb{R}^{2N}$ satisfying $|(K-G)EX| \leq s$ that

$$\dot{V} + 2\kappa V \leq \tilde{X}^\top \Theta_1(\kappa)\tilde{X} + \sum_{n \geq N+1} \Gamma_n z_n^2,$$

where $\Gamma_n = 2\gamma \left(-\lambda_n + q_c + \kappa + \frac{1}{\alpha} \right) + \beta\|\mathcal{R}_N c\|_{L^2}^2 \leq \Theta_3(\kappa)$ for all $n \geq N+1$. Hence the assumptions imply that $\dot{V} + 2\kappa V \leq 0$ for all $X \in \mathbb{R}^{2N}$ is such that $|(K - G)EX| \leq s$.

We now need to give a sufficient condition such that $|(K - G)EX| \leq s$ holds. To do so, consider $X \in \mathbb{R}^{2N}$ and $z \in L^2(0,1)$ such that $V(X, z) \leq 1/\mu$. Applying the Schur complement to $\Theta_2 \succeq 0$, we obtain that $P \succeq \frac{1}{\mu} E^\top (K - G)^\top \mathrm{diag}(s)^{-2}(K - G)E$. This implies that $\|\mathrm{diag}(s)^{-1}(K - G)EX\| \leq 1$, giving in particular that $|(K - G)EX| \leq s$ hence $\dot{V} + 2\kappa V \leq 0$.

From now it is easy to show that, for any initial condition selected such that $(z_0, \hat{z}_0) \in \mathcal{E}_1$ with $z_0 \in D(\mathcal{A})$, we have $V(X(t), z(t, \cdot)) \leq 1/\mu$ and $\dot{V}(X(t), z(t, \cdot)) + 2\kappa V(X(t), z(t, \cdot)) \leq 0$ for all $t \geq 0$. The claimed stability estimate (71) follows from the definition of V. The extension of this result to mild solutions associated with any $(z_0, \hat{z}_0) \in \mathcal{E}_1$ follows from a classical density argument [53, Thm. 6.1.2].

The rest of the proof, which concerns the feasibility of the constraints, is reported in [34]. $\qquad\qquad\qquad\qquad\qquad\qquad\qquad\qquad\qquad\qquad\quad$ □

Stabilization in H^1 norm. The following result deals with the exponential stability of the system trajectories evaluated in H^1-norm.

Theorem 9. *In the context of Theorem 8, we further assume that $q > 0$. Assume that there exist $N \geq N_0 + 1$, a symmetric positive definite $P \in \mathbb{R}^{2N \times 2N}$, $\alpha > 1$, $\beta, \gamma, \mu, \kappa > 0$, a diagonal positive definite $T \in \mathbb{R}^{m \times m}$, and $G \in \mathbb{R}^{m \times N_0}$ such that*

$$\Theta_1(\kappa) \preceq 0, \quad \Theta_2 \succeq 0, \quad \Theta_3(\kappa) \leq 0, \tag{72}$$

where $\Theta_1(\kappa)$ and Θ_2 are defined as in Theorem 8 while

$$\Theta_3(\kappa) = 2\gamma \left\{ -\left(1 - \frac{1}{\alpha}\right) \lambda_{N+1} + q_c + \kappa \right\} + \frac{\beta \|\mathcal{R}_N c\|_{L^2}^2}{\lambda_{N+1}}.$$

Define the ellipsoid

$$\mathcal{E}_2 = \left\{ (z, \hat{z}) \in D(\mathcal{A}) \times \mathbb{R}^N : \Pi(z, \hat{z})^\top P \Pi(z, \hat{z}) + \gamma \|\mathcal{R}_N \mathcal{A}^{1/2} z\|_{L^2}^2 \leq \frac{1}{\mu} \right\}.$$

Then, the origin of the closed-loop system composed of the plant (64) with measured output (65) and the control law (68) is locally exponentially stable in H^1-norm with exponential decay rate κ and with a basin of attraction including $\mathcal{E}_1$. In other words, there exists $M > 0$ such that for any initial condition $(z_0, \hat{z}(0)) \in \mathcal{E}_1$, the solution satisfies

$$\|z(t, \cdot)\|_{H^1}^2 + \sum_{n=1}^N \hat{z}_n(t)^2 \leq M e^{-2\kappa t} \left(\|z_0\|_{H^1}^2 + \sum_{n=1}^N \hat{z}_n(0)^2 \right)$$

for all $t \geq 0$. Moreover, for any fixed $\kappa \in (0, \delta]$, the constraints (72) are always feasible for N large enough.

Proof of Theorem 9. We introduce the Lyapunov functional candidate $V(X, z) = X^\top P X + \gamma \sum_{n \geq N+1} \lambda_n \langle z, \Phi_n \rangle^2$ when $X \in \mathbb{R}^{2N}$ and $z \in D(\mathcal{A})$.

The computation of the time derivative of V along the system solutions to (67) and (69) gives

$$\dot{V} + 2\kappa V = X^\top \left(F^\top P + PF + 2\kappa P\right) X + 2X^\top P\mathcal{L}\zeta$$
$$+ 2X^\top P\mathcal{L}_\phi \phi(K\hat{Z}^{N_0}) + 2\gamma \sum_{n \geq N+1} \lambda_n(-\lambda_n + q_c + \kappa)z_n^2$$
$$+ 2\gamma \sum_{n \geq N+1} \lambda_n z_n L_n^b \tilde{K} X + 2\gamma \sum_{n \geq N+1} \lambda_n z_n L_n^b \phi(K\hat{Z}^{N_0})$$

where $L_n^b = \begin{bmatrix} b_{n,1} & \ldots & b_{n,m} \end{bmatrix}$. Invoking Young's inequality, we obtain for any $\alpha > 0$ and any $w \in \mathbb{R}^m$ that $2\sum_{n \geq N+1} \lambda_n z_n L_n^b w \leq \frac{1}{\alpha} \sum_{n \geq N+1} \lambda_n^2 z_n^2 + \alpha \sum_{k=1}^m \|\mathcal{R}_N b_k\|_{L^2}^2 \|w\|^2$. Let $\tilde{X} = \mathrm{col}(X, \zeta, \phi(K\hat{Z}^{N_0}))$. Proceeding as in the proof of Theorem 8, we deduce that

$$\dot{V} + 2\kappa V \leq \tilde{X}^\top \Theta_1(\kappa)\tilde{X} + \sum_{n \geq N+1} \lambda_n \Gamma_n z_n^2$$

for all $X \in \mathbb{R}^{2N}$ satisfying $|(K - G)EX| \leq s$ and where $\Gamma_n = 2\gamma\left\{-\left(1 - \frac{1}{\alpha}\right)\lambda_n + q_c + \kappa\right\} + \frac{\beta\|\mathcal{R}_N c\|_{L^2}^2}{\lambda_n} \leq \Theta_3(\kappa)$ for all $n \geq N + 1$. The proof now follows similar arguments that the ones employed in the proof of Theorem 8. $\qquad\square$

Remark 6. The conditions in Theorem 8 and in Theorem 9 are nonlinear in the unknown variables. This is due in particular to the product $G^\top T$. Some nonlinearity could be transformed into linear conditions as for the variable α, as discussed in Remark 3. To deal with the particular case $m = 1$, or to deduce convex constraints from these theorems, see [34].

3.5. *Section conclusion*

This section has discussed the topic of output-feedback stabilization of a reaction-diffusion equation by means of in-domain or boundary control inputs. The controllers that are considered in this section are output-feedback laws where the output is defined from a boundary measure or an internal measurement of the state. The control strategy takes the form of a finite-dimensional controller composed of an observer coupled with a finite-dimensional partial state-feedback. The control can be either linear or subject to a saturation map. In the latter scenario, only a local asymptotic stability can be obtained in general along with an estimation of the basin of attraction. The reported stability analysis takes advantage of Lyapunov functionals coupled with the generalized sector condition that has been recalled in Section 2 in the context of finite-dimensional systems. The obtained sets of constraints ensuring the stability of the closed-loop system take an explicit form and have been shown to be feasible when the order of the controller is selected large enough. An explicit subset of the domain of attraction of the closed-loop system has also been derived.

4. Stabilization of wave and KdV equations

Two classes of particular equations are considered in this section: first the wave equation and then the Korteweg-de Vries (KdV) equation. We particularly focus on the boundary stabilization problem. The interest of the first equation is that it gives a transition towards the boundary control of general hyperbolic systems, whereas the second one allows to show perspectives in terms of stabilization of nonlinear partial differential equations, and give a highlighting example of what could be done for boundary control of other classes of hyperbolic PDEs (as considered e.g. in [55, 75]).

For both equations, we solve the common objectives of well-posedness assessment and asymptotic stabilization by means of distributed or boundary control that can be either linear or subject to a nonlinear map (e.g., a saturation).

This section is organized as follows. First, in Section 4.1, the stabilization of the linear wave equation with linear and with nonlinear in-domain control is presented. The topic of boundary control is then considered for the same equation. Finally the nonlinear KdV equation is considered in Section 4.2 with in-domain control. This result is illustrated with some numerical simulations.

4.1. *Wave equation with a bounded control operator*

Motivated by the illustration depicted in Fig. 8, where z stands for the deflection of a membrane with respect to the rest and horizontal axis and that is subject to a distributed force u, we start this section by considering the following wave equation:

$$z_{tt}(t,x) = z_{xx}(t,x) + u(t,x), \ \forall t \geq 0, \ x \in (0,1). \tag{73}$$

We assume that the membrane is clamped at both extremities. This implies the following boundary conditions, for all $t \geq 0$,

$$z(t,0) = 0, \\ z(t,1) = 0. \tag{74}$$

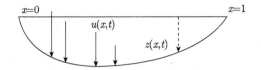

Fig. 8.   Wave equation with bounded control operator

The initial condition is given, for all $x \in (0,1)$, by

$$z(0,x) = z^0(x),$$
$$z_t(0,x) = z^1(x), \tag{75}$$

where z^0 and z^1 stand for the initial deflection and the initial deflection speed, respectively.

Let us note that the function defined by $z(t,x) = 0$, for all (t,x) in $(0,1) \times [0,\infty)$ is a particular solution to (73) and (74) in the uncontrolled scenario $(u = 0)$. Hence the origin is an equilibrium for the studied wave equation. The objective is to render this equilibrium asymptotically stable by designing an adequate feedback control u.

4.1.1. *Internal linear control*

Let us define the linear control by

$$u(t,x) = -az_t(t,x),\ t \geq 0,\ x \in (0,1), \tag{76}$$

and consider

$$V_1 = \frac{1}{2}\int (z_x^2 + z_t^2)\mathrm{d}x. \tag{77}$$

A formal computation gives, along the solutions to (73), (74) and (76),

$$\begin{aligned}
\dot{V}_1 &= \int_0^1 (z_x z_{xt} - az_t^2 + z_t z_{xx})\mathrm{d}x \\
&= -\int_0^1 az_t^2 \mathrm{d}x + [z_t z_x]_{x=0}^{x=1} \\
&= -\int_0^1 az_t^2 \mathrm{d}x.
\end{aligned}$$

Thus, if $a > 0$, V_1 is a (non strict) Lyapunov function.

Using standard technics, such as Lumer-Phillips theorem for the well-posedness (see e.g., [14, Theorem A.4.]) and Huang-Prüss theorem for the exponential stability (see [27] and [58]), we may prove the following result:

H. Lhachemi and C. Prieur

Theorem 10. *For $a > 0$ and $(z^0, z^1) \in H_0^1(0,1) \times L^2(0,1)$, there exists a unique (weak) solution $z \colon [0,\infty) \to H_0^1(0,1) \times L^2(0,1)$ to (73)–(76). Moreover, the origin of $H_0^1(0,1) \times L^2(0,1)$ is an exponentially stable equilibrium, that is there exist two positive values C and $\mu > 0$ such that, for any initial condition $(z^0, z^1) \in H_0^1(0,1) \times L^2(0,1)$, it holds, for all $t \geq 0$,*

$$\|z\|_{H_0^1(0,1)} + \|z_t\|_{L^2(0,1)} \leq Ce^{-\mu t}(\|z^0\|_{H_0^1(0,1)} + \|z^1\|_{L^2(0,1)}).$$

Proof of Theorem 10. Let us first prove the well-posedness. Let A_l be the linear unbounded operator

$$A_l \begin{pmatrix} f \\ g \end{pmatrix} = \begin{pmatrix} g \\ f_{xx} - ag \end{pmatrix}$$

with the domain $D(A_l) = (H^2(0,1) \cap H_0^1(0,1)) \times H_0^1(0,1)$. This domain is dense in $H_0^1(0,1) \times L^2(0,1)$, and the operator A_l is closed. Let us rewrite (73)–(76) as

$$\frac{\partial}{\partial t} \begin{pmatrix} z \\ z_t \end{pmatrix} = A_l \begin{pmatrix} z \\ z_t \end{pmatrix}, \quad \begin{pmatrix} z \\ z_t \end{pmatrix}(t = 0, \cdot) = \begin{pmatrix} z^0 \\ z^1 \end{pmatrix}. \tag{78}$$

Our objective is to prove that A_l generates a contraction semigroup $T_l(t)$, i.e. the solution of (78) is $T_l(t) \begin{pmatrix} z^0 \\ z^1 \end{pmatrix}$ and satisfies

$$\left\| T_l(t) \begin{pmatrix} z^0 \\ z^1 \end{pmatrix} \right\|^2 \leq \left\| \begin{pmatrix} z^0 \\ z^1 \end{pmatrix} \right\|^2, \quad \forall t \geq 0. \tag{79}$$

Informally, one can try to prove (79) by differentiating the right-hand-side with respect to the time. Using $\left\| \begin{pmatrix} z^0 \\ z^1 \end{pmatrix} \right\|^2 = \left\langle \begin{pmatrix} z^0 \\ z^1 \end{pmatrix}, \begin{pmatrix} z^0 \\ z^1 \end{pmatrix} \right\rangle$, we get

$$\frac{d}{dt} \left\| T_l(t) \begin{pmatrix} z^0 \\ z^1 \end{pmatrix} \right\|^2 = \left\langle A_l T_l(t) \begin{pmatrix} z^0 \\ z^1 \end{pmatrix}, \begin{pmatrix} z^0 \\ z^1 \end{pmatrix} \right\rangle + \left\langle \begin{pmatrix} z^0 \\ z^1 \end{pmatrix}, A_l T_l(t) \begin{pmatrix} z^0 \\ z^1 \end{pmatrix} \right\rangle$$

$$= 2\mathrm{Re} \left\langle A_l T_l(t) \begin{pmatrix} z^0 \\ z^1 \end{pmatrix}, \begin{pmatrix} z^0 \\ z^1 \end{pmatrix} \right\rangle$$

where Re denotes the real part. This gives, at time $t = 0$,

$$\frac{d}{dt} \left\| T_l(t) \begin{pmatrix} z^0 \\ z^1 \end{pmatrix} \right\|^2 (t = 0, \cdot) = 2\mathrm{Re} \left\langle A_l \begin{pmatrix} z^0 \\ z^1 \end{pmatrix}, \begin{pmatrix} z^0 \\ z^1 \end{pmatrix} \right\rangle.$$

This formal computation tends to show that in order to obtain (79), a necessary condition is to have $\mathrm{Re} \left\langle A_l \begin{pmatrix} z^0 \\ z^1 \end{pmatrix}, \begin{pmatrix} z^0 \\ z^1 \end{pmatrix} \right\rangle \leq 0$. This condition is one of the two key elements of the Lumer-Phillips theorem which provides a characterization of the unbounded operators generating a contraction semigroup. Specifically, in order to apply the Lumer-Phillips theorem, we need to show that the two following points hold true:

(1) $\operatorname{Re}\left\langle A_l\begin{pmatrix}z^0\\z^1\end{pmatrix},\begin{pmatrix}z^0\\z^1\end{pmatrix}\right\rangle \leq 0$, for all $\begin{pmatrix}z^0\\z^1\end{pmatrix}$ in $D(A_l)$;

(2) there exists $\lambda > 0$ such that $\operatorname{Ran}(I - \lambda A_l) = H_0^1(0,1) \times L^2(0,1)$, where Ran is the range set.

Under these two conditions, the unbounded operator A_l generates a semigroup of contraction and the Cauchy problem (78) is well-posed for strong and weak solutions as considered in Theorem 10.

Even if we do not give here a complete proof of these both properties, note that the interest of the second item is that it replaces the time-dependent Cauchy problem (78) by

$$\forall \begin{pmatrix}f\\g\end{pmatrix} \in H_0^1(0,1) \times L^2(0,1), \text{ find } \begin{pmatrix}\tilde{f}\\\tilde{g}\end{pmatrix} \in D(A_l) \text{ such that}$$

$$(I - \lambda A_l)\begin{pmatrix}\tilde{f}\\\tilde{g}\end{pmatrix} = \begin{pmatrix}f\\g\end{pmatrix}$$

which is a stationary Cauchy problem of a linear ODE with prescribed boundary conditions.

Let us now sketch the proof of the exponential stability. According to Huang-Prüss theorem (see [27] and [58]), it is sufficient to check the two conditions

$$i\mathbb{R} \subset \rho(A), \tag{80}$$

$$\sup_{\beta \in \mathbb{R}} \|(i\beta - A_l)^{-1}\| < \infty. \tag{81}$$

Inspired by [33], let us prove these both properties successively. To prove (80), we argue by contradiction, assuming the existence of an eigenvalue of A_l of the form $i\beta$. Pick $\begin{pmatrix}f\\g\end{pmatrix}$ in $D(A_l) \setminus \{0\}$ such that $(i\beta - A_l)\begin{pmatrix}f\\g\end{pmatrix} = 0$. Then

$$0 = \left\langle (i\beta - A_l)\begin{pmatrix}f\\g\end{pmatrix},\begin{pmatrix}f\\g\end{pmatrix}\right\rangle_{H_0^1(0,1)\times L^2(0,1)}, \tag{82}$$

$$= i\beta \left(\int_0^1 |f'|^2 \mathrm{d}x + \int_0^1 |g|^2 \mathrm{d}x\right) + a\int_0^1 |g|^2 \mathrm{d}x. \tag{83}$$

Thus, inspecting the real part of the previous equation, with $a \neq 0$, we get $g = 0$. Moreover, inspecting the imaginary part, we get $f' = 0$ which gives $f = 0$ using the definition of $D(A_l)$ and the boundary conditions of f. This is a contradiction with $\begin{pmatrix} f \\ g \end{pmatrix} \neq 0$. Therefore (80) holds.

Let us now prove (81), by proceeding again with a contradiction. If (81) is false, then there exist a sequence $(\beta_n)_{n \in \mathbb{N}}$ and a sequence $\begin{pmatrix} f_n \\ g_n \end{pmatrix}_{n \in \mathbb{N}}$ in $D(A_l)$ such that

$$\left\| \begin{pmatrix} f_n \\ g_n \end{pmatrix}_{n \in \mathbb{N}} \right\|_{H_0^1(0,1) \times L^2(0,1)} = 1, \tag{84}$$

$$\beta_n \to_{n \to \infty} +\infty \tag{85}$$

and

$$\left\| \begin{pmatrix} \tilde{f}_n \\ \tilde{g}_n \end{pmatrix} \right\|_{H_0^1(0,1) \times L^2(0,1)} \to_{n \to \infty} 0, \tag{86}$$

where $\begin{pmatrix} \tilde{f}_n \\ \tilde{g}_n \end{pmatrix} = (i\beta_n - A_l) \begin{pmatrix} f_n \\ g_n \end{pmatrix}_{n \in \mathbb{N}}$. We compute

$$\left\langle \begin{pmatrix} \tilde{f}_n \\ \tilde{g}_n \end{pmatrix}, \begin{pmatrix} f_n \\ g_n \end{pmatrix} \right\rangle_{H_0^1(0,1) \times L^2(0,1)}$$

$$= \left\langle (i\beta_n - A_l) \begin{pmatrix} f_n \\ g_n \end{pmatrix}, \begin{pmatrix} f_n \\ g_n \end{pmatrix} \right\rangle_{H_0^1(0,1) \times L^2(0,1)}$$

$$= i\beta_n \left(\int_0^1 |f_n|'^2 \mathrm{d}x + \int_0^1 |g_n|^2 \mathrm{d}x \right) + a \int_0^1 |g_n|^2 \mathrm{d}x.$$

Therefore, with (84) and (86), inspecting the imaginary part in the last equation, we get

$$\beta_n \left(\int_0^1 f_n'^2 \mathrm{d}x + \int_0^1 g_n^2 \mathrm{d}x \right) \to_{n \to \infty} 0,$$

thus with (84), we get $\beta_n \to 0$ which is a contradiction with (85). Therefore (81) holds. This concludes the proof of the exponential stability and of the proof of Theorem 10. □

4.1.2. Internal saturating control

We now study the nonlinear control

$$u(t, x) = -\mathsf{sat}(az_t(t, x)), \quad x \in (0, 1), \ \forall t \geq 0, \tag{87}$$

where sat is the nonlinear function defined in (17) with $m = 1$ and level s_0. Following the terminology of [45], we call this nonlinearity the localized saturated map. The wave equation (73) in closed loop with the control (87) gives the dynamics

$$z_{tt} = z_{xx} - \text{sat}(az_t). \tag{88}$$

A formal computation of the time derivative of V_1 defined by (77) along the solutions to the wave PDE (88) with boundary conditions (74) gives

$$\dot{V}_1 = -\int_0^1 z_t \text{sat}(az_t) \mathrm{d}x.$$

Hence, in order to conclude on the possible stability of the closed-loop system, one needs to handle the nonlinearity $z_t \text{sat}(az_t)$.

Note that other choices of saturation mechanisms can also be considered instead of the *localized* saturation studied in (87). For instance, papers [66] and [32] deal with the L^2 saturation denoted by sat_{L2} and defined for any $\sigma \in L^2(0,1)$ by

$$\text{sat}_{L2}(\sigma)(x) = \begin{cases} \sigma(x), & \text{if } \|\sigma\|_{L^2(0,1)} < 1, \\ \dfrac{\sigma(x)}{\|\sigma\|_{L^2(0,1)}}, & \text{else.} \end{cases} \tag{89}$$

Even if all the different saturation mechanisms are of interest, we focus here on the localized saturation used in (87), which is generally more relevant from a physical point of view and in practical applications.

The well-posedness of the nonlinear PDE (88), which is borrowed from [56], is assessed by the following theorem.

Theorem 11. *For all $a \geq 0$ and (z^0, z^1) in $(H^2(0,1) \cap H_0^1(0,1)) \times H_0^1(0,1)$, there exists a unique solution $z \colon [0, \infty) \to H^2(0,1) \cap H_0^1(0,1)$ to (88) with the boundary conditions (74) and the initial condition (75).*

Proof of Theorem 11. We only provide a sketch of the proof reported in [56]. Consider the nonlinear operator

$$A_1 \begin{pmatrix} f \\ g \end{pmatrix} = \begin{pmatrix} g \\ f_{xx} - \text{sat}(ag) \end{pmatrix}$$

with the domain $D(A_1) = (H^2(0,1) \cap H_0^1(0,1)) \times H_0^1(0,1)$. We want to invoke here a generalization of the Lumer-Phillips theorem, which is the so-called Crandall-Liggett theorem. A precise statement of this theorem can be found in [4]; see also [8] and [49]. To apply this theorem, two conditions need to be checked:

(1) A_1 is dissipative, that is for any two elements of $D(A_1)$,

$$\mathrm{Re}\left(\left\langle A_1\begin{pmatrix} f \\ g \end{pmatrix} - A_1\begin{pmatrix} \tilde{f} \\ \tilde{g} \end{pmatrix}, \begin{pmatrix} f \\ g \end{pmatrix} - \begin{pmatrix} \tilde{f} \\ \tilde{g} \end{pmatrix}\right\rangle\right) \leq 0.$$

(2) For all $\lambda > 0$, $D(A_1) \subset \mathrm{Ran}(I - \lambda A_1)$.

Let us prove the first item. To do that, given $\begin{pmatrix} f \\ g \end{pmatrix}$ and $\begin{pmatrix} \tilde{f} \\ \tilde{g} \end{pmatrix}$ in $H_0^1(0,1) \times L^2(0,1)$, we denote

$$\Delta = \mathrm{Re}\left(\left\langle A_1\begin{pmatrix} f \\ g \end{pmatrix} - A_1\begin{pmatrix} \tilde{f} \\ \tilde{g} \end{pmatrix}, \begin{pmatrix} f \\ g \end{pmatrix} - \begin{pmatrix} \tilde{f} \\ \tilde{g} \end{pmatrix}\right\rangle\right) \leq 0.$$

Let us check that $\Delta \leq 0$. Using the definition of A_1 and of the hermitian product in $H_0^1(0,1) \times L^2(0,1)$, we compute

$$\begin{aligned}
\Delta &= \mathrm{Re}\left(\int_0^1 (g_x(x) - \tilde{g}_x(x))\overline{(f_x(x) - \tilde{f}_x(x))}\mathrm{d}x \right. \\
&\quad + \left. \int_0^1 (f_{xx}(x) - \tilde{f}_{xx}(x))\overline{(g(x) - \tilde{g}(x))}\mathrm{d}x\right) \\
&\quad - \mathrm{Re}\left(\int_0^1 (\mathrm{sat}(a\,g(x)) - \mathrm{sat}(a\,\tilde{g}(x)))\overline{(g(x) - \tilde{g}(x))}\mathrm{d}x\right) \\
&= -\mathrm{Re}\left(\int_0^1 (\mathrm{sat}(a\,g(x)) - \mathrm{sat}(a\,\tilde{g}(x)))\overline{(g(x) - \tilde{g}(x))}\mathrm{d}x\right).
\end{aligned}$$

Note that, for all $a \geq 0$ and for all $(s, \tilde{s})$ in $\mathbb{C} \times \mathbb{C}$,

$$\mathrm{Re}\left((\mathrm{sat}(a\,s) - \mathrm{sat}(a\,\tilde{s}))\overline{(s - \tilde{s})}\right) \geq 0.$$

Thus A_1 is dissipative.

The second item requires to deal with a nonlinear ODE. To be more specific, let $\lambda > 0$ and $\begin{pmatrix} f \\ g \end{pmatrix} \in H_0^1(0,1) \times L^2(0,1)$ be arbitrarily given. Our objective is to find $\begin{pmatrix} \tilde{f} \\ \tilde{g} \end{pmatrix} \in D(A_1)$ such that

$$(I - \lambda A_1)\begin{pmatrix} \tilde{f} \\ \tilde{g} \end{pmatrix} = \begin{pmatrix} f \\ g \end{pmatrix},$$

that is

$$\begin{cases} \tilde{f} - \lambda \tilde{g} = f, \\ \tilde{g} - \lambda(\tilde{f}_{xx} - \mathrm{sat}(a\,\tilde{g})) = g. \end{cases}$$

Using the first identity to express $\tilde{g}$ in function of f and $\tilde{f}$, we only have to find $\tilde{f}$ such that

$$\tilde{f}_{xx} - \frac{1}{\lambda^2}\tilde{f} - \mathsf{sat}\left(\frac{a}{\lambda}(\tilde{f}-f)\right) = -\frac{1}{\lambda}g - \frac{1}{\lambda^2}f,$$
$$\tilde{f}(0) = \tilde{f}(1) = 0$$

holds. The existence of a solution to this nonhomogeneous nonlinear ODE with two boundary conditions is provided by the following lemma.

Lemma 2. *For any $a \geq 0$ and $\lambda > 0$, there exists $\tilde{f}$ solution to*

$$\tilde{f}_{xx} - \frac{1}{\lambda^2}\tilde{f} - \mathsf{sat}\left(\frac{a}{\lambda}(\tilde{f}-f)\right) = -\frac{1}{\lambda}g - \frac{1}{\lambda^2}f, \tag{90}$$
$$\tilde{f}(0) = \tilde{f}(1) = 0.$$

To prove this lemma, let us introduce the mapping:

$$\mathcal{T} : L^2(0,1) \to L^2(0,1),$$
$$y \mapsto z,$$

where $z = \mathcal{T}(y)$ is the unique solution to

$$z_{xx} - \frac{1}{\lambda^2}z = -\frac{1}{\lambda}v - \frac{1}{\lambda^2}u + \mathsf{sat}\left(\frac{a}{\lambda}(y-u)\right),$$
$$z(0) = z(1) = 0.$$

It can be proven that $\mathcal{T}$ is a well defined mapping. Then, it is possible to invoke the Schauder fixed-point theorem (see e.g., [14]) to deduce the existence of y such that $\mathcal{T}(y) = y$. After doing so, we obtain that $\tilde{f} = y$ solves (90). □

After having assessed the well-posedness of the closed-loop system dynamics, we can focus on the study of its stability. The global asymptotic stability of this nonlinear PDE is stated in the following result.

Theorem 12. *For all $a > 0$, the origin of the PDE (88) with the boundary conditions (74) is globally asymptotically stable. More specifically, for all (z^0, z^1) in $(H^2(0,1) \cap H_0^1(0,1)) \times H_0^1(0,1)$, the solution to (88) with the boundary conditions (74) and the initial condition (75) satisfies, $\forall t \geq 0$,*

$$\|z(t,\cdot)\|_{H_0^1(0,1)} + \|z_t(t,\cdot)\|_{L^2(0,1)} \leq \|z^0\|_{H_0^1(0,1)} + \|z^1\|_{L^2(0,1)}$$

together with the attractivity property

$$\|z(t,\cdot)\|_{H_0^1(0,1)} + \|z_t(t,\cdot)\|_{L^2(0,1)} \to 0, \quad as \ t \to \infty.$$

Proof of Theorem 12. Due to Theorem 11, the formal computation of the time derivative of V_1 previously computed is rigorously justified. Hence we have

$$\dot{V}_1 = -\int_0^1 z_t \mathsf{sat}(az_t)\mathrm{d}x.$$

This is a weak Lyapunov function because $\dot{V}_1 \le 0$ which guarantees the stability of the origin. In order to prove the attractivity of the origin, we are going to invoke LaSalle's Invariance Principle [17, Chapter 11] for infinite-dimensional systems. To apply LaSalle's Invariance Principle, we have to check that the set of solutions is precompact. This result can be obtained here by following the approach reported in [19, 20] and relies on the following lemma (see below for a sketch of proof).

Lemma 3. *The canonical embedding from $D(A_1)$, equipped with the graph norm, into $H_0^1(0,1) \times L^2(0,1)$ is compact.*

Using the dissipativity of A_1 and Lemma 3, the trajectory $\begin{pmatrix} z(t,\cdot) \\ z_t(t,\cdot) \end{pmatrix}$ is precompact in $H_0^1(0,1) \times L^2(0,1)$. Moreover the ω-limit set $\omega\left[\begin{pmatrix} z(0,\cdot) \\ z_t(0,\cdot) \end{pmatrix}\right] \subset D(A_1)$ is not empty and is invariant with respect to the nonlinear semigroup $T(t)$ (see [66]). With these elements in hand, we can indeed apply LaSalle's Invariance Principle to show that $\omega\left[\begin{pmatrix} z(0,\cdot) \\ z_t(0,\cdot) \end{pmatrix}\right] = \{0\}$. This shows that the origin of the equation (88) with the boundary conditions (74) is attractive. This concludes the proof of Theorem 12. $\square$

Let us now give the main steps of the proof of Lemma 3.

Proof of Lemma 3. Consider a sequence $\begin{pmatrix} f_n \\ g_n \end{pmatrix}_{n\in\mathbb{N}}$ in $D(A_1)$, which is bounded in graph norm, that is there exists $M > 0$ such that, for all $n \in \mathbb{N}$,

$$\left\|\begin{pmatrix} f_n \\ g_n \end{pmatrix}\right\|_{D(A_1)}^2 := \left\|\begin{pmatrix} f_n \\ g_n \end{pmatrix}\right\|^2 + \left\|A_1\begin{pmatrix} f_n \\ g_n \end{pmatrix}\right\|^2 \le M$$

which means that

$$\int_0^1 (|f_n'|^2 + |g_n|^2 + |g_n'|^2 + |f_n'' - a\mathsf{sat}(g_n)|^2)\mathrm{d}x \le M.$$

From that, we deduce that $\int_0^1(|g_n|^2 + |g_n'|^2)\mathrm{d}x$ and $\int_0^1(|f_n'|^2 + |f_n''|^2)\mathrm{d}x$ are bounded. Hence, with compact injection of $H_0^1(0,1)$ in $L^2(0,1)$, and of

$H^2(0,1)$ in $H_0^1(0,1)$ we infer the existence of a subsequence of $\begin{pmatrix} f_n \\ g_n \end{pmatrix}_{n\in\mathbb{N}}$ which converges in $H_0^1(0,1) \times L^2(0,1)$, giving the precompactness of the set of solutions to equation (88) with the boundary conditions (74). □

4.1.3. *A boundary linear control*

We now consider the wave equation with a boundary control, as depicted in Fig. 9. The system dynamics reads

$$z_{tt}(t,x) = z_{xx}(t,x), \quad \forall x \in (0,1), \quad t \ge 0, \tag{91}$$

with the boundary conditions

$$\begin{aligned} z(t,0) &= 0, \\ z_x(t,1) &= u(t), \end{aligned} \tag{92}$$

for all $t \ge 0$ and with the initial condition

$$\begin{aligned} z(0,x) &= z^0(x), \\ z_t(0,x) &= z^1(x) \end{aligned} \tag{93}$$

for all x in $(0,1)$.

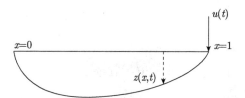

Fig. 9. Wave equation with unbounded control operator

We define the linear control

$$u(t) = -bz_t(t,1), \quad x \in (0,1), \quad \forall t \ge 0 \tag{94}$$

and we consider

$$V_2 = \frac{1}{2}\int (e^{\mu x}(z_t + z_x)^2)\mathrm{d}x + \int(e^{-\mu x}(z_t - z_x)^2)\mathrm{d}x.$$

A formal computation of the time derivative along the solutions to (91), (92) and (94) gives

$$\dot{V}_2 = -\mu V_2 + \frac{1}{2}\left(e^\mu(1-b)^2 - e^{-\mu}(1+b)^2\right)z_t^2(t,1).$$

Assuming that $b > 0$ and letting $\mu > 0$ such that[d] $e^\mu(1-b)^2 \le e^{-\mu}(1+b)^2$, it holds that $\dot{V}_2 \le -\mu V_2$. Hence V_2 is a strict Lyapunov function and thus the origin of (91) with boundary conditions (92) and command (94) is exponentially stable.

4.1.4. *A boundary saturating control*

Let us consider now the nonlinear control $u(t) = -\text{sat}(bz_t(t,1))$, for all $t \ge 0$. The boundary conditions become:

$$z(t,0) = 0, \quad z_x(t,1) = -\text{sat}(bz_t(t,1)). \tag{95}$$

Inspired by [14, Sec. 2.4], we introduce $H^1_{(0)}(0,1) = \{u \in H^1(0,1), u(0) = 0\}$ and $\|u\|_{H^1_{(0)}(0,1)} = \sqrt{\int_0^1 |u'|^2(x)\mathrm{d}x}$ for all $u \in H^1_{(0)}$. We are now in position to state the following well-posedness and asymptotic stability result (see [56] for a complete proof).

Theorem 13. *For all $b > 0$, the origin of the PDE (91) with the boundary conditions (95) is globally asymptotically stable. More specifically, for all $(z^0, z^1) \in \{(f,g) \in H^2(0,1) \times H^1_{(0)}(0,1) : f_x(1) + \text{sat}(bg(1)) = 0, f(0) = 0\}$, there exists a unique solution to (91) with the boundary conditions (95) and the initial condition (75). Moreover it satisfies the following stability property, for all $t \ge 0$,*

$$\|z(t,\cdot)\|_{H^1_{(0)}(0,1)} + \|z_t(t,\cdot)\|_{L^2(0,1)} \le \|z^0\|_{H^1_{(0)}(0,1)} + \|z^1\|_{L^2(0,1)},$$

together with the attractivity property

$$\|z(t,\cdot)\|_{H^1_{(0)}(0,1)} + \|z_t(t,\cdot)\|_{L^2(0,1)} \to 0, \quad \text{as } t \to \infty.$$

Proof of Theorem 13. To prove the well-posedness of the Cauchy problem, we can show that A_2 defined by

$$A_2 \begin{pmatrix} f \\ g \end{pmatrix} = \begin{pmatrix} g \\ f'' \end{pmatrix}$$

with the domain $D(A_2) = \{(f,g) \in H^2(0,1) \times H^1_{(0)}(0,1) : f'(1) + \text{sat}(bg(1)) = 0, f(0) = 0\}$ is a semigroup of contraction by applying Lumer-Phillips theorem. The global stability property is immediately inferred from contraction property (consequence of the dissipativity of A_2). Finally, the global attractivity property comes from the following lemma establishing that the origin of the PDE (91) with the boundary conditions (95) is semi-globally exponentially stable. This completes the proof of the theorem. $\square$

[d]This constraint is always satisfied for $\mu > 0$ small enough by a continuity argument at $\mu = 0$.

Lemma 4. *For all $r > 0$, there exists $\mu > 0$ such that, for all initial condition satisfying*

$$\|(z^0)''\|^2_{L^2(0,1)} + \|z^1\|^2_{H^1_{(0)}(0,1)} \leq r^2, \tag{96}$$

it holds

$$\dot{V}_2 \leq -\mu V_2$$

along the solutions to (91) with the boundary conditions (95).

Proof of Lemma 4. First note that by dissipativity of A_2, it holds that

$$t \mapsto \left\| A_2 \begin{pmatrix} z(t, \cdot) \\ z_t(t, \cdot) \end{pmatrix} \right\| \tag{97}$$

is a non-increasing function. Moreover by continuity of the trace function on $H^1_{(0)}(0, 1)$, it holds

$$|z_t(t, 1)| \leq \|z_{tx}(t, \cdot)\|_{L^2(0,1)} \leq \left\| A_2 \begin{pmatrix} z(t, \cdot) \\ z_t(t, \cdot) \end{pmatrix} \right\| \leq \left\| A_2 \begin{pmatrix} z(0, \cdot) \\ z_t(0, \cdot) \end{pmatrix} \right\|$$

where the decreasing property of the function in (97) has been used for the last inequality. Thus, for all $t \geq 0$,

$$|z_t(t, 1)| \leq \left\| A_2 \begin{pmatrix} z(0, \cdot) \\ z_t(0, \cdot) \end{pmatrix} \right\|. \tag{98}$$

Now, given $r > 0$, for an initial condition satisfying (96), we have $|z_t(t, 1)| \leq r$ and thus there exists $c \neq b$ such that, for all $t \geq 0$,

$$(b - c)|z_t(t, 1)| \leq 1$$

and thus, recalling the definition of the deadzone function ϕ in (20), the local sector condition holds $\phi(\phi + cz_t(t, 1)) \leq 0$, see (21). Let us now go back to the Lyapunov function candidate V_2. Given $b > 0$, using the previous inequality, we infer that

$$\dot{V}_2 = -\mu V_2 + e^{\mu}(\sigma - \mathsf{sat}(b\sigma))^2 - e^{-\mu}(\sigma + \mathsf{sat}(b\sigma))^2$$
$$\leq -\mu V_2 + e^{\mu}((1 - b)\sigma - \phi)^2 - e^{-\mu}((1 + b)\sigma + \phi)^2 - 2\phi(\phi + c\sigma)$$
$$\leq -\mu V_2 + \begin{pmatrix} \sigma \\ \phi \end{pmatrix}^{\mathsf{T}} \mathcal{M} \begin{pmatrix} \sigma \\ \phi \end{pmatrix},$$

where

$$\mathcal{M}(\mu, c) = \begin{bmatrix} e^{\mu}(1 - b)^2 - e^{-\mu}(1 + b)^2 & -e^{\mu}(1 - b) - e^{-\mu}(1 + b) + c \\ \star & -2 + e^{\mu} - e^{-\mu} \end{bmatrix}.$$

In particular we have at $\mu = 0$ that

$$\mathcal{M}(0,c) = \begin{bmatrix} -4b & -2+c \\ \star & -2 \end{bmatrix}.$$

We have to select c close to b such that $\mathcal{M}(0,c)$ is symmetric semi-definite negative. Of course, $c = b$ is not convenient since $\mathcal{M}(0,c)$ is not semi-definite negative (moreover the choice $c = b$ would yield the global section condition which does not hold, confirming that the choice $c = b$ is not suitable for c). But $c < b$ and close to b exists such that $\det(\mathcal{M}) > 0$. Thus $\mathcal{M} < 0$.

Given $r > 0$, we consider initial condition such that $\left\| A_2 \begin{pmatrix} z(\cdot,0) \\ z_t(\cdot,0) \end{pmatrix} \right\| \leq r$. This implies, for a suitable c ensuring that $(b-c)|z_t(t,1)| \leq 1$ for all $t \geq 0$, that $\dot{V}_2 \leq -\mu V_2$. The semi-global exponential stability follows. $\qquad\square$

Note here that the exponential stability is only achieved on bounded sets of initial conditions. An open question is whether we have (or not) the global exponential stability of the origin of the PDE (91) with the boundary conditions (95).

4.2. KdV equation with a bounded control operator

In the previous section we reviewed two classes of controllers for the linear wave equation with linear and nonlinear feedback. Different methods for proving asymptotic stability have been reported, one using LaSalle's Invariance Principle and another one establishing semi-global exponential stability based on a local sector condition. In this section, we move to a control problems for a nonlinear PDE. Specifically, let us consider the following nonlinear Korteweg-de Vries (KdV) PDE:

$$\begin{cases} z_t + z_x + z_{xxx} + zz_x + u = 0, \ x \in [0,L], \ t \geq 0, \\ z(t,0) = z(t,L) = z_x(t,L) = 0, \ t \geq 0, \\ z(0,x) = z^0(x), \ x \in [0,L], \end{cases} \tag{99}$$

where z stands for the state and u for the control.

As shown in [59], in the uncontrolled scenario ($u = 0$) and for a length L of the spatial domain such that

$$L \in \left\{ 2\pi\sqrt{\frac{k^2 + kl + l^2}{3}} \ / \ k,l \in \mathbb{N}^* \right\}, \tag{100}$$

there exist solutions of the linearized version of (99) for which the $L^2(0,L)$ norm of the state does not decay to zero. This can be observed, for instance, in the particular case for the first critical length $L_1 = 2\pi$ (obtained by letting $k = l = 1$ in (100)) by considering the initial condition $z^0(x) = 1 - \cos(x)$ for all $x \in [0, L]$. Let us denote the second critical by $L_2 = 2\pi\sqrt{\frac{7}{3}}$ (obtained by letting $k = l = 1$ in (100)). We refer the reader to the papers [10, 11, 59] for controllability results of (99) and the role of the so-called critical lengths (100).

In these notes we are interested in the stabilization problem of the origin of the KdV. We refer the reader to [69] for the stabilization of the origin of the linearized KdV equation with anti-diffusion. In [9] and [54], localized damping is considered for the linearized KdV equation. Specifically, when setting linear control $u = a(x)z$, for a non-negative continuous function $a : [0, 1] \to \mathbb{R}$, we obtain

$$\begin{cases} z_t + z_x + z_{xxx} + a(x)z = 0, \ x \in [0, L], \ t \geq 0, \\ z(t, 0) = z(t, L) = z_x(t, L) = 0, \ t \geq 0, \\ z(0, x) = z^0(x), \ x \in [0, L]. \end{cases} \tag{101}$$

The following theorem is proven in [54].

Theorem 14. *The following results hold true for* (101).

- *When L is not a critical length (i.e., (100) does not hold) and $a \equiv 0$, the origin of (101) is asymptotically stable. To be more specific, there exist M and μ such that*

$$\|z(t)\|_{L^2} \leq Me^{-\mu t}\|z(0)\|_{L^2}.$$

- *When $a > 0$ on a non-empty subset of $[0, L]$, then the same conclusion holds.*

Let us now shortly review the stabilization results of the origin of the nonlinear KdV PDE (99) when using a control given by $u = a(x)z$. The papers [13, 54, 68] consider the following closed-loop dynamics:

$$\begin{cases} z_t + z_x + z_{xxx} + zz_x + a(x)z = 0, \ x \in [0, L], \ t \geq 0, \\ z(t, 0) = z(t, L) = z_x(t, L) = 0, \ t \geq 0, \\ z(0, x) = z^0(x), \ x \in [0, L]. \end{cases} \tag{102}$$

The following theorem summarizes some of the contributions contained in these papers (see [13] and [68] for the proof of the first item, respectively for $L = L_1$ and $L = L_2$, and see [54] for the proof of the second item).

Theorem 15. *The following results hold true for* (102).

- *When $L = L_1$ or $L = L_2$ and $a \equiv 0$, the origin of (99) is locally asymptotically stable. More precisely[e], there exist $r > 0$, $M > 0$, and $\mu > 0$ such that the solutions to (102) issuing from $z(0)$ with $\|z(0)\|_{L^2} \le r$ satisfy*

$$\|z(t)\|_{L^2} \le Me^{-\mu t}\|z(0)\|_{L^2}.$$

- *For all $L > 0$, when $a > 0$ on a non-empty subset of $[0, L]$, then the origin of (102) is globally asymptotically stable. More precisely[f], for all $r > 0$, there exist $M > 0$, and $\mu > 0$ such that*

$$\|z(t)\|_{L^2} \le Me^{-\mu t}\|z(0)\|_{L^2}.$$

4.2.1. *Saturating control for KdV*

Let us now consider the case of a saturating control. To simplify the presentation, we will consider the case where the function $a(x)$ in (102) is a constant denoted by a. The localized control is subject to a saturation map. To be more specific, let the KdV equation controlled by a saturated distributed control be described by

$$\begin{cases} z_t + z_x + z_{xxx} + zz_x + \mathsf{sat}(az) = 0, \\ z(t,0) = z(t,L) = 0, \\ z_x(t,L) = 0, \\ z(0,x) = z^0(x), \end{cases} \tag{103}$$

where sat is the saturation map defined in (17) with $m = 1$, and with level s. The corresponding nonlinear equation (103) is studied in [45]. The case of L^2 saturation, defined in (89), is also considered. In these notes we focus on the nonlinear equation (103), but some numerical simulations will also be performed with the L^2 saturation in the next numerical example.

The well-posedness result is proven in [45] by proving first existence of solution for small time following the approach of [12, 60], and then removing the smallness property of the time existence using *a priori estimates*. It yields the following theorem.

Theorem 16. *For any initial condition $z^0 \in L^2(0, L)$, there exists a unique solution $z \in C([0,T]; L^2(0, L)) \cap L^2(0, T; H^1(0, L))$ to (103).*

The global asymptotic stability of the origin, which is also proven in the same paper, can be stated as follows.

---

[e]This property is the definition of the global exponential stability of the origin.
[f]This property is the definition of the semi-global asymptotic stability of the origin.

Theorem 17. *The origin of* (103) *is globally asymptotically stable. More precisely there exist $\mu > 0$ and a class $\mathcal{K}$ function[g] $\alpha : [0, \infty) \to [0, \infty)$ such that for any $z^0 \in L^2(0, L)$, any solution z to* (103) *satisfies, for all $t \geq 0$,*

$$\|z(t)\|_{L^2(0,L)} \leq \alpha(\|z^0\|_{L^2(0,L)})e^{-\mu t}.$$

This result is proved by following the approaches of [9, 60] by showing that the origin of (103) is semi-globally exponentially stable.

Proposition 1. *For any given $r > 0$, there exist positive values C and μ such that for all initial condition z^0 satisfying $\|z^0\|_{L^2(0,L)} \leq r$, the solution to* (103) *satisfies, for all $t \geq 0$,*

$$\|z(t)\|_{L^2(0,L)} \leq C\|z^0\|_{L^2(0,L)}e^{-\mu t}.$$

Proof of Proposition 1. To prove this proposition, a key result is the following claim.

Claim 3. *For all $T > 0$ and $r > 0$, there exists $C > 0$ such that for any solution z to* (103) *starting from $z^0 \in L^2(0, L)$ with $\|z^0\|_{L^2(0,L)} \leq r$, it holds*

$$\|z^0\|_{L^2(0,L)}^2 \leq C\left(\int_0^T |z_x(t,0)|^2 \mathrm{d}t + 2\int_0^T \int_0^L \mathbf{sat}(az)z\mathrm{d}t\mathrm{d}x\right). \quad (104)$$

Assume Claim 3 holds for the time being. Then with (104) it holds

$$\|z(t,\cdot)\|_{L^2(0,L)}^2 = \|z^0\|_{L^2(0,L)}^2 - \int_0^T |z_x(t,0)|^2\mathrm{d}t$$
$$- 2\int_0^T \int_0^L \mathbf{sat}(az)z\mathrm{d}x\mathrm{d}t,$$

we get

$$\|z(\cdot,kT)\|_{L^2(0,L)}^2 \leq \gamma^k\|z^0\|_{L^2(0,L)}^2, \qquad \forall k \geq 0,$$

where $\gamma \in (0,1)$. From the dissipativity property, we have $\|z(t,\cdot)\|_{L^2(0,L)} \leq \|z(\cdot,kT)\|_{L^2(0,L)}$ for $kT \leq t \leq (k+1)T$. Thus we obtain, for all $t \geq 0$,

$$\|z(t,\cdot)\|_{L^2(0,L)}^2 \leq \frac{1}{\gamma}\|z^0\|_{L^2(0,L)}e^{\frac{\log \gamma}{T}t}.$$

We conclude the proof of the semi-global exponential stability, as stated in Proposition 1. □

Let us now prove Claim 3 that has been used in the proof of Proposition 1.

---

[g]A class $\mathcal{K}$ function is a continuous and increasing function that is zero at zero.

Proof of the Claim 3. We prove (104) by contradiction. Assume that there exists a sequence of solution z^n to (103) with

$$\|z^n(\cdot,0)\|_{L^2(0,L)} \le r \qquad (105)$$

and such that

$$\lim_{n\to+\infty} \frac{\|z^n\|^2_{L^2(0,T;L^2(0,L))}}{\int_0^T |z_x^n(t,0)|^2 dt + 2\int_0^T \int_0^L \mathbf{sat}(az^n(t,x))z^n(t,x)dtdx} = +\infty. \qquad (106)$$

By dissipativity property, there exists $\beta > 0$ such that

$$\sup_{t\in[0,T]} \|z^n(t,\cdot)\|_{L^2(0,L)} \le r \,, \qquad \sup_{x\in[0,L]} \int_0^T |z^n(t,x)|^2 dt \le \beta. \qquad (107)$$

Now let us define $\Omega_i := \left\{ t\in[0,T], \sup_{x\in[0,L]} |z(t,x)| > i \right\} \subset [0,T]$. We have

$$\beta \ge \int_0^T \sup_{x\in[0,L]} |z^n(t,x)|^2 dt \ge \int_{\Omega_i} \sup_{x\in[0,L]} |z^n(t,x)|^2 dt \ge i^2 \nu(\Omega_i).$$

Therefore, denoting the Lebesgue mesure by ν, and the complementary set of Ω_i^c by $\nu(\Omega_i^c)$, we obtain, with (107), $\nu(\Omega_i) \le \frac{\beta}{i^2}$, and thus $\nu(\Omega_i^c) \ge \max\left(T - \frac{\beta}{i^2}, 0\right)$.

Let us note that denoting, $k(i) = \min(\frac{s}{ai}, 1)$, for each i in $\mathbb{N}$, it holds for all z in Ω_i^c, $|z| \le i$, and thus[h]

$$(\mathbf{sat}(az) - k(i)az)z \ge 0. \qquad (108)$$

Moreover, using again the local sector condition, we have

$$\int_0^T \int_0^L \mathbf{sat}(az^n)z^2 dtdx = \int_{\Omega_i} \int_0^L \mathbf{sat}(az^n)z^n dtdx + \int_{\Omega_i^c} \int_0^L \mathbf{sat}(az^n)z^n dtdx$$

$$\ge 0 + \int_{\Omega_i^c} \int_0^L ak(i)(z^n)^2 dtdx, \qquad (109)$$

where (108) has been used in the last inequality. Thus, with (105), for all i in $\mathbb{N} \setminus \{0\}$,

$$\|z^n(t,\cdot)\|^2_{L^2(0,L)} \le \|z^n(\cdot,0)\|^2_{L^2(0,L)} - \int_0^T |z_x^n(t,0)|^2 dt$$

$$- 2\int_{\Omega_i^c} \int_0^L ak(i)(z^n)^2 dtdx.$$

---

[h]To prove (108), assume first that $ai \le s$, then $k(i) = 1$, and $\mathbf{sat}(az) = az$, which gives (108). Second, if $ai > s$ and $\mathbf{sat}(az) = az$, then $1 - k(i) > 0$ and $(\mathbf{sat}(az) - k(i)az) = (1 - k(i))az$, which gives (108). Third, if $ai > s$ and $az > s$, then $(\mathbf{sat}(az) - k(i)az)z = s - \frac{s}{ai}az = s(1 - \frac{i}{z}) \ge 0$, which gives (108). The fourth case $ai > s$ and $az < -s$ is studied in a similar way as the third one.

Let $\lambda^n := \|z^n\|_{L^2(0,T;L^2(0,L))}$ and $v^n(t,x) = \frac{z^n(t,x)}{\lambda^n}$. Due to (105), up to extracting a subsequence, we may assume that $\lambda^n \to \lambda \geq 0$. Due to (106) and (109), we have, for all $i \in \mathbb{N}$,

$$\int_0^T |v_x^n(t,0)|^2 \mathrm{d}t + 2 \int_{\Omega_i^c} \int_0^L ak(i)(v^n)^2 \mathrm{d}t\mathrm{d}x \to 0. \qquad (110)$$

Using Aubin-Lions lemma in [65], we get $\{v^n\}_{n\in\mathbb{N}}$ converges strongly in $L^2(0,T;L^2(0,L))$. Thus, with (110), we have, for all $i \in \mathbb{N}$,

$$v_x(t,0) = 0, \ \forall t \in (t,0) \text{ and } v(t,x) = 0, \ \forall x \in [0,L], \forall t \in \Omega_i^c.$$

We know that $\nu\left(\bigcup_{i\in\mathbb{N}} \Omega_i^c\right) = T$. We get a contradiction with $\|v\|_{L^2(0,T;L^2(0,L))} = 1$. This concludes the proof of Claim 3. $\qquad \square$

Example 7. Let us discretize (103) and illustrate Theorem 17. Moreover we will discretize this equation using the saturation map $\mathbf{sat}_{L2}$ instead of $\mathbf{sat}$ and without any saturation map (the equation becomes (102)).

```
1  """
2  Discretizing KdV equation with saturating control.
3  Use of central difference in space and forward
4  Euler in time schemes.
5  Code originally written by S. Marx for
6  [S. Marx et al, SIAM J. Control Opt., 2017]
7  """
8  import numpy as np
9  import matplotlib as mpl
10 import matplotlib.pyplot as plt
11 from mpl_toolkits.mplot3d import Axes3D
12
13 # Parameters of the PDE
14 L=2*np.pi;
15 a=1.0
16
17 # Space discretization
18 Nx = 30
19 x= np.linspace(0,L,Nx+1)
20 dx = L/Nx;
21
22 # Time discretization
23 dt = 0.06; tfinal=6
24 Nt= np.floor(tfinal/dt).astype(int)
25
26 # Saturation level
27 s0=0.5
28
29 # Set initial condition
30 z0=[]
```

```
31 for ii in range(Nx+1):
32     z0.append(1-np.cos(x[ii]))
```

On line 14, we set the first critical length $L = 2\pi$ and the initial condition $z^0(x) = 1 - \cos(x)$ is chosen on line 32 so that its energy is constant along the linearized KdV equation, without any control. The function a is chosen as the constant value 1 on line 15, and the level of the saturation map is set at 0.5 on line 27. The space and time discretization steps are selected respectively at lines 20 and 23.

```
33
34 t = 0 # current time
35 j = 0 # current time index
36
37 # pointwise saturation function
38 def sat(u):
39     m=np.size(u)
40     sigma=u
41     for i in range(m):
42         if np.absolute(u[i])>s0:
43             sigma[i]=s0*np.sign(u[i])
44     return sigma
45
46 # L2 saturation function
47 def sat2(u):
48     L2=np.linalg.norm(u)*np.sqrt(dx)
49     sigma=u
50     if not L2<s0:
51         for ii in range(Nx):
52             sigma[ii]=s0*u[ii]/L2
53     return sigma
54
55 L2norm=[] # L2norm of the sol with sat
56 L2normNoSat=[] # L2norm of the sol without sat
57 ztot=np.zeros((Nx+1,Nt+1)) # to save the solution
58 ztotNoSat=np.zeros((Nx+1,Nt+1)) # to save the solution without sat
59
60 ztot[:,0]=z0
61 ztotNoSat[:,0]=z0
62 L2norm.append(np.linalg.norm(ztot[:,0])*np.sqrt(dx))
63 L2normNoSat.append(np.linalg.norm(ztotNoSat[:,0])*np.sqrt(dx))
```

The localized saturation sat and the L^2 saturation sat_{L2} are defined after line 37 and line 46 respectively. Between lines 55 and 63 it shows the initialization of the state and of its norm for both the linear control (thus with (102)) and the saturated control (thus with (103) with either the saturation map sat or with sat_{L2}.

```
65  def discretNoSat(z,dx,dt,a):
66      """
67      discretization of the nonlinear KdV using
68      [Pazoto, et al, Numer. Math., 2010]
69      method without saturation
70      """
71      n=len(z)
72      n1=n-1
73      Dm=1/dx*np.identity(n1)
74      Dp=-Dm
75      for i in range(n1-1):
76          Dp[i,i+1]=-Dp[i,i]
77          Dm[i+1,i]=-Dm[i,i]
78      D=1/2*(Dm+Dp)
79      I=np.identity(n1)
80      A=np.dot(np.dot(Dp,Dp),Dm)+D
81      C=I+dt*A
82      NS=np.zeros((n1,n1))
83      NS[n1-1,n1-1]=C[n1-1,n1-1]
84
85      #Fixed-point method
86      NIter = 100 # number of iterations
87      J =[]
88      J.append(z[:-1])
89      tmp=J[-1]
90      for k in range(NIter):
91          tmp=np.linalg.solve(C-NS,z[:-1]-dt/2*np.dot(D,np.multiply(
            tmp,tmp))-np.dot(dt*a,tmp))
92          J.append(tmp)
93      return tmp
94
95  def discret(z,dx,dt,a):
96      """
97      discretization of the nonlinear KdV using
98      [Pazoto, et al, Numer. Math., 2010]
99      method with saturation (select sat or sat2 function)
100     """
101     n=len(z)
102     n1=n-1
103     Dm=1/dx*np.identity(n1)
104     Dp=-Dm
105     for i in range(n1-1):
106         Dp[i,i+1]=-Dp[i,i]
107         Dm[i+1,i]=-Dm[i,i]
108     D=1/2*(Dm+Dp)
109     I=np.identity(n1)
110     A=np.dot(np.dot(Dp,Dp),Dm)+D
111     C=I+dt*A
112     NS=np.zeros((n1,n1))
113     NS[n1-1,n1-1]=C[n1-1,n1-1]
```

156 H. Lhachemi and C. Prieur

```
114
115    #Fixed-point method
116    NIter = 100 # number of iterations
117    J =[]
118    J.append(z[:-1])
119    tmp=J[-1]
120    for k in range(NIter):
121        tmp=np.linalg.solve(C-NS,z[:-1]-dt/2*np.dot(D,np.multiply(
       tmp,tmp))-dt*sat(np.dot(a,tmp)))
122        J.append(tmp)
123    return tmp
```

To discretize (102) and (103), we follow the approach of [52], and solve, at each time-step, a fixed point problem. No proof of convergence of the numerical scheme is guaranteed in the context of (103), since another non-linearity is considered in [52]. In particular the term z_{xxx} is discretized as follows

$$D_+ D_+ D_- z^i,$$

where z^i is the discretized version of z, and where Dp and Dm are the matrices defined by

$$Dp = \frac{1}{dx} \begin{bmatrix} -1 & 1 & 0 & \dots & 0 \\ 0 & -1 & 1 & \dots & 0 \\ \vdots & \ddots & \ddots & \ddots & \vdots \\ 0 & 0 & 0 & \dots & -1 \end{bmatrix}, \quad Dm = \frac{1}{dx} \begin{bmatrix} 1 & -1 & 0 & \dots & 0 \\ 0 & 1 & -1 & \dots & 0 \\ \vdots & \ddots & \ddots & \ddots & \vdots \\ 0 & 0 & 0 & \dots & 1 \end{bmatrix}.$$

It yields two discretizations, for the equations (102) and (103), between lines 66–93 and lines 95–123 respectively. It asks to solve a fixed point problem that is solved using a iteration scheme with 100 steps (see after lines 85 and 115). The choice of the saturation map (either sat or sat$_{L2}$) is made on line 121. In the python code given here, sat is considered.

```
125  # making a loop until t > tfinal
126  while t<tfinal-dt:
127      #Forward Euler step
128      ztotNoSat[:-1,j+1]=discretNoSat(ztotNoSat[:,j],dx,dt,a)
129      ztot[:-1,j+1]=discret(ztot[:,j],dx,dt,a)
130      t+=dt
131      j+=1
132      L2normNoSat.append(np.linalg.norm(ztotNoSat[:,j])*np.sqrt(dx))
133      L2norm.append(np.linalg.norm(ztot[:,j])*np.sqrt(dx))
134
135  # plotting the figures
136  space= np.linspace(0,np.pi,Nx+1)
137  t=np.linspace(0,tfinal,Nt+1)
138
```

```
139  fig , ax= plt.subplots()
140  ax.plot(t,L2normNoSat, label='without saturation')
141  ax.plot(t,L2norm, label='with saturation')
142  ax.set_xlabel('t')
143  ax.set_ylabel('L2 norm')
144  ax.legend()
145  plt.savefig('pde-l2norm.png',bbox_inches='tight')
146
147  mpl.rcParams['legend.fontsize'] = 10
148  fig = plt.figure(); ax = fig.add_subplot(111, projection='3d')
149  SX, ST = np.meshgrid(space, t)
150  ax.plot_surface(SX, ST, ztotNoSat.T, cmap='jet')
151  ax.set_xlabel('x')
152  ax.set_ylabel('t')
153  ax.set_zlabel('z(t,x)')
154  ax.set_title('Solution without saturation')
155  ax.view_init(elev=15, azim=20) # adjust view so it is easy to see
156  plt.savefig('pde-3dNoSat.png')
157
158  mpl.rcParams['legend.fontsize'] = 10
159  fig = plt.figure(); ax = fig.add_subplot(111, projection='3d')
160  ax.plot_surface(SX, ST, ztot.T, cmap='jet')
161  ax.set_xlabel('x')
162  ax.set_ylabel('t')
163  ax.set_zlabel('z(t,x)')
164  ax.set_title('Solution with saturation')
165  ax.view_init(elev=15, azim=20) # adjust view so it is easy to see
166  plt.savefig('pde-3d.png')
```

The discretization in time is done after line 125, where an Euler scheme
is used. The figures are drawn after line 135. It yields Fig. 10 and Fig. 11
where the time-evolutions of the solutions to (103) and to (102) are respec-

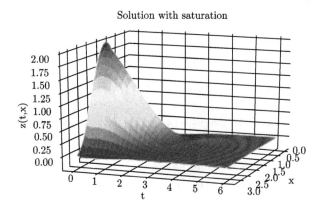

Fig. 10.   Solution to (103) with the saturation map sat

tively given. It is observed that the solutions converge to the origin, as predicted in Theorem 17 and the second item of Theorem 15.

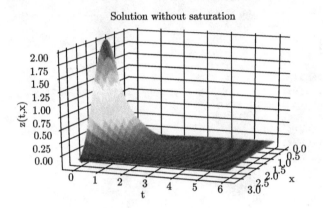

Fig. 11.   Solution to (102)

On Fig. 12 the corresponding L^2 norms are compared where it is checked that, as expected, the L^2 norm decreases faster along the solution to (102) than along the solution to (103) with the chosen initial condition.

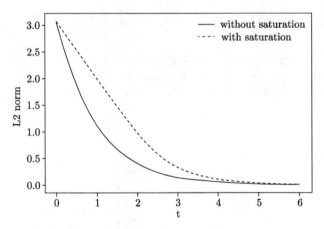

Fig. 12.   Comparison of the time-evolutions of the L^2 of the solutions to (103), with the saturation map **sat**, and to (102)

Selecting the saturation map $\mathbf{sat}_{L2}$ gives the Fig. 13 and Fig. 14 with analogous conclusions on the time-evolution of the solution to (103) and on the L^2 norm.

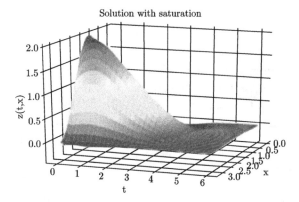

Fig. 13.   Solution to (103) with the saturation map sat_{L2}

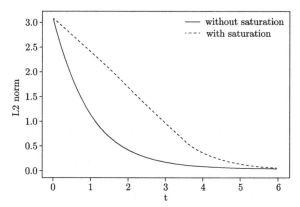

Fig. 14.   Comparison of the time-evolutions of the L^2 of the solutions to (103), with the saturation map sat_{L2}, and to (102)

4.3. *Conclusion so far*

In this section we have reviewed the well-posedness and the asymptotic stability of the origin of the wave equation and of the Korteweg-de Vries equation in presence of (possibly saturating) control. Different proofs have been provided for the attractivity, using either direct Lyapunov method or a LaSalle's Invariance Principle or a contradiction argument.

Let us emphasize that the approaches presented in this section are also useful for certain other classes of equations such as hyperbolic systems. See [5] for the stabilization of linear and quasilinear hyperbolic systems. See also [64] for the stabilization of hyperbolic systems using saturated control.

In [46] an output-feedback control has been computed for the linearized KdV equation. It would be relevant to evaluate the impact of the saturation map on the obtained result.

Finally, in addition to the stabilization control problem, the impact of disturbances could be studied. It would be relevant to obtain Input-to-State Stability results in the context of this section (see [28, 47] for introductory presentations on this subject for infinite-dimensional systems).

5. Conclusion

This chapter has reviewed some recent results on stability analysis of distributed parameter systems as those modeled by parabolic partial differential equations, or the wave equation or the Korteweg-de Vries equation. The suggested approach succeeds to design boundary stabilizing controllers, possibly subject to amplitude constraint, ensuring an asymptotic stability of the closed-loop equation. The constructive approach is based on Lyapunov function, and numerically tractable conditions. Some simulations have illustrated our results and design methods. More recent works follow the present chapter as the control of reaction-diffusion equation coupled with ordinary differential equations (see [35]), or control of such partial differential equation by means of delayed control (see [42]) to cite just a few. As far as hyperbolic system are considered, nonlinear controllers could be also designed as done in [56, 72]. Finally, let us cite the papers [36, 37] dealing with regulation problems, that could be seen as generalizations of stabilization problems for both the parabolic equations and the wave equation.

References

[1] P. J. Antsaklis and A. N. Michel. *Linear systems.* Springer Science & Business Media, 2006.
[2] A. Bacciotti and L. Rosier. *Lyapunov functions and stability in control theory.* Springer Science & Business Media, 2005.
[3] M. J. Balas. Finite-dimensional controllers for linear distributed parameter systems: exponential stability using residual mode filters. *Journal of Mathematical Analysis and Applications*, 133(2):283–296, 1988.
[4] V. Barbu. *Nonlinear semigroups and differential equations in Banach spaces.* Editura Academiei, 1976.
[5] G. Bastin and J.-M. Coron. *Stability and Boundary Stabilization of 1-D Hyperbolic Systems*, volume 88 of *Progress in Nonlinear Differential Equations and Their Applications.* Springer, 2016.

[6] S. P. Boyd, L. El Ghaoui, E. Feron, and V. Balakrishnan. *Linear matrix inequalities in system and control theory*, volume 15. SIAM, 1994.

[7] F. Boyer. Controllability of linear parabolic equations and systems. Technical Report hal-02470625, Hal, 2020.

[8] H. Brezis. *Opérateurs maximaux monotones et semi-groupes de contractions dans les espaces de Hilbert.* North-Holland, 1973.

[9] E. Cerpa. Control of a Korteweg-de Vries equation: a tutorial. *Math. Control Relat. Fields*, 4(1):45–99, 2014.

[10] E. Cerpa. Exact controllability of a nonlinear Korteweg-de Vries equation on a critical spatial domain. *SIAM Journal on Control and Optimization*, 46(3):877–899, 2007.

[11] E. Cerpa and E. Crépeau. Boundary controllability for the nonlinear Korteweg-de Vries equation on any critical domain. *Annales de l'IHP Analyse non linéaire*, 26(2):457–475, 2009.

[12] M. Chapouly. Global controllability of a nonlinear Korteweg-de Vries equation. *Communications in Contemporary Mathematics*, 11(3):495–521, 2009.

[13] J. Chu, J.-M. Coron, and P. Shang. Asymptotic stability of a nonlinear Korteweg-de Vries equation with critical lengths. *Journal of Differential Equations*, 259(8):4045–4085, 2015.

[14] J.-M. Coron. *Control and Nonlinearity*, volume 136 of *Mathematical Surveys and Monographs*. American Mathematical Society, Providence, RI, 2007.

[15] J.-M. Coron and E. Trélat. Global steady-state controllability of one-dimensional semilinear heat equations. *SIAM Journal on Control and Optimization*, 43(2):549–569, 2004.

[16] J.-M. Coron and E. Trélat. Global steady-state stabilization and controllability of 1D semilinear wave equations. *Commun. Contemp. Math.*, 8(4):535–567, 2006.

[17] R. Curtain and H. Zwart. *Introduction to infinite-dimensional systems theory: a state-space approach*, volume 71. Springer Nature, 2020.

[18] R. Curtain. Finite-dimensional compensator design for parabolic distributed systems with point sensors and boundary input. *IEEE Transactions on Automatic Control*, 27(1):98–104, 1982.

[19] C. M. Dafermos and M. Slemrod. Asymptotic behavior of nonlinear contraction semigroups. *Journal of Functional Analysis*, 13(1):97–106, 1973.

[20] B. d'Andréa Novel, F. Boustany, F. Conrad, and B. P. Rao. Feedback stabilization of a hybrid PDE-ODE system: application to an overhead crane. *Mathematics of Control, Signals, and Systems*, 7(1):1–22, 1994.

[21] E. D. Klerk. *Aspects of semidefinite programming: interior point algorithms and selected applications*, volume 65. Springer Science & Business Media, 2006.

[22] P. Gahinet, A. Nemirovski, A. J. Laub, and M. Chilali. LMI control toolbox. *The Math Works Inc*, 1996.

[23] J. M. Gomes da Silva Jr and S. Tarbouriech. Anti-windup design with guaranteed regions of stability: an LMI-based approach. *IEEE Transactions on Automatic Control*, 50(1):106–111, 2005.

[24] W. Hahn. *Stability of Motion*, volume 138. Springer, 1967.

[25] C. Harkort and J. Deutscher. Finite-dimensional observer-based control of linear distributed parameter systems using cascaded output observers. *International Journal of Control*, 84(1):107–122, 2011.

[26] J. P. Hespanha. *Linear systems theory*. Princeton university press, 2018.

[27] F.-L. Huang. Characteristic conditions for exponential stability of linear dynamical systems in Hilbert spaces. *Ann. Differential Equations*, 1(1):43–56, 1985.

[28] I. Karafyllis and M. Krstic. *Input-to-State Stability for PDEs*. Communications and Control Engineering. Springer, 2019.

[29] R. Katz and E. Fridman. Constructive method for finite-dimensional observer-based control of 1-D parabolic PDEs. *Automatica*, 122:109285, 2020.

[30] H. K. Khalil. *Nonlinear Systems*. Prentice-Hall, 3rd edition, 2002.

[31] M. Krstic and A. Smyshlyaev. *Boundary control of PDEs: A course on backstepping designs*. SIAM, 2008.

[32] I. Lasiecka and T. I. Seidman. Strong stability of elastic control systems with dissipative saturating feedback. *Systems & Control Letters*, 48(3-4):243–252, 2003.

[33] P. L. Gall, C. Prieur, and L. Rosier. Exact controllability and output feedback stabilization of a bimorph mirror. *ESAIM Proc.*, 25:19–28, 2008.

[34] H. Lhachemi and C. Prieur. Local output feedback stabilization of a reaction-diffusion equation with saturated actuation. *IEEE Transactions on Automatic Control*, 68 (1):564–571, 2023.

[35] H. Lhachemi and C. Prieur. Stability analysis of reaction-diffusion PDEs coupled at the boundaries with an ODE. *Automatica*, 144:110465, 2022.

[36] H. Lhachemi, C. Prieur, and E. Trélat. Proportional integral regulation of a one-dimensional semilinear wave equation. *SIAM Journal on Control and Optimization*, 60(1):1–21, 2022.

[37] H. Lhachemi, C. Prieur, and E. Trélat. PI regulation of a reaction-diffusion equation with delayed boundary control. *IEEE Transactions on Automatic Control*, 66(4):1573–1587, 2021.

[38] H. Lhachemi and C. Prieur. Boundary output feedback stabilization of reaction-diffusion PDEs with delayed boundary measurement. *International Journal of Control*, to appear, 2022.

[39] H. Lhachemi and C. Prieur. Global output feedback stabilization of semilinear reaction-diffusion PDEs. 4th IFAC Workshop on Control of Systems Governed by Partial Differential Equations (CPDE'22), Kiel, Germany, 2022.

[40] H. Lhachemi and C. Prieur. Nonlinear boundary output feedback stabilization of reaction diffusion equations. *Systems and Control Letters*, 166:105301, 2022.

[41] H. Lhachemi and C. Prieur. Finite-dimensional observer-based boundary stabilization of reaction–diffusion equations with either a Dirichlet or Neumann boundary measurement. *Automatica*, 135:109955, 2022.

[42] H. Lhachemi and C. Prieur. Predictor-based output feedback stabilization of an input delayed parabolic PDE with boundary measurement. *Automatica*, 137:110115, 2022.

[43] H. Lhachemi, C. Prieur, and R. Shorten. An LMI condition for the robust-ness of constant-delay linear predictor feedback with respect to uncertain time-varying input delays. *Automatica*, 109:108551, 2019.

[44] H. Lhachemi and R. Shorten. Boundary output feedback stabilization of state delayed reaction-diffusion pdes. *arXiv*: 2105.15056, 2021.

[45] S. Marx, E. Cerpa, C. Prieur, and V. Andrieu. Global stabilization of a Korteweg-de Vries equation with saturating distributed control. *SIAM Journal on Control and Optimization*, 55(3):1452–1480, 2017.

[46] S. Marx and E. Cerpa. Output feedback stabilization of the Korteweg-de Vries equation. *Automatica*, 87:210–217, 2018.

[47] A. Mironchenko and C. Prieur. Input-to-state stability of infinite-dimensional systems: recent results and open questions. *SIAM Review*, 62(3):529–614, 2020.

[48] A. Mironchenko, C. Prieur, and F. Wirth. Local stabilization of an unstable parabolic equation via saturated controls. *IEEE Transactions on Automatic Control*, 66(5):2162–2176, 2021.

[49] I. Miyadera. *Nonlinear Semigroups*. Translations of mathematical mono-graphs. American Mathematical Society, 1992.

[50] Y. V. Orlov. Discontinuous unit feedback control of uncertain infinite-dimensional systems. *IEEE Transactions on Automatic Control*, 45(5):834–843, 2000.

[51] Y. Orlov. On general properties of eigenvalues and eigenfunctions of a Sturm–Liouville operator: comments on "ISS with respect to boundary dis-turbances for 1-D parabolic PDEs". *IEEE Transactions on Automatic Control*, 62(11):5970–5973, 2017.

[52] A. F. Pazoto, M. Sepúlveda, and O. V. Villagrán. Uniform stabilization of numerical schemes for the critical generalized Korteweg-de Vries equation with damping. *Numerische Mathematik*, 116(2):317–356, 2010.

[53] A. Pazy. *Semigroups of linear operators and applications to partial differential equations*. Applied mathematical sciences. Springer-Verlag, 1983.

[54] G. P. Menzala, C. F. Vasconcellos, and E. Zuazua. Stabilization of the Korteweg-de Vries equation with localized damping. *Quarterly of applied Mathematics*, 60(1):111–129, 2002.

[55] C. Prieur and J. de Halleux. Stabilization of a 1-D tank containing a fluid modeled by the shallow water equations. *Systems & Control Letters*, 52(3-4):167–178, 2004.

[56] C. Prieur, S. Tarbouriech, and J. M. Gomes da Silva Jr. Wave equation with cone-bounded control laws. *IEEE Transactions on Automatic Control*, 61(11):3452–3463, 2016.

[57] C. Prieur and E. Trélat. Feedback stabilization of a 1D linear reaction-diffusion equation with delay boundary control. *IEEE Transactions on Automatic Control*, 64(4):1415–1425, 2019.

[58] J. Prüss. On the spectrum of C_0-semigroups. *Trans. Amer. Math. Soc.*, 284(2):847–857, 1984.

[59] L. Rosier. Exact boundary controllability for the Korteweg-de Vries equa-tion on a bounded domain. *ESAIM: Control, Optimisation and Calculus of Variations*, 2:33–55, 1997.

[60] L. Rosier and B.-Y. Zhang. Global stabilization of the generalized Korteweg-de Vries equation posed on a finite domain. *SIAM Journal on Control and Optimization*, 45(3):927–956, 2006.

[61] D. L. Russell. Controllability and stabilizability theory for linear partial differential equations: recent progress and open questions. *SIAM Review*, 20(4):639–739, 1978.

[62] Y. Sakawa. Feedback stabilization of linear diffusion systems. *SIAM Journal on Control and Optimization*, 21(5):667–676, 1983.

[63] R. Sepulchre, M. Janković, and P. V. Kokotović. *Constructive nonlinear control*. Communications and Control Engineering Series. Springer-Verlag, Berlin, 1997.

[64] S. Shreim, F. Ferrante, and C. Prieur. Design of saturated boundary control for hyperbolic systems. In *1st Virtual IFAC World Congress*, Berlin, Germany, 2020.

[65] J. Simon. Compact sets in the space $l^p(o, t; b)$. *Annali di Matematica pura ed applicata*, 146(1):65–96, 1986.

[66] M. Slemrod. Feedback stabilization of a linear control system in Hilbert space with an a priori bounded control. *Mathematics of Control, Signals, and Systems*, 2(3):265–285, 1989.

[67] E. D. Sontag, H. J. Sussmann, and Y. D. Yang. A general result on the stabilization of linear systems using bounded controls. *IEEE Trans. Automat. Control*, 39(12):2411–2424, 1994.

[68] S. Tang, J. Chu, P. Shang, and J.-M. Coron. Asymptotic stability of a Korteweg-de Vries equation with a two-dimensional center manifold. *Advances in Nonlinear Analysis*, 7(4):497–515, 2018.

[69] S. Tang and M. Krstic. Stabilization of linearized Korteweg-de Vries systems with anti-diffusion. In *2013 American Control Conference*, pages 3302–3307. IEEE, 2013.

[70] S. Tarbouriech, G. Garcia, J.-M. Gomes da Silva Jr, and I. Queinnec. *Stability and Stabilization of Linear Systems with Saturating Actuators*. Springer, London, 2011.

[71] A. L. Tits and Y. Yang. Globally convergent algorithms for robust pole assignment by state feedback. *IEEE Transactions on Automatic Control*, 41(10):1432–1452, 1996.

[72] N. Vanspranghe, F. Ferrante, and C. Prieur. Velocity stabilization of a wave equation with a nonlinear dynamic boundary condition. *IEEE Transactions on Automatic Control*, 67(12):6786–6793, 2022.

[73] W. Wonham. On pole assignment in multi-input controllable linear systems. *IEEE Transactions on Automatic Control*, 12(6):660–665, 1967.

[74] L. Zaccarian and A. R. Teel. *Modern Anti-windup Synthesis: Control Augmentation for Actuator Saturation*. Princeton Series in Applied Mathematics. Princeton University Press, 2011.

[75] L. Zhang and C. Prieur. Stochastic stability of Markov jump hyperbolic systems with application to traffic flow control. *Automatica*, 86:29–37, 2017.

State Observation for Stochastic Partial Differential Equations*

Qi Lü

School of Mathematics, Sichuan University, Chengdu, China, 610064
lu@scu.edu.cn

Abstract. The main purpose of this lecture note is to give a basic introduction to some recent results on state observation problems for stochastic hyperbolic equations, stochastic parabolic equations and stochastic Schrödinger equations. The main tool to solve those problems is Carleman estimate.

1. Introduction

Many important systems in reality (such as those in the microelectronics industry, in the atmospheric motion, in communication and transportation, also in chemistry, biology, microelectronics industry, pharmaceutical industry, and so on) exhibit very complicated dynamics, including substantial model uncertainty, actuator and state constraints, and high dimensionality (usually even infinite). These systems are often best described by stochastic partial differential equations (SPDEs for short) (e.g., [1–6]). This leads to a major requirement for the study on the control theory of SPDEs (e.g., [7, 8]). The operation and control of such systems usually require precise information on the state of the underlying mathematical model. Usually, the information is only available through observations made on a suitable subset of the domain where the system is evolving on. As a result, it is desired to study state observation problems of SPDEs. Besides its own interest, the state observation problems are also related to some other important problems, such as data assimilation and Kalman filtering for SPDEs (e.g., [9]).

---

*This work was partially supported by NSF of China under grants 12025105, 11971334, 11931011, and by the Science Development Project of Sichuan University under grants 2020SCUNL101 and 2020SCUNL201.

This lecture note mainly focuses on the observation problems just mentioned above, that is, whether one can determine the state of the underlying system by a given observation.

We do not attempt to cover all recent results for the state observation problems for SPDEs, which is virtually hopeless. Rather, with admitted bias, we choose subjects that are undergoing rapid change and require new approaches to meet the challenges and opportunities.

A word of caution is necessary concerning the references. No attempt will be made to provide a complete bibliography for the topic of this note, which would only tend to make the narrative very disjointed. Since it depends on personal views and author's taste, the selection of the bibliography is not impartial.

The main technique we employ in this note is the Carleman type estimate, which traces its origin to early work on uniqueness of the solution to elliptic PDEs by T. Carleman ([10]). In that paper, he introduced an energy estimate with an exponential weight to prove a strong unique continuation property for some elliptic PDEs in $\mathbb{R}^2$. This type of weighted energy estimate, now referred to as Carleman estimate, has become one of the major tools in the study of unique continuation property (e.g., [11, 12]), inverse problems (e.g., [13, 14]) and control theory of PDEs (e.g., [15, 16]). However, it is only in recent years that the power of the global Carleman estimate in the context of control theory of SPDEs came to be realized (e.g., [17–35]).

In this note, we consider state observation problems for SPDEs driven by a standard one dimensional Brownian motion. The system is completely observable (meaning that the controller is able to observe the system state completely). The reason for these settings is that we would like to show readers some fundamental structure and properties of state observation problems for SPDEs in a clean and clear way, and avoid technicalities caused by more complicated models.

Let us now sketch the main contents of this lecture note. In Section 2, some preliminary materials on stochastic calculus are presented. It also unifies terminologies and notations used in the rest of this note. These materials are mainly for beginners. They also serve as a quick reference for knowledgeable readers. Section 3 is devoted to the formulation of state observation problems which will be studied in the rest of this lecture. In Sections 4–7, we address the state observation problems for stochastic hyperbolic equations and stochastic heat equations, respectively.

2. Some preliminary results on Probability Theory and Stochastic Analysis

In this section, we recall some basic knowledge of Probability Theory and Stochastic Analysis which will be used in this lecture note. More details can be found in [28].

2.1. *Probability, random variables and expectation*

Definition 1. A *probability space* is a measure space $(\Omega, \mathcal{F}, \mathbb{P})$ for which $\mathbb{P}(\Omega) = 1$. In this case, $\mathbb{P}$ is called a *probability measure* or a *probability*.

Any $\omega \in \Omega$ is called a *sample point*; any $A \in \mathcal{F}$ is called an *event*, and $\mathbb{P}(A)$ represents the *probability* of the event A.

If an event $A \in \mathcal{F}$ satisfies that $\mathbb{P}(A) = 1$, then we may alternatively say that A holds, $\mathbb{P}$-a.s., or simply A holds a.s. (if the probability $\mathbb{P}$ is clear from the context).

Let $\mathcal{X}$ be a Banach space and $(\Omega, \mathcal{F}, \mathbb{P})$ be a probability space. Denote by $\mathcal{B}(\mathcal{X})$ the Borel σ-algebra of $\mathcal{X}$. Each $\mathcal{X}$-valued, strongly measurable function $f : \Omega \to \mathcal{X}$ is called an ($\mathcal{X}$-valued) *random variable*. Clearly, $f^{-1}(\mathcal{B}(\mathcal{X}))$ is a sub-σ-field of $\mathcal{F}$, which is called the σ-field generated by f and denoted by $\sigma(f)$. For a given index set Λ and a family of $\mathcal{X}$-valued, random variables $\{f_\lambda\}_{\lambda \in \Lambda}$ (defined on $(\Omega, \mathcal{F})$), we denote by $\sigma(f_\lambda; \lambda \in \Lambda)$ the σ-field generated by $\cup_{\lambda \in \Lambda} \sigma(f_\lambda)$.

If f is Bochner integrable w.r.t. the measure $\mathbb{P}$, then we denote the integral by $\mathbb{E}f$ and call it the *mean* or *mathematical expectation* of f.

Next, we introduce an important notation, *independence*, which distinguishes probability theory from the usual measure theory.

Definition 2. Let $(\Omega, \mathcal{F}, \mathbb{P})$ be a probability space.

(1) We say that two events A and B are independent if $\mathbb{P}(A \cap B) = \mathbb{P}(A)\mathbb{P}(B)$.

(2) Let $\mathcal{K}_1$ and $\mathcal{K}_2$ be two subsets of $\mathcal{F}$. We say that $\mathcal{K}_1$ and $\mathcal{K}_2$ are independent if $\mathbb{P}(A \cap B) = \mathbb{P}(A)\mathbb{P}(B)$ for any $A \in \mathcal{K}_1$ and $B \in \mathcal{K}_2$.

(3) Let f and g be two random variables defined on $(\Omega, \mathcal{F})$ (with possibly different ranges), and $\mathcal{K} \subset \mathcal{F}$. We say that f and g are independent if $\sigma(f)$ and $\sigma(g)$ are independent. We say that f and $\mathcal{K}$ are independent if $\sigma(f)$ and $\mathcal{K}$ are independent.

(4) Let $\{A_\lambda\}_{\lambda \in \Lambda} \subset \mathcal{F}$. We say that $\{A_\lambda\}_{\lambda \in \Lambda}$ is a class of mutually

independent events if

$$\mathbb{P}(A_{\lambda_1} \cap \cdots \cap A_{\lambda_n}) = \mathbb{P}(A_{\lambda_1}) \cdots \mathbb{P}(A_{\lambda_n})$$

for any $n \in \mathbb{N}$, $\lambda_1, \cdots, \lambda_n \in \Lambda$ satisfying $\lambda_j \neq \lambda_k$ whenever $j \neq k$, $j, k = 1, \cdots, n$.

2.2. *Distribution, density and Gaussian random variable*

Let $(\Omega, \mathcal{F}, \mathbb{P})$ be a probability space, and $X : \Omega \to \mathcal{X}$ be a random variable. Then, X induces a probability measure $\mathbb{P}_X$ on $(\mathcal{X}, \mathcal{B}(\mathcal{X}))$ via

$$\mathbb{P}_X(A) \triangleq \mathbb{P}(X^{-1}(A)), \quad \forall A \in \mathcal{B}(\mathcal{X}). \tag{2.1}$$

We call $\mathbb{P}_X$ the *distribution* of X. If X is Bochner integrable w.r.t. $\mathbb{P}$, then

$$\mathbb{E}X = \int_{\mathcal{X}} \eta \, d\mathbb{P}_X(\eta).$$

When $\mathcal{X} = \mathbb{R}^m$ ($m \in \mathbb{N}$), $\mathbb{P}_X$ can be uniquely determined by the following function:

$$F(x) = F(x_1, \cdots, x_m) \triangleq \mathbb{P}\{X_j \leq x_j, \ 1 \leq j \leq m\}, \tag{2.2}$$

where $x = (x_1, \cdots, x_m)$ and $X = (X_1, \cdots, X_m)$. We call F the *distribution function* of X. If F is absolutely continuous w.r.t. the Lebesgue measure in $\mathbb{R}^m$, then there exists a (nonnegative) function $f \in L^1(\mathbb{R}^m)$ such that

$$F(x) = \int_{-\infty}^{x_1} \cdots \int_{-\infty}^{x_m} f(\xi_1, \cdots, \xi_m) d\xi_1 \cdots d\xi_m.$$

The function $f(\cdot)$ is called the *density* of X.

If

$$f(x) = [(2\pi)^m \det Q]^{-1/2} \exp\left\{ -\frac{1}{2}(x - \lambda)Q^{-1}(x - \lambda)^\top \right\}, \quad x \in \mathbb{R}^m,$$

for some $\lambda \in \mathbb{R}^m$ and $Q \in \mathbb{R}^{m \times m}$ with $Q^\top = Q > 0$, then we say that X has a *normal distribution* with parameter (λ, Q), denoted by $X \sim \mathcal{N}(\lambda, Q)$ or X is normal distributed with parameter (λ, Q). Normal distributed random variables appear in many physical models (e.g., [36]).

2.3. *Stochastic processes*

Definition 3. Let $\mathcal{I} = [0, T]$ with $T > 0$. A family of $\mathcal{X}$-valued random variables $\{X(t)\}_{t \in \mathcal{I}}$ is called a *stochastic process*. For any $\omega \in \Omega$, the map $t \mapsto X(t, \omega)$ is called a *sample path* (of X).

We will interchangeably use $\{X(t)\}_{t\in\mathcal{I}}$, $X(\cdot)$ or even X to denote a (stochastic) process if there is no confusion.

Definition 4. An ($\mathcal{X}$-valued) process $X(\cdot)$ is said to be *continuous* if there is a $\mathbb{P}$-null set $N \in \mathcal{F}$, such that for any $\omega \in \Omega \setminus N$, the sample path $X(\cdot, \omega)$ is continuous in $\mathcal{X}$.

Definition 5. Two ($\mathcal{X}$-valued) processes $X(\cdot)$ and $\overline{X}(\cdot)$ are said to be *stochastically equivalent* if $\mathbb{P}(\{X(t) = \overline{X}(t)\}) = 1$ for each $t \in \mathcal{I}$. In this case, one is said to be a *modification* of the other.

Definition 6. We call a family of sub-σ-fields $\{\mathcal{F}_t\}_{t\in\mathcal{I}}$ of $\mathcal{F}$ a *filtration* if $\mathcal{F}_{t_1} \subset \mathcal{F}_{t_2}$ for all $t_1, t_2 \in \mathcal{I}$ with $t_1 \leq t_2$. For every $t \in \mathcal{I}$, we put

$$\mathcal{F}_{t+} \overset{\Delta}{=} \bigcap_{s\in(t,+\infty)\cap\mathcal{I}} \mathcal{F}_s, \qquad \mathcal{F}_{t-} \overset{\Delta}{=} \bigcup_{s\in[0,t)\cap\mathcal{I}} \mathcal{F}_s.$$

We call $\{\mathcal{F}_t\}_{t\in\mathcal{I}}$ *right (resp. left) continuous* if $\mathcal{F}_{t+} = \mathcal{F}_t$ (resp. $\mathcal{F}_{t-} = \mathcal{F}_t$).

In the sequel, for simplicity, we write $\mathbf{F} = \{\mathcal{F}_t\}_{t\in\mathcal{I}}$ unless we need to emphasize what $\mathcal{F}_t$ or $\mathcal{I}$ exactly is. We call $(\Omega, \mathcal{F}, \mathbf{F}, \mathbb{P})$ a *filtered probability space*.

Definition 7. We say that $(\Omega, \mathcal{F}, \mathbf{F}, \mathbb{P})$ satisfies the *usual condition* if $(\Omega, \mathcal{F}, \mathbb{P})$ is complete, $\mathcal{F}_0$ contains all $\mathbb{P}$-null sets in $\mathcal{F}$, and $\mathbf{F}$ is right continuous.

Definition 8. Let $X(\cdot)$ be an $\mathcal{X}$-valued process.

(1) $X(\cdot)$ is said to be *measurable* if the map $(t, \omega) \mapsto X(t, \omega)$ is strongly $(\mathcal{B}(\mathcal{I}) \times \mathcal{F})/\mathcal{B}(\mathcal{X})$-measurable;

(2) $X(\cdot)$ is said to be $\mathbf{F}$-*adapted* if it is measurable, and for each $t \in \mathcal{I}$, the map $\omega \mapsto X(t, \omega)$ is strongly $\mathcal{F}_t/\mathcal{B}(\mathcal{X})$-measurable;

(3) $X(\cdot)$ is said to be $\mathbf{F}$-*progressively measurable* if for each $t \in \mathcal{I}$, the map $(s, \omega) \mapsto X(s, \omega)$ from $[0, t] \times \Omega$ to $\mathcal{X}$ is strongly $(\mathcal{B}([0, t]) \times \mathcal{F}_t)/\mathcal{B}(\mathcal{X})$-measurable.

Definition 9. A set $A \in \mathcal{I} \times \Omega$ is called *progressively measurable* w.r.t. $\mathbf{F}$ if the process $\chi_A(\cdot)$ is progressive. The class of all progressively measurable sets is a σ-field, called the *progressive σ-field* w.r.t. $\mathbf{F}$, denoted by $\mathbb{F}$.

Proposition 1. *An ($\mathcal{X}$-valued) process $\varphi : [0, T] \times \Omega \to \mathcal{X}$ is $\mathbf{F}$-progressively measurable if and only if it is strongly $\mathbb{F}$-measurable.*

It is clear that if $X(\cdot)$ is **F**-progressively measurable, it must be **F**-adapted. Conversely, it can be proved that, for any **F**-adapted process $X(\cdot)$, there is an **F**-progressively measurable process $\widetilde{X}(\cdot)$ which is stochastically equivalent to $X(\cdot)$. For this reason, in the sequel, by saying that a process $X(\cdot)$ is **F**-adapted, we mean that it is **F**-progressively measurable.

For any $p, q \in [1, \infty)$, write

$$L_{\mathbb{F}}^p(\Omega; L^q(0, T; \mathcal{X}))$$

$$\triangleq \left\{ \varphi : (0, T) \times \Omega \to \mathcal{X} \,\middle|\, \varphi(\cdot) \text{ is } \mathbf{F}\text{-adapted and } \mathbb{E}\left(\int_0^T |\varphi(t)|_{\mathcal{X}}^q dt \right)^{\frac{p}{q}} < \infty \right\},$$

$$L_{\mathbb{F}}^q(0, T; L^p(\Omega; \mathcal{X}))$$

$$\triangleq \left\{ \varphi : (0, T) \times \Omega \to \mathcal{X} \,\middle|\, \varphi(\cdot) \text{ is } \mathbf{F}\text{-adapted and } \int_0^T (\mathbb{E}|\varphi(t)|_{\mathcal{X}}^p)^{\frac{q}{p}} dt < \infty \right\}.$$

Similarly, we can define (for $1 \leq p, q < \infty$)

$$\begin{cases} L_{\mathbb{F}}^\infty(\Omega; L^q(0, T; \mathcal{X})), & L_{\mathbb{F}}^p(\Omega; L^\infty(0, T; \mathcal{X})), & L_{\mathbb{F}}^\infty(\Omega; L^\infty(0, T; \mathcal{X})), \\ L_{\mathbb{F}}^\infty(0, T; L^p(\Omega; \mathcal{X})), & L_{\mathbb{F}}^q(0, T; L^\infty(\Omega; \mathcal{X})), & L_{\mathbb{F}}^\infty(0, T; L^\infty(\Omega; \mathcal{X})). \end{cases}$$

All the above spaces are Banach spaces (with the canonical norms). We shall simply denote $L_{\mathbb{F}}^p(\Omega; L^p(0, T; \mathcal{X})) \equiv L_{\mathbb{F}}^p(0, T; L^p(\Omega; \mathcal{X}))$ by $L_{\mathbb{F}}^p(0, T; \mathcal{X})$; and further simply denote $L_{\mathbb{F}}^p(0, T; \mathbb{R})$ by $L_{\mathbb{F}}^p(0, T)$.

For any $p \in [1, \infty)$, set

$$L_{\mathbb{F}}^p(\Omega; C([0, T]; \mathcal{X}))$$

$$\triangleq \left\{ \varphi : [0, T] \times \Omega \to \mathcal{X} \,\middle|\, \varphi(\cdot) \text{ is continuous, } \mathbf{F}\text{-adapted and} \right.$$

$$\mathbb{E}\left(|\varphi(\cdot)|_{C([0,T];\mathcal{X})}^p \right) < \infty \Big\}.$$

One can show that $L_{\mathbb{F}}^p(\Omega; C([0, T]; \mathcal{X}))$ is a Banach space with the norm

$$|\varphi(\cdot)|_{L_{\mathbb{F}}^p(\Omega; C([0,T];\mathcal{X}))} = \left(\mathbb{E}(|\varphi(\cdot)|_{C([0,T];\mathcal{X})}^p) \right)^{1/p}.$$

If $\mathcal{X}$ is a Fréchet space, then all the spaces given above can be similarly defined and they are all Fréchet space with canonical norms.

Definition 10. Let $(\Omega, \mathcal{F}, \mathbf{F}, \mathbb{P})$ (with $\mathbf{F} = \{\mathcal{F}_t\}_{t \in \mathcal{I}}$) be a filtered probability space. A continuous **F**-adapted process $W(\cdot)$ is called a 1-*dimensional Brownian motion* (over $\mathcal{I}$), if for all $s, t \in \mathcal{I}$ with $0 \leq s < t < T$, $W(t) - W(s)$ is independent of $\mathcal{F}_s$, and normally distributed with mean 0 and variance $t - s$. In addition, if $\mathbb{P}(W(0) = 0) = 1$, then $W(\cdot)$ is called a 1-*dimensional standard Brownian motion*.

From now on, we fix a filtered probability space $(\Omega, \mathcal{F}, \mathbf{F}, \mathbb{P})$ satisfying the usual condition, on which a standard Brownian motion is defined. In the rest of this note, *unless other stated, we omit the argument $\omega\ (\in \Omega)$ in any random variable or stochastic process defined on the probability space* $(\Omega, \mathcal{F}, \mathbb{P})$. Assume $\{W(t)\}_{t\in[0,T]}$ is a 1-dimensional standard Brownian motion defined on $(\Omega, \mathcal{F}, \mathbf{F}, \mathbb{P})$.

2.4. *Itô integral and its properties*

Let H be a Hilbert space and $T > 0$. Note that one cannot define

$$\int_0^T f(t)dW(t) \tag{2.3}$$

as a Lebesgue-Stieltjes type integral by regarding ω as a parameter, due to the fact that the map $t(\in [0,T]) \mapsto W(t, \cdot)$ is not of bounded variation, $\mathbb{P}$-a.s.

Write $\mathcal{L}_0$ for the set of $f \in L^2_{\mathbb{F}}(0,T;H)$ as

$$f(t,\omega) = \sum_{j=0}^n f_j(\omega)\chi_{[t_j,t_{j+1})}(t), \quad (t,\omega) \in [0,T] \times \Omega, \tag{2.4}$$

where $n \in \mathbb{N}$, $0 = t_0 < t_1 < \cdots < t_{n+1} = T$, f_j is $\mathcal{F}_{t_j}$-measurable and

$$\sup\left\{|f_j(\omega)|_H \,\middle|\, j \in \{0, \cdots, n\}, \omega \in \Omega\right\} < \infty.$$

For $f \in \mathcal{L}_0$, we set

$$I(f)(t) = \sum_{j=0}^n f_j(\omega)[W(t \wedge t_{j+1}) - W(t \wedge t_j)]. \tag{2.5}$$

It is easy to show that $I(f)(t) \in L^2_{\mathcal{F}_t}(\Omega; H)$ and the following Itô isometry holds:

$$|I(f)(t)|_{L^2_{\mathcal{F}_t}(\Omega;H)} = |f|_{L^2_{\mathbb{F}}(0,t;H)}. \tag{2.6}$$

For $f \in L^2_{\mathbb{F}}(0,T;H)$, one can find a sequence of $\{f_k\}_{k=1}^\infty \subset \mathcal{L}_0$ such that

$$\lim_{k\to\infty} |f_k - f|_{L^2_{\mathbb{F}}(0,T;H)} = 0.$$

Since

$$|I(f_j)(t) - I(f_l)(t)|_{L^2_{\mathcal{F}_t}(\Omega;H)} = |f_j - f_l|_{L^2_{\mathbb{F}}(0,t;H)}, \quad \forall j,l \in \mathbb{N},$$

one gets that $\{I(f_k)(t)\}_{k=1}^\infty$ is a Cauchy sequence in $L^2_{\mathcal{F}_t}(\Omega; H)$ and therefore, it converges to a unique $X \in L^2_{\mathcal{F}_t}(\Omega; H)$, which is independent of the particular choice of $\{f_k\}_{k=1}^\infty$. We call this element the *Itô integral* of f (w.r.t.

the Brownian motion $W(\cdot)$) on $[0,t]$ and denote it by $X(t)$ to emphasize the time variable t.

For $0 \le s < t \le T$, we call $X(t) - X(s)$ the *Itô integral* of $f \in L_{\mathbb{F}}^2(0,T;H)$ (w.r.t. the Brownian motion $W(\cdot)$) on $[s,t]$ and denote it by $\int_s^t f(\tau)dW(\tau)$ or simply by $\int_s^t f dW$.

The Itô integral has the following properties.

Theorem 1. *Let* $f,g \in L_{\mathbb{F}}^2(0,T;H)$ *and* $a,b \in L_{\mathcal{F}_s}^2(\Omega)$, $0 \le s < t \le T$. *Then*

(1) $\displaystyle\int_s^t f dW \in L_{\mathbb{F}}^p(\Omega; C([s,t];H))$;

(2) $\displaystyle\int_s^t (af + bg)dW = a \int_s^t f dW + b \int_s^t g dW$, $\mathbb{P}$*-a.s.*;

(3) $\displaystyle\mathbb{E}\left(\int_s^t f dW \right) = 0$;

(4) $\displaystyle\mathbb{E}\left(\left\langle \int_s^t f dW, \int_s^t g dW \right\rangle_H \right) = \mathbb{E}\int_s^t \langle f(r,\cdot), g(r,\cdot) \rangle_H dr$.

Definition 11. An H-valued, $\mathbf{F}$-adapted, continuous process $X(\cdot)$ is called an (H-valued) *Itô process* if there exist two H-valued stochastic processes $\phi(\cdot), \Phi(\cdot) \in L_{\mathbb{F}}^2(0,T;H)$ such that for any $t \in [0,T]$,

$$X(t) = X(0) + \int_0^t \phi(s)ds + \int_0^t \Phi(s)dW(s), \quad \mathbb{P}\text{-a.s.} \qquad (2.7)$$

The following fundamental result is known as *Itô's formula.*

Theorem 2. *Let* $X(\cdot)$ *be given by* (2.7). *Let* $F : [0,T] \times H \to \mathbb{R}$ *be a function such that its partial derivatives* F_t, F_x *and* F_{xx} *are uniformly continuous on any bounded subset of* $[0,T] \times H$. *Then for any* $t \in [0,T]$,

$$
\begin{aligned}
F(t, X(t)) - F(0, X(0)) = \int_0^t & \Big[F_t(s, X(s)) + \langle F_x(s, X(s)), \phi(s) \rangle_H \\
& + \frac{1}{2} \langle F_{xx}(s, X(s))\Phi(s), \Phi(s) \rangle_H \Big] ds \\
& + \int_0^t F_x(s, X(s))\Phi(s)dW(s), \quad \mathbb{P}\text{-a.s.} \quad (2.8)
\end{aligned}
$$

Remark 1. Usually, people write the formula (2.8) in the following differential form:

$$
\begin{aligned}
dF(t, X(t)) = & F_t(t, X(t))dt + \langle F_x(t, X(t)), \phi(t) \rangle_H dt \\
& + \frac{1}{2} \langle F_{xx}(t, X(t))\Phi(t), \Phi(t) \rangle_H dt + F_x(t, X(t))\Phi(t)dW(t).
\end{aligned}
$$

Theorem 2 works well for Itô processes in the (strong) form (2.7). However, usually it is too restrictive in the study of stochastic evolution equations in infinite dimensions. Indeed, in the infinite dimensional setting sometimes one has to handle Itô processes in a weaker form, to be presented below.

Let $\mathcal{V}$ be a Hilbert space such that the embedding $\mathcal{V} \subset H$ is continuous and dense. Denote by $\mathcal{V}^*$ the dual space of $\mathcal{V}$ w.r.t. the pivot space H. Then, $\mathcal{V} \subset H = H^* \subset \mathcal{V}^*$, continuously and densely and

$$\langle z, v \rangle_{\mathcal{V}, \mathcal{V}^*} = \langle z, v \rangle_H, \qquad \forall\, v \in H,\ z \in \mathcal{V}.$$

We have the following Itô's formula for a weak form of Itô process.

Theorem 3. *Suppose that* $X_0 \in L^2_{\mathcal{F}_0}(\Omega; H)$, $\phi(\cdot) \in L^2_{\mathbb{F}}(0, T; \mathcal{V}^*)$ *and* $\Phi(\cdot) \in L^2_{\mathbb{F}}(0, T; H)$. *Let*

$$X(t) = X_0 + \int_0^t \phi(s)ds + \int_0^t \Phi(s)dW(s), \ \ t \in [0, T].$$

If $X \in L^2_{\mathbb{F}}(0, T; \mathcal{V})$, *then* $X(\cdot) \in C([0, T]; H)$, $\mathbb{P}$*-a.s., and for any* $t \in [0, T]$,

$$|X(t)|^2_H = |X_0|^2_H + 2 \int_0^t \langle \phi(s), X(s) \rangle_{\mathcal{V}^*, \mathcal{V}} ds + \int_0^t |\Phi(s)|^2_H ds$$
$$+ 2 \int_0^t \langle \Phi(s), X(s) \rangle_H dW(s), \ \ \mathbb{P}\text{-}a.s. \tag{2.9}$$

As an immediate corollary of Theorem 3, we have the following result.

Corollary 1. *Suppose that* $X_0, Y_0 \in L^2_{\mathcal{F}_0}(\Omega; H)$, $\phi_1(\cdot)$, $\phi_2(\cdot) \in L^2_{\mathbb{F}}(0, T; \mathcal{V}^*)$, *and* $\Phi_1(\cdot), \Phi_2(\cdot) \in L^p_{\mathbb{F}}(\Omega; L^2(0, T; H))$. *Let*

$$X(t) = X_0 + \int_0^t \phi_1(s)ds + \int_0^t \Phi_1(s)dW(s), \ \ t \in [0, T]$$

and

$$Y(t) = Y_0 + \int_0^t \phi_2(s)ds + \int_0^t \Phi_2(s)dW(s), \ \ t \in [0, T].$$

If $X(\cdot), Y(\cdot) \in L^2_{\mathbb{F}}(0, T; \mathcal{V})$, *then* $X(\cdot), Y(\cdot) \in C([0, T]; H)$, $\mathbb{P}$*-a.s., and for any* $t \in [0, T]$,

$$\langle X(t), Y(t) \rangle_H - \langle X_0, Y_0 \rangle_H$$
$$= \int_0^t \big(\langle \phi_2(s), X(s) \rangle_{\mathcal{V}^*, \mathcal{V}} + \langle \phi_1(s), Y(s) \rangle_{\mathcal{V}^*, \mathcal{V}} \big) ds + \int_0^t \langle \Phi_1(s), \Phi_2(s) \rangle_H ds$$
$$+ \int_0^t \big(\langle \Phi_2(s), X(s) \rangle_H + \langle \Phi_1(s), Y(s) \rangle_H \big) dW(s), \ \ \mathbb{P}\text{-}a.s. \tag{2.10}$$

Remark 2. For simplicity, we usually write the formula (2.9) in the following differential form:

$$d|X(t)|_H^2 = 2\langle\phi(t), X(t)\rangle_{V^*,V} + |\Phi(t)|_H^2 dt + 2\langle\Phi(t), X(t)\rangle_H dW(t)$$

and denote $|\Phi(t)|_H^2 dt$ by $|dX|_H^2$. Similarly, we write the formula (2.10) in the following differential form:

$$\begin{aligned}
d\langle X(t), Y(t)\rangle_H &= \langle\phi_2(t), X(t)\rangle_{V^*,V} + \langle\phi_1(t), Y(t)\rangle_{V^*,V} \\
&\quad + \langle\Phi_1(t), \Phi_2(t)\rangle_H dt + \langle\Phi_2(t), X(t)\rangle_H dW(t) \\
&\quad + \langle\Phi_1(t), Y(t)\rangle_H dW(t)
\end{aligned}$$

and denote $\langle\Phi_1(t), \Phi_2(t)\rangle_H dt$ by $\langle dX(t), dY(t)\rangle_H$ for simplicity.

2.5. *Stochastic evolution equations*

In what follows, we shall always assume that H is a separable Hilbert space, and A generates a contraction semigroup $\{S(t)\}_{t\geq 0}$ on H. Consider the following stochastic evolution equation:

$$\begin{cases} dX(t) = [AX(t) + F(t, X(t))]dt + \widetilde{F}(t, X(t))dW(t) & \text{in } (0, T], \\ X(0) = X_0. \end{cases} \quad (2.11)$$

Here $X_0 \in L^2_{\mathcal{F}_0}(\Omega; H)$, and $F(\cdot, \cdot)$ and $\widetilde{F}(\cdot, \cdot)$ are measurable functions from $[0, T] \times \Omega \times H$ to H, satisfying

$$\begin{cases} F(\cdot, 0) \in L^2_{\mathbb{F}}(\Omega; L^1(0, T; H)), \quad \widetilde{F}(\cdot, 0) \in L^2_{\mathbb{F}}(0, T; H), \\ |F(t, y) - F(t, z)|_H \leq L_1(t)|y - z|_H, \\ |\widetilde{F}(t, y) - \widetilde{F}(t, z)|_H \leq L_2(t)|y - z|_H, \\ \qquad\qquad\qquad \forall\, y, z \in H, \text{ a.e. } t \in [0, T], \mathbb{P}\text{-a.s.}, \end{cases} \quad (2.12)$$

for some $L_1(\cdot) \in L^1(0, T)$ and $L_2(\cdot) \in L^2(0, T)$.

First, we give the notion of strong solution to the equation (2.11).

Definition 12. A process $X(\cdot) \in L^2_{\mathbb{F}}(\Omega; C([0, T]; H))$ is called a *strong solution* to (2.11) if $X(t, \omega) \in D(A)$ for a.e. $(t, \omega) \in [0, T] \times \Omega$, $AX(\cdot) \in L^1(0, T; H)$, $\mathbb{P}$-a.s., and for all $t \in [0, T]$,

$$X(t) = X_0 + \int_0^t \big(AX(s) + F(s, X(s))\big)ds + \int_0^t \widetilde{F}(s, X(s))dW(s), \quad \mathbb{P}\text{-a.s.}$$

One needs very restrictive conditions to guarantee the existence of a strong solution. Thus, people introduce two types of "weak" solutions.

Definition 13. A process $X(\cdot) \in L^2_{\mathbb{F}}(\Omega; C([0,T]; H))$ is called a *weak solution* to (2.11) if for any $t \in [0,T]$ and $\xi \in D(A^*)$,

$$\langle X(t), \xi \rangle_H = \langle X_0, \xi \rangle_H + \int_0^t \left(\langle X(s), A^*\xi \rangle_H + \langle F(s, X(s)), \xi \rangle_H \right) ds$$

$$+ \int_0^t \langle \widetilde{F}(s, X(s)), \xi \rangle_H dW(s), \quad \mathbb{P}\text{-a.s.}$$

Definition 14. A process $X(\cdot) \in L^2_{\mathbb{F}}(\Omega; C([0,T]; H))$ is called a *mild solution* to (2.11) if for any $t \in [0,T]$,

$$X(t) = S(t)X_0 + \int_0^t S(t-s)F(s, X(s))ds$$

$$+ \int_0^t S(t-s)\widetilde{F}(s, X(s))dW(s), \quad \mathbb{P}\text{-a.s.}$$

It is easiest to show the well-posedness of (2.11) in the framework of mild solution among the above three kinds of solutions. Indeed, we have the following result.

Theorem 4. *The equation* (2.11) *admits a unique mild solution* $X(\cdot) \in L^2_{\mathbb{F}}(\Omega; C([0,T]; H))$. *Moreover,*

$$|X(\cdot)|_{L^2_{\mathbb{F}}(\Omega; C([0,T]; H))}$$
$$\leq \mathcal{C}\left(|X_0|_{L^2_{\mathcal{F}_0}(\Omega; H)} + |F(\cdot, 0)|_{L^2_{\mathbb{F}}(\Omega; L^1(0,T;H))} + |\widetilde{F}(\cdot, 0)|_{L^2_{\mathbb{F}}(0,T;H)} \right).$$

The following result indicates the space smoothing effect of mild solutions to a class of stochastic evolution equations, such as the stochastic parabolic equation.

Theorem 5. *Assume that* A *is a self-adjoint, negative definite* (*unbounded linear*) *operator on* H. *Then, the equation* (2.11) *admits a unique mild solution*

$$X(\cdot) \in L^2_{\mathbb{F}}(\Omega; C([0,T]; H)) \cap L^2_{\mathbb{F}}(0,T; D((-A)^{\frac{1}{2}})).$$

Moreover,

$$|X(\cdot)|_{L^2_{\mathbb{F}}(\Omega; C([0,T]; H))} + |X(\cdot)|_{L^2_{\mathbb{F}}(0,T; D((-A)^{\frac{1}{2}}))}$$
$$\leq \mathcal{C}\left(|X_0|_{L^2_{\mathcal{F}_0}(\Omega; H)} + |F(\cdot, 0)|_{L^2_{\mathbb{F}}(\Omega; L^1(0,T;H))} + |\widetilde{F}(\cdot, 0)|_{L^2_{\mathbb{F}}(0,T;H)} \right).$$

The next result reveals the relationship between mild and weak solutions to (2.11).

Theorem 6. *Each weak solution to* (2.11) *is also a mild solution to* (2.11) *and vice versa.*

In many applications, the mild solution does not have enough regularity. For example, to establish the pointwise identity for Carleman estimate, we need the functions to be twice differentiable in the sense of weak derivative w.r.t. the spatial variable. These problems can be solved by the following strategy:

(1) Introduce a family of equations with strong solutions such that the limit of these strong solutions is the mild or weak solution to the original equation.
(2) Obtain the desired properties for these strong solutions.
(3) Utilize the density argument to establish the desired properties for the mild/weak solutions.

Let us introduce an approximation equation of (2.11) as follows:

$$\begin{cases} dX^\lambda(t) = AX^\lambda(t)dt + R(\lambda)F(t, X^\lambda(t))dt \\ \qquad\quad +R(\lambda)\widetilde{F}(t, X^\lambda(t))dW(t) \qquad \text{in } (0,T], \qquad (2.13) \\ X^\lambda(0) = R(\lambda)X_0 \in D(A). \end{cases}$$

Here $\lambda \in \rho(A)$, the resolvent set of A, and $R(\lambda) \triangleq \lambda(\lambda I - A)^{-1}$ with I being the identity operator on H.

Theorem 7. *For any $X_0 \in L^2_{\mathcal{F}_0}(\Omega; H)$ and $\lambda \in \rho(A)$, the equation (2.13) admits a unique strong solution $X^\lambda(\cdot)$. Moreover, as $\lambda \to \infty$, the solution $X^\lambda(\cdot)$ converges to $X(\cdot)$ in $L^2_{\mathbb{F}}(\Omega; C([0,T]; H))$, where $X(\cdot)$ solves (2.11) in the sense of the mild solution.*

For many results in this note, rigorous proofs should employ the above processes (1)–(3). However, writing it down every time could be quite cumbersome. Hence, we omit it.

3. Formulation of state observation problems

Let U be a Hilbert space and $\mathfrak{C} : D(A) \subset H \to U$ be a bounded linear operator. Let X be the mild solution to (2.11).

Put

$$(\Psi_t X_0)(s) = \begin{cases} \mathfrak{C}X(s), & \text{if } s \in [0,t], \\ 0, & \text{if } s \in (t,T]. \end{cases}$$

Definition 15. The operator $\mathfrak{C} \in \mathcal{L}(D(A); U)$ is called an admissible observation operator for (2.11) if for some $t_0 > 0$, Ψ_{t_0} has a continuous extension to H.

Equivalently, $\mathfrak{C} \in \mathcal{L}(D(A); U)$ is an *admissible observation operator* for (2.11) if and only if for some $t_0 > 0$, there exists a constant $\mathcal{C}_{t_0} > 0$ such that

$$\mathbb{E} \int_0^{t_0} |\mathfrak{C}X(s)|_U^2 ds \le \mathcal{C}_{t_0}^2 |X_0|_H^2, \quad \forall X_0 \in H. \tag{3.1}$$

If (3.1) holds for some $t_0 > 0$, then for any $t > 0$, there is a constant $\mathcal{C}_t > 0$ such that

$$\mathbb{E} \int_0^t |\mathfrak{C}X(s)|_U^2 ds \le \mathcal{C}_t^2 |X_0|_H^2, \quad \forall X_0 \in H. \tag{3.2}$$

Obviously, every $\mathfrak{C} \in \mathcal{L}(H; U)$ is admissible for (2.11). If $\mathfrak{C}$ is an admissible observation operator for (2.11), then we denote the (unique) extension of Ψ_t to H by the same symbol. Clearly, the norm of the extended operator, $|\Psi_t|_{\mathcal{L}(H; L_\mathbb{F}^2(0,t;U))}$, is the smallest constant $\mathcal{C}_t$ for which (3.2) holds.

Now we give the formulation of the *state observation problem* for (2.11).

- **Identifiability.** Is the solution $X \in L_\mathbb{F}^2(\Omega; C([0,T]; H))$ determined uniquely by the observation $\mathfrak{C}X$?

- **Stability.** Assume that two solutions X and $\widehat{X}$ are given. Let $\mathfrak{C}X$ and $\mathfrak{C}\widehat{X}$ be the corresponding observations. Can we find a positive constant C such that

$$|X - \widehat{X}| \le C \|\mathfrak{C}X - \mathfrak{C}\widehat{X}\|, \tag{3.3}$$

  with appropriate norms in both sides?

- **Reconstruction.** Is it possible to reconstruct $X \in L_\mathbb{F}^2(\Omega; C([0,T]; H))$, in some sense, from the observation $\mathfrak{C}X$?

If one can prove the stability, one also get the identifiability. In this note, we focus on the first two problems, which are concerned with whether one can reconstruct the state of (2.11) in a unique fashion from the observation. If (3.3) holds, then there exists a continuous reconstruction operator R_0

$$R_0 : \mathcal{R}(\mathfrak{C}) \to H, \quad R_0 \mathfrak{C}X = X_0.$$

However, it is very difficult to construct the operator R_0.

Next, we present the formulation of observability for linear stochastic evolution equations, which is closely related to state observation problems for the same equation.

Consider the following linear stochastic evolution equation:

$$\begin{cases} d\widetilde{X}(t) = [A + \widetilde{J}(t)]\widetilde{X}(t)dt + \widetilde{K}(t)\widetilde{X}(t)dW(t) & \text{in } (0,T], \\ \widetilde{X}(0) = \widetilde{X}_0. \end{cases} \tag{3.4}$$

In (3.4), $\widetilde{X}_0 \in H$, $\widetilde{J} \in L_{\mathbb{F}}^\infty(0, T; \mathcal{L}(H))$ and $\widetilde{K} \in L_{\mathbb{F}}^\infty(0, T; \mathcal{L}(H))$.

From Theorem 4, the equation (3.4) admits a unique solution $\widetilde{X}(\cdot) \in L_{\mathbb{F}}^2(\Omega; C([0, T]; H))$ which reads

$$
\begin{aligned}
\widetilde{X}(t) = S(t)\widetilde{X}_0 &+ \int_0^t S(t - s)\widetilde{J}(t)\widetilde{X}(s)ds \\
&+ \int_0^t S(t - s)\widetilde{K}(t)\widetilde{X}(s)dW(s), \quad \mathbb{P}\text{-a.s., } \forall t \in [0, T].
\end{aligned}
\tag{3.5}
$$

Definition 16. (Continuously initially observable on $[0, T]$.)

We say that (3.4) is *continuously initially observable* on $[0, T]$ if there exists $\mathcal{C} > 0$, such that

$$
|\widetilde{X}_0|_H \le \mathcal{C}|\mathfrak{C}X|_{L_{\mathbb{F}}^2(0, T; \widetilde{U})}.
\tag{3.6}
$$

The inequality (3.6) is called the *initial time observability estimate* for (3.4).

Definition 17. (Initially observable on $[0, T]$.)

We say that (3.4) is *initially observable* on $[0, T]$ if

$$
Ker(\Psi_T) = \{0\}.
$$

Initially observability is also called *unique continuation property* of (3.4).

Definition 18. (Continuously finally observable on $[0, T]$.)

We say that (3.4) is *continuously finally observable* on $[0, T]$ if there exists $\mathcal{C} > 0$, such that

$$
|\widetilde{X}(T)|_{L_{\mathcal{F}_T}^2(\Omega; H)} \le \mathcal{C}|\mathfrak{C}\widetilde{X}|_{L_{\mathbb{F}}^2(0, T; \widetilde{U})}.
\tag{3.7}
$$

The inequality (3.7) is called the *final time observability estimate* for (3.4).

Remark 3. Inequality (3.6) (*resp.* (3.7)) is referred to as observability estimate since it provides a quantitative estimate of the norm of the initial data (*resp.* final data) in terms of the observed quantity, by means of the observability constant $\mathcal{C}$. This sort of inequality is strongly relevant to controllability problems and state observation problems for stochastic evolution equations.

When $\widetilde{J}$ is deterministic, i.e., $\widetilde{J} \in L^\infty(0, T; \mathcal{L}(H))$, then the continuously initial observability problem of (3.4) can be reduced to the continuously initially observability problem of the following deterministic evolution equation:

$$
\begin{cases}
\dfrac{d\widehat{X}(t)}{dt} = \left(A + \widetilde{J}\right)\widehat{X}(t) \quad \text{in } (0, T], \\
\widehat{X}(0) = \widetilde{X}_0.
\end{cases}
\tag{3.8}
$$

Recall that the equation (3.8) is continuously initially observable on $[0, T]$ for the observation operator $\mathfrak{C}$ if

$$|\widetilde{X}_0|_H^2 \leq C \int_0^T |\mathfrak{C}\widehat{X}(t)|_{\widehat{U}}^2 dt. \tag{3.9}$$

We have the following result.

Proposition 2. *The equation (3.4) is continuously initially observable on $[0, T]$ for the observation operator $\mathfrak{C}$, provided that the equation (3.8) is continuously initially observable on $[0, T]$ for the observation operator $\mathfrak{C}$.*

Proof. Let $\widetilde{X}(\cdot)$ be a solution to (3.4). Then, $\mathbb{E}\widetilde{X}(\cdot)$ solves (3.8). If (3.8) is continuously initially observable on $[0, T]$, then

$$|X_0|_H^2 \leq C \int_0^T |\mathfrak{C}\mathbb{E}\widetilde{X}(t)|_{\widetilde{U}}^2 dt.$$

This, together with Hölder's inequality, implies (3.6). $\qquad\square$

The above argument does not work for other types of observability problems.

4. Global state observation for stochastic hyperbolic equations

This section is devoted to global state observation problems for stochastic hyperbolic equations. We consider two different problems. The first one is to determine the initial state of the system by means of a boundary observation. The second one is to determine both the initial state and a function in the diffusion term via a boundary observation. The main contents of this section are taken from [23, 37]. We do some minor modifications.

4.1. *Formulation of the problem*

Let $T > 0$ and $G \in \mathbb{R}^n$ ($n \in \mathbb{N}$) be a given bounded domain with the C^2 boundary Γ. Let Γ_0 be a suitable chosen nonempty subset of Γ, whose definition will be given later. Put

$$Q \overset{\triangle}{=} (0, T) \times G, \quad \Sigma \overset{\triangle}{=} (0, T) \times \Gamma, \quad \Sigma_0 \overset{\triangle}{=} (0, T) \times \Gamma_0.$$

Let $(b^{jk})_{1 \leq j,k \leq n} \in C^1(G; \mathbb{R}^{n \times n})$ satisfying that $b^{jk} = b^{kj}$ for $j, k = 1, 2, \cdots, n$ and

$$\sum_{j,k=1}^n b^{jk}\xi_j\xi_k \geq s_0|\xi|^2, \quad \forall (x, \xi) \overset{\triangle}{=} (x, \xi_1, \cdots, \xi_n) \in G \times \mathbb{R}^n, \tag{4.1}$$

for some constant $s_0 > 0$.

Let

$$F(\eta, \varrho, \zeta) : \mathbb{R} \times \mathbb{R} \times \mathbb{R}^n \to \mathbb{R}$$

and

$$K(\eta) : \mathbb{R} \to \mathbb{R}$$

be two nonlinear functions satisfying that

$$|F(\eta_1, \varrho_1, \zeta_1) - F(\eta_2, \varrho_2, \zeta_2)| \leq L(|\eta_1 - \eta_2| + |\varrho_1 - \varrho_2| + |\zeta_1 - \zeta_2|_{\mathbb{R}^n}),$$
$$\forall \, (\eta_i, \varrho_i, \zeta_i) \in \mathbb{R} \times \mathbb{R} \times \mathbb{R}^n, \, i = 1, 2,$$
$$|F(0, \varrho, \zeta)| \leq L(|\varrho| + |\zeta|_{\mathbb{R}^n}), \, \forall \, (\varrho, \zeta) \in \mathbb{R} \times \mathbb{R}^n,$$
$$|K(0)| \in L_{\mathbb{F}}^2(0, T), \, |K(\eta_1) - K(\eta_2)| \leq L|\eta_1 - \eta_2|, \, \forall \, \eta_1, \eta_2 \in \mathbb{R}$$

for some constant $L > 0$.

Consider the following semilinear stochastic hyperbolic equation:

$$\begin{cases} dy_t - \displaystyle\sum_{j,k=1}^n (b^{jk} y_{x_j})_{x_k} dt = F(y, y_t, \nabla y)dt + K(y)dW(t), & \text{in } Q, \\ y = 0, & \text{on } \Sigma, \\ y(0) = y_0, \quad y_t(0) = y_1, & \text{in } G, \end{cases} \tag{4.2}$$

where the initial data $(y_0, y_1) \in L_{\mathcal{F}_0}^2(\Omega; H_0^1(G) \times L^2(G))$ are unknown random variables.

By Theorem 4, we know that the equation (4.2) admits a unique (mild) solution

$$y \in \mathbb{H}_T \stackrel{\triangle}{=} L_{\mathbb{F}}^2(\Omega; C([0, T]; H_0^1(G))) \cap L_{\mathbb{F}}^2(\Omega; C^1([0, T]; L^2(G))).$$

By Proposition 4 below, we know $\frac{\partial y}{\partial \nu} \in L_{\mathbb{F}}^2(0, T; L^2(\Gamma_0))$. Let us define a map $\mathcal{M}_1$ as follows:

$$\begin{cases} \mathcal{M}_1 : L_{\mathcal{F}_0}^2(\Omega; H_0^1(G) \times L^2(G)) \to L_{\mathbb{F}}^2(0, T; L^2(\Gamma_0)), \\ \mathcal{M}_1(y_0, y_1) = \dfrac{\partial y}{\partial \nu}\Big|_{(0,T) \times \Gamma_0}, \end{cases}$$

where y solves the equation (4.2).

Our first problem is as follows:

(P1) *Does there exist a constant $C > 0$ such that for any initial data* $(y_0, y_1), (\hat{y}_0, \hat{y}_1) \in L_{\mathcal{F}_0}^2(\Omega; H_0^1(G) \times L^2(G))$,

$$\begin{aligned} &|(y_0 - \hat{y}_0, y_1 - \hat{y}_1)|_{L_{\mathcal{F}_0}^2(\Omega; H_0^1(G) \times L^2(G))} \\ &\leq C|\mathcal{M}_1(y_0, y_1) - \mathcal{M}_1(\hat{y}_0, \hat{y}_1)|_{L_{\mathbb{F}}^2(0, T; L^2(\Gamma_0))}? \end{aligned} \tag{4.3}$$

Next, we introduce the second problem. Consider the following linear stochastic hyperbolic equation:

$$
\begin{cases}
dz_t - \displaystyle\sum_{j,k=1}^{n} (b^{jk} z_{x_j})_{x_k} dt = \big(b_1 z_t + b_2 \cdot \nabla z + b_3 z + f\big) dt \\
\qquad\qquad\qquad\qquad\qquad + (b_4 z + g) dW(t), & \text{in } Q, \qquad (4.4) \\
z = 0, & \text{on } \Sigma, \\
z(0) = z_0,\ z_t(0) = z_1, & \text{in } G.
\end{cases}
$$

Here the initial data $(z_0, z_1) \in L^2_{\mathcal{F}_0}(\Omega; H^1_0(G) \times L^2(G))$, the coefficients b_j $(1 \leq j \leq 4)$ satisfy that

$$
\begin{aligned}
b_1 &\in L^\infty_{\mathbb{F}}(0,T; L^\infty(G)), & b_2 &\in L^\infty_{\mathbb{F}}(0,T; L^\infty(G; \mathbb{R}^n)), \\
b_3 &\in L^\infty_{\mathbb{F}}(0,T; L^p(G))\ (p \in [n,\infty]), & b_4 &\in L^\infty_{\mathbb{F}}(0,T; L^\infty(G)),
\end{aligned}
\qquad (4.5)
$$

and the nonhomogeneous terms

$$
f \in L^2_{\mathbb{F}}(0,T; L^2(G)), \quad g \in L^2_{\mathbb{F}}(0,T; L^2(G)). \qquad (4.6)
$$

By Theorem 4, we know that the equation (4.4) admits a unique mild solution $z \in \mathbb{H}_T$.

The random force $\int_0^t g\, dW$ is assumed to cause the random vibration starting from some initial state (z_0, z_1). Our aim is to determine the unknown random force intensity g and the unknown initial displacement z_0 and initial velocity z_1 from the (partial) boundary observation $\left.\frac{\partial z}{\partial \nu}\right|_{(0,T) \times \Gamma_0}$ and the measurement on the terminal displacement $z(T)$. More precisely, we are concerned with the following problem:

(P2) *Assume that $f = 0$. Do $\left.\dfrac{\partial z}{\partial \nu}(z_0, z_1, g)\right|_{(0,T) \times \Gamma_0} = 0$ and $z(z_0, z_1, g)(T) = 0$ in G, $\mathbb{P}$-a.s. imply that $g = 0$ in Q and $z_0 = z_1 = 0$ in G, $\mathbb{P}$-a.s.?*

Before presenting the main results, we first give some assumptions on $(b^{jk})_{1 \leq j,k \leq n}$, T and Γ_0.

Condition 4.1. There exists a positive function $\psi(\cdot) \in C^2(\overline{G})$ satisfying the following:

(1) For some constant $\mu_0 > 0$, it holds that

$$
\sum_{j,k=1}^{n} \left\{ \sum_{j',k'=1}^{n} \left[2 b^{jk'} (b^{j'k} \psi_{x_{j'}})_{x_{k'}} - b^{jk}_{x_{k'}} b^{j'k'} \psi_{x_{j'}} \right] \right\} \xi_j \xi_k
$$

$$
\geq \mu_0 \sum_{j,k=1}^{n} b^{jk} \xi_j \xi_k, \quad \forall (x, \xi_1, \cdots, \xi_n) \in \overline{G} \times \mathbb{R}^n. \qquad (4.7)
$$

(2) There is no critical point of $\psi(\cdot)$ in $\overline{G}$, i.e.,

$$\min_{x \in \overline{G}} |\nabla \psi(x)| > 0. \tag{4.8}$$

Remark 4. If $(b^{jk})_{1 \le j,k \le n}$ is the identity matrix I, then $\psi(x) = |x - x_0|^2$ satisfies Condition 4.1, where x_0 is any point in $\mathbb{R}^n \setminus \overline{G}$. On the other hand, if $c \in C^1(\overline{G})$ such that $c(x) \ne 0$ for any $x \in \overline{G}$ and

$$(x - x_0) \cdot \nabla c^{-2}(x) \ge \mu_1, \; \forall x \in \overline{G},$$

for some $\mu_1 > 0$, then Condition 4.1 also holds for $\psi(x) = |x - x_0|^2$ when $(b^{jk})_{1 \le j,k \le n} = c^2 I$ (e.g., [38, Theorem 1.10.2]).

Remark 5. Condition 4.1 is a kind of pseudo-convex condition (e.g., [11, Section 28.3]). A detailed study of this condition is given in [39].

Denote by $\nu(x) = (\nu_1(x), \cdots, \nu_n(x))$ the unit outward normal vector of Γ at point x. Let

$$\Gamma_0 \triangleq \left\{ x \in \Gamma \; \middle| \; \sum_{j,k=1}^n b^{jk}(x)\psi_{x_j}(x)\nu_k(x) > 0 \right\}. \tag{4.9}$$

It is easy to check that if $\psi(\cdot)$ satisfies Condition 4.1, then for any given constants $a \ge 1$ and $b \in \mathbb{R}$, the function $\tilde{\psi} = a\psi + b$ also works well with μ_0 replaced by $a\mu_0$. Therefore we may choose ψ, μ_0, $c_0 > 0$, $c_1 > 0$ and T such that the following condition holds:

Condition 4.2.

(1) $\dfrac{1}{4} \displaystyle\sum_{j,k=1}^n b^{jk}(x)\psi_{x_j}(x)\psi_{x_k}(x) \ge R_1^2 \triangleq \max_{x \in \overline{G}} \psi(x) \ge R_0^2 \triangleq \min_{x \in \overline{G}} \psi(x),$

$$\forall x \in \overline{G}. \tag{4.10}$$

(2) $T > T_0 \triangleq 2R_1$.

(3) $\left(\dfrac{2R_1}{T}\right)^2 < c_1 < \dfrac{2R_1}{T}$.

(4) $\mu_0 - 4c_1 - c_0 > 0$.

Remark 6. As we have explained, Condition 4.2 could be satisfied obviously. We put it here just to emphasize the relationship between $0 < c_0 < c_1 < 1$, μ_0 and T.

We have the following result.

Theorem 8. *Let Conditions* 4.1 *and* 4.2 *be satisfied. There exists a constant* $C > 0$ *such that for any initial data* $(y_0, y_1), (\hat{y}_0, \hat{y}_1) \in L^2_{\mathcal{F}_0}(\Omega; H^1_0(G) \times L^2(G))$, *the inequality* (4.3) *holds.*

Remark 7. Theorem 8 indicates that the state $y(t)$ of (4.2) (for $t \in [0, T]$) can be uniquely determined from the observation $\left. \dfrac{\partial y}{\partial \nu} \right|_{(0,T) \times \Gamma_0}$, $\mathbb{P}$-a.s., and continuously depends on it. Therefore, it provides positive answers to the first and second questions for the state observation problem of the system (4.2).

To solve Problem **(P2)**, we need furthermore the following condition:

Condition 4.3.

$$
\begin{cases}
(1) \ \mu_0 - 4c_1 - c_0 > 0, \\
(2) \ \dfrac{\mu_0}{(8c_1 + c_0)} \displaystyle\sum_{j,k=1}^{n} b^{jk} \psi_{x_j} \psi_{x_k} > 4c_1^2 T^2 > \displaystyle\sum_{j,k=1}^{n} b^{jk} \psi_{x_j} \psi_{x_k}.
\end{cases}
$$

Theorem 9. *Assume that* $f = 0$. *Let Conditions* 4.1 *and* 4.3 *hold. Assume that the solution* $z \in \mathbb{H}_T$ *of* (4.4) *satisfies that* $z(T) = 0$ *in* G, $\mathbb{P}$-*a.s., then we have*

$$
|(z_0, z_1)|_{L^2_{\mathcal{F}_0}(\Omega; H^1_0(G) \times L^2(G))} + |\sqrt{T - t}g|_{L^2_{\mathbb{F}}(0,T; L^2(G))} \leq C \left| \frac{\partial z}{\partial \nu} \right|_{L^2_{\mathbb{F}}(0,T; L^2(\Gamma_0))}.
$$

Remark 8. Similar to Problem **(P2)**, and stimulated by Theorem 9, it seems natural and reasonable to expect a similar result for the following equation:

$$
\begin{cases}
dz_t - \displaystyle\sum_{j,k=1}^{n} (b^{jk} z_{x_j})_{x_k} dt = (b_1 z_t + b_2 \cdot \nabla z + b_3 z + f) \, dt + b_4 z dW(t), & \text{in } Q, \\
z = 0, & \text{on } \Sigma, \\
z(0) = z_0, \ z_t(0) = z_1, & \text{in } G,
\end{cases}
$$

in which z_0, z_1 and f are unknowns and one expects to determine them through the boundary observation $\left. \dfrac{\partial z}{\partial \nu} \right|_{(0,T) \times \Gamma_0}$ and the terminal measurement $z(T)$. However the same conclusion as that in Theorem 9 does NOT hold true even for deterministic wave equations. To see this, let us choose a $y \in C_0^\infty(Q)$ so that it does not vanish in some proper nonempty subdomain of Q. Put $f = y_{tt} - \Delta y$. Then, it is easy to see that y solves the following wave equation:

$$\begin{cases} y_{tt} - \Delta y = f, & \text{in } Q, \\ y = 0, & \text{on } \Sigma, \\ y(0) = 0, \ y_t(0) = 0, & \text{in } G. \end{cases}$$

One can show that $y(T) = 0$ in G and $\dfrac{\partial y}{\partial \nu} = 0$ on Σ. However, it is clear that f does not vanish in Q. This counterexample shows that the formulation of the state observation problem for SPDEs may differ considerably from its deterministic counterpart.

4.2. *Some preliminaries*

In this subsection, we present some preliminary results. First, we give a hidden regularity property of solutions to the equations (4.2) and (4.4). To this end, we recall the following result about the existence of a special vector field.

Proposition 3. *There exists a vector field* $\xi = (\xi_1, \cdots, \xi_n) \in C^1(\mathbb{R}^n; \mathbb{R}^n)$ *such that* $\xi = \nu$ *on* Γ.

Proposition 3 is a well-known result (e.g., [40, Lemma 3.1]). We provide a proof below for the convenience of beginners.

Proof. Since Γ is C^2, for every fixed $x_0 \in \Gamma$, there is an open neighborhood V of x_0 in $\mathbb{R}^n$ and a function $\varphi : V \to \mathbb{R}$ of class C^2 such that $\nabla\varphi(x) \neq 0$ for $x \in V$ and $\varphi(x) = 0$ for $x \in V \cap \Gamma$. Replacing φ by $-\varphi$ if necessary, we may assume that $\nu(x_0) \cdot \nabla\varphi(x_0) > 0$. Choosing V sufficiently small we may assume also that $V \cap \Gamma$ is connected. Then the function $\eta : V \to \mathbb{R}^n$ defined by $\eta \overset{\triangle}{=} \nabla\varphi/|\nabla\varphi|$ is of class C^1 and $\eta = \nu$ on $V \cap \Gamma$.

Since Γ is compact, it can be covered by a finite number of neighborhoods $V_1, \cdots, V_m$, which are constructed as the above V. Denote by $\eta_1, \cdots, \eta_m$ the corresponding functions constructed as the above η. We have

$$\begin{cases} \Gamma \subset \displaystyle\bigcup_{j=1}^{m} V_j, \\ \eta_j = \nu \text{ on } V_j \cap \Gamma, & \text{for } j = 1, \cdots, m. \end{cases}$$

Let us choose a $V_0 \subset\subset G$ such that

$$\overline{G} \subset \bigcup_{j=0}^{m} V_j$$

and define $\eta_0 : V_0 \to \mathbb{R}^n$ by $\eta_0(x) = 0$ for $x \in V_0$.

Let $\alpha_0, \cdots, \alpha_m$ be a partition of unity of class C^2, corresponding to the covers $V_0, \cdots, V_m$ of G. Then, the vector field $\xi = \sum_{j=0}^{m} \alpha_j \eta_j$ is as desired. $\square$

Put

$$\begin{cases} r_1 = |b_1|_{L^\infty_{\mathbb{F}}(0,T;L^\infty(G))} + |b_2|_{L^\infty_{\mathbb{F}}(0,T;L^\infty(G;\mathbb{R}^n))} + |b_4|_{L^\infty_{\mathbb{F}}(0,T;L^\infty(G))}, \\ r_2 = |b_3|_{L^\infty_{\mathbb{F}}(0,T;L^p(G))}. \end{cases} \quad (4.11)$$

Proposition 4. *For any solution of the equation* (4.4), *it holds that*

$$\left| \frac{\partial z}{\partial \nu} \right|_{L^2_{\mathbb{F}}(0,T;L^2(\Gamma_0))}$$

$$\leq e^{\mathcal{C}(r_1^2 + r_2^2 + 1)} \Big(|(z_0, z_1)|_{L^2_{\mathcal{F}_0}(\Omega;H_0^1(G) \times L^2(G))} + |f|_{L^2_{\mathbb{F}}(0,T;L^2(G))} \quad (4.12)$$

$$+ |g|_{L^2_{\mathbb{F}}(0,T;L^2(G))} \Big).$$

Remark 9. By means of Proposition 4, we know that $\left| \frac{\partial y}{\partial \nu} \right|^2_{L^2_{\mathbb{F}}(0,T;L^2(\Gamma_0))}$ makes sense. Compared with Theorem 8, Proposition 4 tells us that $\left| \frac{\partial z}{\partial \nu} \right|^2_{L^2_{\mathbb{F}}(0,T;L^2(\Gamma_0))}$ can be bounded by the initial datum and non-homogenous terms. This result is the converse of Theorem 8 in some sense.

Proof. For any

$$\eta \stackrel{\triangle}{=} (\eta_1, \cdots, \eta_n) \in C^1(\mathbb{R}_t \times \mathbb{R}_x^n; \mathbb{R}^n),$$

by direct computation, we can show that

$$-\sum_{k=1}^{n} \left[2(\eta \cdot \nabla z) \sum_{j=1}^{n} b^{jk} z_{x_j} + \eta_k \left(z_t^2 - \sum_{j,k'=1}^{n} b^{jk'} z_{x_j} z_{x_{k'}} \right) \right]_{x_k} dt$$

$$= 2 \Big\{ \Big[dz_t - \sum_{j,k=1}^{n} \left(b^{jk} z_{x_j} \right)_{x_k} dt \Big] \eta \cdot \nabla z - d(z_t \eta \cdot \nabla z) + z_t \eta_t \cdot \nabla z \, dt$$

$$- \sum_{j,k,k'=1}^{n} b^{jk} z_{x_j} z_{x_{k'}} \eta_{k',x_k} dt \Big\} - (\nabla \cdot \eta) z_t^2 dt + \sum_{j,k=1}^{n} z_{x_k} z_{x_j} \nabla \cdot (b^{jk} \eta) dt.$$

$$(4.13)$$

Since $\Gamma \in C^2$, by Proposition 3, one can find a vector field $\xi = (\xi_1, \cdots, \xi_n) \in C^1(\mathbb{R}^n; \mathbb{R}^n)$ such that $\xi = \nu$ on Γ. Setting $\eta = \xi$ in the equality (4.13), integrating (4.13) in Q, taking expectation in Ω and integrating by parts, we get the inequality (4.12) immediately. □

As an immediate corollary of Proposition 2, we have the following result.

Corollary 2. *For any solution of the equation (4.2), it holds that*

$$\left|\frac{\partial y}{\partial \nu}\right|_{L_{\mathcal{F}}^2(0,T;L^2(\Gamma_0))} \leq C|(y_0,y_1)|_{L_{\mathcal{F}_0}^2(\Omega;H_0^1(G) \times L^2(G))}. \qquad (4.14)$$

Next, we give an energy estimate for the equation (4.4).

Proposition 5. *For any z solving the equation (4.4), it holds that*

$$\mathbb{E}\int_G \left(|z_t(t,x)|^2 + |\nabla z(t,x)|^2\right)dx$$
$$\leq e^{C\left(r_1^2 + r_2^{\frac{1}{2-n/p}} + 1\right)T}\mathbb{E}\int_G \left(|z_t(s,x)|^2 + |\nabla z(s,x)|^2\right)dx \qquad (4.15)$$
$$+ C\mathbb{E}\int_0^T\int_G \left[f(\tau,x)^2 + g(\tau,x)^2\right]dxd\tau,$$

for any $0 \leq s, t \leq T$.

Proof. Without loss of generality, we assume that $t \leq s$. Let

$$\mathcal{E}(t) = \mathbb{E}\int_G \left(|z_t(t,x)|^2 + |\nabla z(t,x)|^2 + r_2^{\frac{2}{2-n/p}}|z(t,x)|^2\right)dx.$$

It follows from Poincaré's inequality that

$$\mathbb{E}\int_G \left(|z_t(t)|^2 + |\nabla z(t)|^2\right)dx$$
$$\leq \mathcal{E}(t) \leq C\left(r_2^{\frac{2}{2-n/p}} + 1\right)\mathbb{E}\int_G \left(|z_t(t)|^2 + |\nabla z(t)|^2\right)dx. \qquad (4.16)$$

By Itô's formula, we have

$$d(z_t^2) = 2z_t dz_t + |dz_t|^2,$$

which implies that

$$\mathbb{E}\int_G \left(|z_t(s,x)|^2 + r_2^{\frac{2}{2-n/p}}|z(s,x)|^2\right)dx$$
$$- \mathbb{E}\int_G \left(|z_t(t,x)|^2 + r_2^{\frac{2}{2-n/p}}|z(t,x)|^2\right)dx$$

$$= -2\mathbb{E}\int_t^s \int_G \sum_{j,k=1}^n b^{jk} z_{x_j} z_{x_k t}\,dx d\tau \tag{4.17}$$

$$+2\mathbb{E}\int_t^s \int_G z_t\big(b_1 z_t + b_2 \cdot \nabla z + b_3 z + f\big)dx d\tau$$

$$+\mathbb{E}\int_t^s \int_G \big|b_4 z + g\big|^2 dx d\tau + 2r_2^{\frac{2}{2-n/p}}\mathbb{E}\int_t^s \int_G z_t z\,dx d\tau.$$

Therefore, we obtain that

$$\mathbb{E}\int_G \Big[|z_t(s,x)|^2 + \sum_{j,k=1}^n b^{jk}(x) z_{x_j}(s,x) z_{x_k}(s,x)\Big]dx$$

$$-\mathbb{E}\int_G \Big[|z_t(t,x)|^2 + \sum_{j,k=1}^n b^{jk}(x) z_{x_j}(t,x) z_{x_k}(t,x)\Big]dx$$

$$= \mathbb{E}\int_t^s \int_G z_t\big(b_1 z_t + b_2 \cdot \nabla z + b_3 z + f\big)dx d\tau$$

$$+\mathbb{E}\int_t^s \int_G \big|b_4 z + g\big|^2 dx d\tau + 2r_2^{\frac{2}{2-n/p}}\mathbb{E}\int_t^s \int_G z_t z\,dx d\tau \tag{4.18}$$

$$\leq C(r_1^2+1)\mathbb{E}\int_t^s \int_G \big(z_t^2 + |\nabla z|^2 + z^2\big)dx d\tau + 2r_2^{\frac{2}{2-n/p}}\mathbb{E}\int_t^s \int_G z_t z\,dx d\tau$$

$$+\mathbb{E}\int_t^s \int_G b_3 z z_t\,dx d\tau + 2\mathbb{E}\int_t^s \int_G \big(f^2 + g^2\big)dx d\tau.$$

Put $p_1 = \frac{2p}{n-2}$ and $p_2 = \frac{2p}{p-n}$. It is easy to check that

$$\frac{1}{p} + \frac{1}{p_1} + \frac{1}{p_2} + \frac{1}{2} = 1$$

and

$$\frac{1}{2(n/p)^{-1}} + \frac{1}{2(1-n/p)^{-1}} + \frac{1}{2} = 1.$$

By Hölder's inequality and Sobolev's embedding theorem, we find that

$$\Big|\mathbb{E}\int_G b_3(\tau,x) z(\tau,x) z_t(\tau,x)\,dx\Big|$$

$$\leq \mathbb{E}\int_G |b_3(\tau,x)||z(\tau,x)|^{\frac{n}{p}}|z(\tau,x)|^{1-\frac{n}{p}}|z_t(\tau,x)|\,dx$$

$$\leq r_2\mathbb{E}\Big(\big||z(\tau,\cdot)|^{\frac{n}{p}}\big|_{L^{p_1}(G)}\big||z(\tau,\cdot)|^{1-\frac{n}{p}}\big|_{L^{p_2}(G)}\big|z_t(\tau,\cdot)\big|_{L^2(G)}\Big) \tag{4.19}$$

$$= r_2\mathbb{E}\Big(\big|z(\tau,\cdot)\big|^{\frac{n}{p}}_{L^{\frac{n}{n-2}}(G)}\big|z(\tau,\cdot)\big|^{1-\frac{n}{p}}_{L^2(G)}\big|z_t(\tau,\cdot)\big|_{L^2(G)}\Big)$$

$$= r_2^{\frac{1}{2-n/p}}\mathbb{E}\Big(\big|z(\tau,\cdot)\big|^{\frac{n}{p}}_{L^{\frac{n}{n-2}}(G)} r_2^{\frac{1-n/p}{2-n/p}}\big|z(\tau,\cdot)\big|^{1-\frac{n}{p}}_{L^2(G)}\big|z_t(\tau,\cdot)\big|_{L^2(G)}\Big).$$

Since

$$|z(\tau,\cdot)|_{L^{\frac{n}{n-2}}(G)}^{\frac{n}{p}}$$
$$\leq \left[\int_G \left(|z_t(\tau,x)|^2 + |\nabla z(\tau,x)|^2 + r_2^{\frac{2}{2-n/p}}|z(\tau,x)|^2\right)dx\right]^{\frac{n}{2p}},$$
$$r_2^{\frac{1-n/p}{2-n/p}}|z(\tau,\cdot)|_{L^2(G)}^{1-\frac{n}{p}}$$
$$\leq \left[\int_G \left(|z_t(\tau,x)|^2 + |\nabla z(\tau,x)|^2 + r_2^{\frac{2}{2-n/p}}|z(\tau,x)|^2\right)dx\right]^{\frac{1}{2}-\frac{n}{2p}},$$

and

$$|z_t(\tau,\cdot)|_{L^2(G)} \leq \left[\int_G \left(|z_t(\tau,x)|^2 + |\nabla z(\tau,x)|^2 + r_2^{\frac{2}{2-n/p}}|z(\tau,x)|^2\right)dx\right]^{\frac{1}{2}},$$

from (4.19), we get that

$$\left|\mathbb{E}\int_G b_3(\tau,x)z(\tau,x)z_t(\tau,x)dx\right| \leq r_2^{\frac{1}{2-n/p}}\mathcal{E}(\tau). \tag{4.20}$$

By a similar argument, we obtain that

$$r_2^{\frac{2}{2-n/p}}\mathbb{E}\int_G z(\tau,x)z_t(\tau,x)dx$$
$$\leq \frac{1}{2}r_2^{\frac{1}{2-n/p}}\mathbb{E}\int_G \left(r_2^{\frac{2}{2-n/p}}z(\tau,x)^2 + z_t(\tau,x)^2\right)dx \tag{4.21}$$
$$\leq \frac{1}{2}r_2^{\frac{1}{2-n/p}}\mathcal{E}(\tau).$$

From (4.20), (4.21) and (4.1), we find that

$$\mathcal{E}(t) \leq C\left[\mathcal{E}(s) + \left(r_1^2 + r_2^{\frac{1}{2-n/p}} + 1\right)\int_t^s \mathcal{E}(\tau)d\tau \right.$$
$$\left. + \mathbb{E}\int_t^s \int_G (f^2 + g^2)dxdt\right]. \tag{4.22}$$

This, together with backward Gronwall's inequality, implies that

$$\mathcal{E}(t) \leq e^{C\left(r_1^2 + r_2^{\frac{1}{2-n/p}} + 1\right)(s-t)}\mathcal{E}(s) + C\mathbb{E}\int_t^s \int_G (f(\tau,x)^2 + g(\tau,x)^2)dxdt. \tag{4.23}$$

By (4.16) and (4.23), we get

$$\mathbb{E}\int_G \left(|z_t(t,x)|^2 + |\nabla z(t,x)|^2\right)dx$$
$$\leq Ce^{C\left(r_1^2 + r_2^{\frac{1}{2-n/p}} + 1\right)(s-t)}\mathbb{E}\int_G \left(|z_t(s,x)|^2 + |\nabla z(s,x)|^2\right)dx$$
$$+ C\mathbb{E}\int_t^s \int_G (f^2 + g^2)dxdt,$$

which leads to the inequality (4.15) immediately. $\qquad\square$

We shall use the stochastic version of Carleman estimate for stochastic hyperbolic equations to solve our state observation problem. To this end, we introduce the following known result.

Lemma 1. [26, Lemma 2.1] *Let $\phi \in C^1((0,T) \times \mathbb{R}^n)$, $p^{jk} = p^{kj} \in C^1(\mathbb{R}^n)$ for $j, k = 1, 2, \cdots, n$, and $\ell, \Psi \in C^2((0,T) \times \mathbb{R}^n)$. Assume u is an $H^2(G)$-valued $\mathbb{F}$-adapted process such that u_t is an $L^2(G)$-valued Itô process. Set $\theta = e^\ell$ and $v = \theta u$. Then, for a.e. $x \in G$ and $\mathbb{P}$-a.s. $\omega \in \Omega$,[a]*

$$\theta\left(-2\phi\ell_t v_t + 2\sum_{j,k=1}^n p^{jk}\ell_{x_j}v_{x_k} + \Psi v\right)\left[\phi du_t - \sum_{j,k=1}^n (p^{jk}u_{x_j})_{x_k}dt\right]$$

$$+\sum_{j,k=1}^n\left[\sum_{j',k'=1}^n\left(2p^{jk}p^{j'k'}\ell_{x_{j'}}v_{x_j}v_{x_{k'}} - p^{jk}p^{j'k'}\ell_{x_j}v_{x_{j'}}v_{x_{k'}}\right)\right.$$

$$\left.-2p^{jk}\ell_t v_{x_j}v_t + \phi p^{jk}\ell_{x_j}v_t^2 + \Psi p^{jk}v_{x_j}v - \left(\mathcal{A}\ell_{x_j} + \frac{1}{2}\Psi_{x_j}\right)p^{jk}v^2\right]_{x_k}dt$$

$$+d\left[\phi\sum_{j,k=1}^n p^{jk}\ell_t v_{x_j}v_{x_k} - 2\phi\sum_{j,k=1}^n p^{jk}\ell_{x_j}v_{x_k}v_t + \phi^2\ell_t v_t^2 - \phi\Psi v_t v\right.$$

$$\left.+\left(\phi\mathcal{A}\ell_t + \frac{1}{2}(\phi\Psi)_t\right)v^2\right] \tag{4.24}$$

$$= \left\{\left[(\phi^2\ell_t)_t + \sum_{j,k=1}^n (\phi p^{jk}\ell_{x_j})_{x_k} - \phi\Psi\right]v_t^2\right.$$

$$-2\sum_{j,k=1}^n [(\phi p^{jk}\ell_{x_k})_t + p^{jk}(\phi\ell_t)_{x_k}]v_{x_j}v_t + \sum_{j,k=1}^n c^{jk}v_{x_j}v_{x_k}$$

$$\left.+Bv^2 + \left(-2\phi\ell_t v_t + 2\sum_{j,k=1}^n p^{jk}\ell_{x_j}v_{x_k} + \Psi v\right)^2\right\}dt + \phi^2\theta^2\ell_t(du_t)^2,$$

where

$$\begin{cases} \mathcal{A} \overset{\triangle}{=} \phi(\ell_t^2 - \ell_{tt}) - \sum_{j,k=1}^n (p^{jk}\ell_{x_j}\ell_{x_k} - p_{x_k}^{jk}\ell_{x_j} - p^{jk}\ell_{x_jx_k}) - \Psi, \\[2mm] \mathcal{B} \overset{\triangle}{=} \mathcal{A}\Psi + (\phi\mathcal{A}\ell_t)_t - \sum_{j,k=1}^n (\mathcal{A}p^{jk}\ell_{x_j})_{x_k} + \frac{1}{2}\left[(\phi\Psi)_{tt} - \sum_{j,k=1}^n (p^{jk}\Psi_{x_j})_{x_k}\right], \\[2mm] c^{jk} = (\phi p^{jk}\ell_t)_t + \sum_{j',k'=1}^n \left[2p^{jk'}(p^{j'k}\ell_{x_{j'}})_{x_{k'}} - (p^{jk}p^{j'k'}\ell_{x_{j'}})_{x_{k'}}\right] + \Psi p^{jk}. \end{cases}$$

$$\tag{4.25}$$

---

[a]See Remark 2 for the notation $(du_t)^2$.

Proof of Lemma 1. By $v(t,x) = \theta(t,x)u(t,x)$, we have

$$u_t = \theta^{-1}\big(v_t - \ell_t v\big), \quad u_{x_j} = \theta^{-1}\big(v_{x_j} - \ell_{x_j} v\big), \quad j = 1, 2, \cdots, n.$$

Then we have

$$\phi du_t = \phi d\big[\theta^{-1}(v_t - \ell_t v)\big] = \phi\theta^{-1}\big[dv_t - 2\ell_t v_t dt + \big(\ell_t^2 - \ell_{tt}\big)v dt\big]. \quad (4.26)$$

Moreover, we find that

$$\sum_{j,k=1}^{n} \big(p^{jk}u_{x_j}\big)_{x_k}$$

$$= \sum_{j,k=1}^{n} \Big[p^{jk}\theta^{-1}\big(v_{x_j} - \ell_{x_j} v\big)\Big]_{x_k}$$

$$= \theta^{-1}\sum_{j,k=1}^{n} \Big[\big(p^{jk}v_{x_j}\big)_{x_k} - 2p^{jk}\ell_{x_j}v_{x_k} + \big(p^{jk}\ell_{x_j}\ell_{x_k} - p^{jk}_{x_k}\ell_{x_j} - p^{jk}\ell_{x_j x_k}\big)v\Big].$$

$$(4.27)$$

As an immediate result of (4.26) and (4.27), we have that

$$\phi du_t - \sum_{j,k=1}^{n} \big(p^{jk}u_{x_j}\big)_{x_k} dt$$

$$= \theta^{-1}\bigg\{\Big[\phi dv_t - \sum_{j,k=1}^{n} \big(p^{jk}v_{x_j}\big)_{x_k} dt\Big] + \Big(-2\phi\ell_t v_t + 2\sum_{j,k=1}^{n} p^{jk}\ell_{x_j}v_{x_k}\Big)dt$$

$$+ \Big[\phi\big(\ell_t^2 - \ell_{tt}\big) - \sum_{j,k=1}^{n} \big(p^{jk}\ell_{x_j}\ell_{x_k} - p^{jk}_{x_k}\ell_{x_j} - p^{jk}\ell_{x_j x_k}\big)\Big]v dt\bigg\}. \quad (4.28)$$

Therefore, by (4.28) and the definition of $\mathcal{A}$ in (4.25), we get that

$$\theta\Big(-2\phi\ell_t v_t + 2\sum_{j,k=1}^{n} p^{jk}\ell_{x_j}v_{x_k} + \Psi v\Big)\Big[\phi du_t - \sum_{j,k=1}^{n} \big(p^{jk}u_{x_j}\big)_{x_k} dt\Big]$$

$$= \Big(-2\phi^2\ell_t v_t + 2\phi\sum_{j,k=1}^{n} p^{jk}\ell_{x_j}v_{x_k} + \phi\Psi v\Big)dv_t \quad (4.29)$$

$$+ \Big(-2\phi\ell_t v_t + 2\sum_{j,k=1}^{n} p^{jk}\ell_{x_j}v_{x_k} + \Psi v\Big)\Big[-\sum_{j,k=1}^{n} \big(p^{jk}v_{x_j}\big)_{x_k} + \mathcal{A}v\Big]dt$$

$$+ \Big(-2\phi\ell_t v_t + 2\sum_{j,k=1}^{n} p^{jk}\ell_{x_j}v_{x_k} + \Psi v\Big)^2 dt.$$

Noting that

$$-2\phi^2 \ell_t v_t dv_t = d\left(-\phi^2 \ell_t v_t^2\right) + \phi^2 \ell_t \left(dv_t\right)^2 + \left(\phi^2 \ell_t\right)_t v_t^2 dt,$$

$$2\phi \sum_{j,k=1}^{n} p^{jk} \ell_{x_j} v_{x_k} dv_t = d\left(2\phi v_t \sum_{j,k=1}^{n} p^{jk} \ell_{x_j} v_{x_k}\right) - 2 \sum_{j,k=1}^{n} \left(\phi p^{jk} \ell_{x_j}\right)_t v_{x_k} v_t dt$$

$$-2\phi \sum_{j,k=1}^{n} p^{jk} \ell_{x_j} v_{x_j t} v_t dt,$$

and

$$\phi \Psi v dv_t = d\left(\Psi \phi v v_t\right) - \left(\phi \Psi\right)_t v v_t dt - \phi \Psi v_t^2 dt,$$

we get that

$$\left(-2\phi \ell_t v_t + 2 \sum_{j,k=1}^{n} p^{jk} \ell_{x_j} v_{x_k} + \Psi v\right) \phi dv_t$$

$$= d\left[-\phi^2 \ell_t v_t^2 + 2\phi v_t \sum_{j,k=1}^{n} p^{jk} \ell_{x_j} v_{x_k} + \phi \Psi v v_t - \frac{1}{2}(\phi \Psi)_t v^2\right] \qquad (4.30)$$

$$-\left\{\sum_{j,k=1}^{n} \left(\phi p^{jk} \ell_{x_j} v_t^2\right)_{x_k} - \left[(\phi \ell_t)_t + \sum_{j,k=1}^{n} \left(\phi p^{jk} \ell_{x_j}\right)_{x_k} - \phi \Psi\right] v_t^2\right.$$

$$\left. + 2 \sum_{j,k=1}^{n} \left(\phi p^{jk} \ell_{x_j}\right)_t v_{x_k} v_t - \frac{1}{2}(\phi \Psi)_{tt} v^2\right\} dt + \phi^2 \ell_t \left(dv_t\right)^2.$$

In a similar manner, for the second term in the right-hand side of (4.29), we find that

$$-2\phi \ell_t v_t \left[-\sum_{j,k=1}^{n} \left(p^{jk} v_{x_j}\right)_{x_k} + \mathcal{A}v\right]$$

$$= 2\left[\sum_{j,k=1}^{n} \left(\phi p^{jk} \ell_t v_{x_j} v_t\right)_{x_k} - \sum_{j,k=1}^{n} p^{jk} \left(\phi \ell_t\right)_{x_k} v_{x_j} v_t\right] \qquad (4.31)$$

$$+ \sum_{j,k=1}^{n} \left(\phi p^{jk} \ell_t\right)_t v_{x_j} v_{x_k} - \left(\phi \sum_{j,k=1}^{n} p^{jk} \ell_t v_{x_j} v_{x_k} + \phi \mathcal{A} \ell_t v^2\right)_t + \left(\phi \mathcal{A} \ell_t\right)_t v^2,$$

$$2 \sum_{j,k=1}^{n} p^{jk} \ell_{x_j} v_{x_k} \left[-\sum_{j,k=1}^{n} \left(p^{jk} v_{x_j}\right)_{x_k} + \mathcal{A}v\right]$$

$$= -\sum_{j,k=1}^{n} \left[\sum_{j',k'=1}^{n} \left(2 p^{jk} p^{j'k'} \ell_{x_{j'}} v_{x_j} v_{x_{k'}} - p^{jk} p^{j'k'} \ell_{x_j} v_{x_{j'}} v_{x_{k'}}\right) - \mathcal{A} p^{jk} \ell_{x_j} v^2\right]_{x_k}$$

$$+ \sum_{j,k,j',k'=1}^{n} \left[2p^{jk'}(p^{j'k}\ell_{x_{j'}})_{x_{k'}} - (p^{jk}p^{j'k'}\ell_{x_{j'}})_{x_{k'}} \right] v_{x_j} v_{x_k} \qquad (4.32)$$

$$- \sum_{j,k=1}^{n} (Ap^{jk}\ell_{x_j})_{x_k} v^2,$$

and

$$\Psi v \left[- \sum_{j,k=1}^{n} (p^{jk} v_{x_j})_{x_k} + Av \right]$$

$$= - \sum_{j,k=1}^{n} \left(\Psi p^{jk} v v_{x_j} - \frac{1}{2} \Psi_{x_j} p^{jk} v^2 \right)_{x_k} + \Psi \sum_{j,k=1}^{n} p^{jk} v_{x_j} v_{x_k} \qquad (4.33)$$

$$+ \left[- \frac{1}{2} \sum_{j,k=1}^{n} (p^{jk} \Psi_{x_j})_{x_k} + A\Psi \right] v^2.$$

Finally, from (4.29) to (4.33), we arrive at the desired equality (4.24).
$\square$

4.3. *State observation for unknown initial data*

We first give a boundary observability estimate for the equation (4.4).

Theorem 10. *Let Conditions 4.1 and 4.2 be satisfied. Then any solution to the equation (4.4) satisfies that*

$$|(z_0, z_1)|_{L^2_{\mathcal{F}_0}(\Omega; H^1_0(G) \times L^2(G))}$$

$$\leq e^{C\left(r_1^2 + r_2^{\frac{1}{3/2-n/p}} + 1\right)} \left(\left| \frac{\partial z}{\partial \nu} \right|_{L^2_{\mathbb{F}}(0,T;L^2(\Gamma_0))} + |f|_{L^2_{\mathbb{F}}(0,T;L^2(G))} \right. \qquad (4.34)$$

$$\left. + |g|_{L^2_{\mathbb{F}}(0,T;L^2(G))} \right).$$

Remark 10. An internal observability estimate for the equation (4.4) was established in [41] by a technical proof.

In what follows, for $\lambda \in \mathbb{R}$, we use $O(\lambda^r)$ to denote a function of order λ^r for large λ.

Proof of Theorem 10. We divide the proof into three steps.

Step 1. Let

$$\ell(t, x) = \lambda \left[\psi(x) - c_1 \left(t - \frac{T}{2} \right)^2 \right], \quad (t, x) \in Q.$$

On the one hand, from Condition 4.2, we know that there is an $\varepsilon_1 \in (0, 1/2)$ such that

$$\ell(t, x) \leq \lambda \left(\frac{R_1^2}{2} - \frac{cT^2}{8} \right) < 0,$$

$$\forall (t, x) \in \left[\left(0, \frac{T}{2} - \varepsilon_1 T \right) \bigcup \left(\frac{T}{2} + \varepsilon_1 T, T \right) \right] \times G. \tag{4.35}$$

On the other hand, since

$$\ell \left(\frac{T}{2}, x \right) = \lambda \psi(x) \geq \lambda R_0^2, \qquad \forall x \in G,$$

we can find an $\varepsilon_0 \in (0, \varepsilon_1)$ such that

$$\ell(t, x) \geq \frac{\lambda R_0^2}{2}, \quad \forall (t, x) \in \left(\frac{T}{2} - \varepsilon_0 T, \frac{T}{2} + \varepsilon_0 T \right) \times G. \tag{4.36}$$

Choose a $\chi \in C_0^\infty[0, T]$ satisfying

$$\chi = 1 \text{ in } \left(\frac{T}{2} - \varepsilon_1 T, \frac{T}{2} + \varepsilon_1 T \right). \tag{4.37}$$

Let $u = \chi z$ with z solving the equation (4.4). Then u solves the following equation:

$$\begin{cases} du_t - \displaystyle\sum_{j,k=1}^n (b^{jk} u_{x_j})_{x_k} dt = \left(b_1 u_t + b_2 \cdot \nabla u + b_3 u + \chi f + \alpha \right) dt \\ \qquad\qquad\qquad\qquad\qquad + \left(b_4 u + \chi g \right) dW(t), & \text{in } Q, \\ u = 0, & \text{on } \Sigma, \\ u(0) = u(T) = 0, \ u_t(0) = u_t(T) = 0, & \text{in } G. \end{cases} \tag{4.38}$$

Here $\alpha = \chi_{tt} z + 2 \chi_t z_t - b_1 \chi_t z$.

Step 2. We apply Lemma 1 to the solution of the equation (4.38). In the present case, we choose

$$\phi \equiv 1, \quad p^{jk} = b^{jk}, \quad \Psi = \ell_{tt} + \sum_{j,k=1}^n \left(b^{jk} \ell_{x_j} \right)_{x_k} - c_0 \lambda, \tag{4.39}$$

and then estimate the terms in (4.24) one by one.

We first analyze the terms which stand for the "energy" of the solution. The point is to compute the order of λ in the coefficients of $|v_t|^2$, $|\nabla v|^2$ and $|v|^2$. Clearly, the term for $|v_t|^2$ reads

$$\left[\ell_{tt} + \sum_{j,k=1}^n (b^{jk} \ell_{x_j})_{x_k} - \Psi \right] v_t^2 = c_0 \lambda v_t^2. \tag{4.40}$$

Noting that b^{jk} $(1 \leq j, k \leq n)$ are independent of t and $\ell_{tx_j} = \ell_{x_j t} = 0$, we get that

$$\sum_{j,k=1}^{n} \left[(b^{jk}\ell_{x_j})_t + b^{jk}\ell_{tx_j} \right] v_{x_k} v_t = 0. \tag{4.41}$$

By Conditions 4.1 and 4.2, we have that

$$\sum_{j,k=1}^{n} \left\{ (b^{jk}\ell_t)_t + \sum_{j',k'=1}^{n} \left[2b^{jk'}(b^{j'k}\ell_{x_{j'}})_{x_{k'}} - (b^{jk}b^{j'k'}\ell_{x_{j'}})_{x_{k'}} \right] + \Psi b^{jk} \right\} v_{x_j} v_{x_k}$$

$$= \sum_{j,k=1}^{n} \left\{ 2b^{jk}\ell_{tt} - b^{jk}\lambda c_0 + \sum_{j',k'=1}^{n} \left[2b^{jk'}(b^{j'k}\ell_{x_{j'}})_{x_{k'}} - b^{jk}_{x_{k'}} b^{j'k'}\ell_{x_{j'}} \right] \right\} v_{x_j} v_{x_k}$$

$$\geq \lambda(\mu_0 - 4c_1 - c_0) \sum_{j,k=1}^{n} b^{jk} v_{x_j} v_{x_k}. \tag{4.42}$$

Now we compute the coefficients of v^2. From (4.25) and (4.39), we have that

$$\mathcal{A} = \ell_t^2 - \ell_{tt} - \sum_{j,k=1}^{n} \left[b^{jk}\ell_{x_j}\ell_{x_k} - (b^{jk}\ell_{x_j})_{x_k} \right] - \Psi$$

$$= c_1^2 \lambda^2 (2t - T)^2 + 4c_1 \lambda + c_0 \lambda - \sum_{j,k=1}^{n} b^{jk}\ell_{x_j}\ell_{x_k}$$

$$= \lambda^2 \left[c_1^2 (2t - T)^2 - \sum_{j,k=1}^{n} b^{jk}\psi_{x_j}\psi_{x_k} \right] + O(\lambda).$$

This, together with (4.25) and (4.39), implies that

$$\mathcal{B} = \mathcal{A}\Psi + (\mathcal{A}\ell_t)_t - \sum_{j,k=1}^{n} (\mathcal{A}b^{jk}\ell_{x_j})_{x_k} + \frac{1}{2}\sum_{j,k=1}^{n} \left[\Psi_{tt} - (b^{jk}\Psi_{x_j})_{x_k} \right]$$

$$= 2\mathcal{A}\ell_{tt} - \lambda c_0 \mathcal{A} - \sum_{j,k=1}^{n} b^{jk}\ell_{x_j}\mathcal{A}_{x_k} + \mathcal{A}_t \ell_t$$

$$- \frac{1}{2}\sum_{j,k=1}^{n}\sum_{j',k'=1}^{n} \left[b^{jk}(b^{j'k'}\ell_{x_{j'}})_{x_{k'}x_j} \right]_{x_k}$$

$$= 2\lambda^3 \left[-2c_1^3(2t - T)^2 + 2c_1 \sum_{j,k=1}^{n} b^{jk}\psi_{x_j}\psi_{x_k} \right] - \lambda^3 c_0 c_1^2 (2t - T)^2 \tag{4.43}$$

$$+ \lambda^3 c_0 \sum_{j,k=1}^{n} b^{jk}\psi_{x_j}\psi_{x_k} + \lambda^3 \sum_{j,k=1}^{n}\sum_{j',k'=1}^{n} b^{jk}\psi_{x_j}(b^{j'k'}\psi_{x_{j'}}\psi_{x_{k'}})_{x_k}$$

$$-4\lambda^3 c_1^3 (2t - T)^2 + O(\lambda^2)$$

$$= (4c_1 + c_0)\lambda^3 \sum_{j,k=1}^{n} b^{jk}\psi_{x_j}\psi_{x_k} + \lambda^3 \sum_{j,k=1}^{n} \sum_{j',k'=1}^{n} b^{jk}\psi_{x_j}\left(b^{j'k'}\psi_{x_{j'}}\psi_{x_{k'}}\right)_{x_k}$$

$$-(8c_1^3 + c_0 c_1^2)\lambda^3 (2t - T)^2 + O(\lambda^2).$$

Now we estimate $\sum_{j,k=1}^{n} \sum_{j',k'=1}^{n} b^{jk}\psi_{x_j}\left(b^{j'k'}\psi_{x_{j'}}\psi_{x_{k'}}\right)_{x_k}$. From Condition 4.1, we get that

$$\mu_0 \sum_{j,k=1}^{n} b^{jk}\psi_{x_j}\psi_{x_k}$$

$$\leq \sum_{j,k=1}^{n} \sum_{j',k'=1}^{n} \left[2b^{jk'}\left(b^{j'k}\psi_{x_{j'}}\right)_{x_{k'}} - b^{jk}_{x_{k'}} b^{j'k'}\psi_{x_{j'}}\right]\psi_{x_j}\psi_{x_k}$$

$$= \sum_{j,k=1}^{n} \sum_{j',k'=1}^{n} \left(2b^{jk'}b^{j'k}_{x_{k'}}\psi_{x_{j'}} + 2b^{jk'}b^{j'k}\psi_{x_{j'}x_{k'}} - b^{jk}_{x_{k'}} b^{j'k'}\psi_{x_{j'}}\right)\psi_{x_j}\psi_{x_k}$$

$$= \sum_{j,k=1}^{n} \sum_{j',k'=1}^{n} \left(2b^{jk'}b^{j'k}_{x_{k'}}\psi_{x_{j'}}\psi_{x_j}\psi_{x_k} + 2b^{jk'}b^{j'k}\psi_{x_{j'}x_k}\psi_{x_j}\psi_{x_k}\right. \tag{4.44}$$

$$\left. -b^{jk}_{x_{k'}} b^{j'k'}\psi_{x_{j'}}\psi_{x_j}\psi_{x_k}\right)$$

$$= \sum_{j,k=1}^{n} \sum_{j',k'=1}^{n} \left(b^{j'k'}b^{jk}_{x_{k'}}\psi_{x_{j'}}\psi_{x_j}\psi_{x_k} + b^{jk}b^{j'k'}\psi_{x_{j'}x_k}\psi_{x_j}\psi_{x_{k'}}\right.$$

$$\left. +b^{jk}b^{j'k'}\psi_{x_{k'}x_k}\psi_{x_j}\psi_{x_{j'}}\right)$$

$$= \sum_{j,k=1}^{n} \sum_{j',k'=1}^{n} b^{jk}\psi_{x_j}\left(b^{j'k'}\psi_{x_{j'}}\psi_{x_{k'}}\right)_{x_k}.$$

By (4.43), (4.44) and Condition 4.2, we obtain that

$$\mathcal{B} \geq (4c_1 + c_0)\lambda^3 \sum_{j,k=1}^{n} b^{jk}\psi_{x_j}\psi_{x_k} + \mu_0\lambda^3 \sum_{j,k=1}^{n} b^{jk}\psi_{x_j}\psi_{x_k}$$

$$-(8c_1^3 + 2c_0 c_1^2)\lambda^3 (2t - T)^2 + O(\lambda^2)$$

$$\geq (4c_1 + c_0)\lambda^3 \sum_{j,k=1}^{n} b^{jk}\psi_{x_j}\psi_{x_k} + \mu_0\lambda^3 \sum_{j,k=1}^{n} b^{jk}\psi_{x_j}\psi_{x_k}$$

$$-2c_1^2(4c_1 + c_0)\lambda^3 T^2 + O(\lambda^2)$$

$$\geq 2(4c_1 + c_0)\lambda^3 \left(\sum_{j,k=1}^{n} b^{jk}\psi_{x_j}\psi_{x_k} - c_1^2 T^2\right) + O(\lambda^2)$$

$$\geq 2(4c_1 + c_0)\lambda^3 (4R_1^2 - c_1^2 T^2) + O(\lambda^2).$$

Then we know that there exists a $\lambda_1 > 0$ such that for any $\lambda \geq \lambda_1$, we have that

$$\mathcal{B}v^2 \geq 8c_1 (4R_1^2 - c_1^2 T^2) \lambda^3 v^2. \tag{4.45}$$

Since

$$v(0, x) = \theta(0, x)u(0, x) = 0$$

and

$$v_t(0, x) = \theta_t(0, x)u(0, x) + \theta(0, x)u_t(0, x) = 0,$$

at time $t = 0$, we have that

$$\sum_{j,k=1}^{n} b^{jk} \ell_t v_{x_j} v_{x_k} - 2 \sum_{j,k=1}^{n} b^{jk} \ell_{x_j} v_{x_k} v_t$$
$$+\ell_t v_t^2 - \Psi v_t v + \left(A\ell_t + \frac{\Psi_t}{2} \right) v^2 = 0.$$

Likewise, we see that at time $t = T$,

$$\sum_{j,k=1}^{n} b^{jk} \ell_t v_{x_j} v_{x_k} - 2 \sum_{j,k=1}^{n} b^{jk} \ell_{x_j} v_{x_k} v_t$$
$$+\ell_t v_t^2 - \Psi v_t v + \left(A\ell_t + \frac{\Psi_t}{2} \right) v^2 = 0.$$

Step 3. Integrating (4.24) in Q, taking expectation in Ω and by the argument above, we obtain that

$$\mathbb{E} \int_Q \theta \left\{ \left(-2\ell_t v_t + 2 \sum_{j,k=1}^{n} b^{jk} \ell_{x_j} v_{x_k} + \Psi v \right) \right.$$
$$\left. \times \left[du_t - \sum_{j,k=1}^{n} (b^{jk} u_{x_j})_{x_k} dt \right] - \theta \ell_t |du_t|^2 \right\} dx \tag{4.46}$$
$$+\lambda \mathbb{E} \int_\Sigma \sum_{j,k=1}^{n} \sum_{j',k'=1}^{n} \left(2b^{jk} b^{j'k'} \psi_{x_{j'}} v_{x_j} v_{x_{k'}} - b^{jk} b^{j'k'} \psi_{x_j} v_{x_{j'}} v_{x_{k'}} \right) \nu_k d\Sigma$$
$$\geq C\mathbb{E} \int_Q \theta^2 \left[(\lambda u_t^2 + \lambda |\nabla u|^2) + \lambda^3 u^2 \right] dx dt$$
$$+\mathbb{E} \int_Q \left(-2\ell_t v_t + 2 \sum_{j,k=1}^{n} b^{jk} \ell_{x_j} v_{x_k} + \Psi v \right)^2 dx dt.$$

Since $u = v = 0$ on Σ, $\mathbb{P}$-a.s., from (4.9), we have that

$$\mathbb{E}\int_\Sigma \sum_{j,k=1}^n \sum_{j',k'=1}^n \left(2b^{jk}b^{j'k'}\psi_{x_{j'}}v_{x_j}v_{x_{k'}} - b^{jk}b^{j'k'}\psi_{x_j}v_{x_{j'}}v_{x_{k'}}\right)\nu_k d\Sigma$$

$$= \mathbb{E}\int_\Sigma \sum_{j,k=1}^n \sum_{j',k'=1}^n \left(2b^{jk}b^{j'k'}\psi_{x_{j'}}\frac{\partial v}{\partial\nu}\nu_j\frac{\partial v}{\partial\nu}\nu_{k'}\right.$$

$$\left. - b^{jk}b^{j'k'}\psi_{x_j}\frac{\partial v}{\partial\nu}\nu_{j'}\frac{\partial v}{\partial\nu}\nu_{k'}\right)\nu_k d\Sigma$$

$$= \mathbb{E}\int_\Sigma \left(\sum_{j,k=1}^n b^{jk}\nu_j\nu_k\right)\left(\sum_{j',k'=1}^n b^{j'k'}\psi_{x_{j'}}\nu_{k'}\right)\left|\frac{\partial v}{\partial\nu}\right|^2 d\Sigma \qquad (4.47)$$

$$= \mathbb{E}\int_\Sigma \left(\sum_{j,k=1}^n b^{jk}\nu_j\nu_k\right)\left(\sum_{j',k'=1}^n b^{j'k'}\psi_{x_{j'}}\nu_{k'}\right)\left|\theta\frac{\partial u}{\partial\nu} + u\frac{\partial\theta}{\partial\nu}\right|^2 d\Sigma$$

$$= \mathbb{E}\int_\Sigma \left(\sum_{j,k=1}^n b^{jk}\nu_j\nu_k\right)\left(\sum_{j',k'=1}^n b^{j'k'}\psi_{x_{j'}}\nu_{k'}\right)\theta^2\left|\frac{\partial u}{\partial\nu}\right|^2 d\Sigma$$

$$\leq \mathbb{E}\int_{\Sigma_0} \left(\sum_{j,k=1}^n b^{jk}\nu_j\nu_k\right)\left(\sum_{j',k'=1}^n b^{j'k'}\psi_{x_{j'}}\nu_{k'}\right)\theta^2\left|\frac{\partial u}{\partial\nu}\right|^2 d\Sigma.$$

By (4.46) and (4.47), we obtain that

$$\mathbb{E}\int_Q \theta\left\{\left(-2\ell_t v_t + 2\sum_{j,k=1}^n b^{jk}\ell_{x_j}v_{x_k} + \Psi v\right)\left[du_t - \sum_{j,k=1}^n \left(b^{jk}u_{x_j}\right)_{x_k} dt\right]\right.$$

$$\left. -\theta\ell_t|du_t|^2\right\}dx + \lambda\mathbb{E}\int_{\Sigma_0} \theta^2\left(\sum_{j,k=1}^n b^{jk}\nu_j\nu_k\right)\left(\sum_{j',k'=1}^n b^{j'k'}\psi_{x_{j'}}\nu_{k'}\right)\left|\frac{\partial u}{\partial\nu}\right|^2 d\Sigma$$

$$\geq C\mathbb{E}\int_Q \left[\theta^2\left(\lambda u_t^2 + \lambda|\nabla u|^2\right) + \lambda^3\theta^2 u^2\right]dxdt \qquad (4.48)$$

$$+ \mathbb{E}\int_Q \left(-2\ell_t v_t + 2\sum_{j,k=1}^n b^{jk}\ell_{x_j}v_{x_k} + \Psi v\right)^2 dxdt.$$

Since u solves the equation (4.38), we know that

$$\mathbb{E}\int_Q \theta\left\{\left(-2\ell_t v_t + 2\sum_{j,k=1}^n b^{jk}\ell_{x_j}v_{x_k} + \Psi v\right)\left[du_t - \sum_{j,k=1}^n \left(b^{jk}u_{x_j}\right)_{x_k} dt\right]\right.$$

$$\left. -\theta\ell_t(du_t)^2\right\}dx$$

$$= \mathbb{E}\int_Q \theta\left\{\left(-2\ell_t v_t + 2\sum_{j,k=1}^n b^{jk}\ell_{x_j}v_{x_k} + \Psi v\right)\left[b_1 u_t + b_2\cdot\nabla u + b_3 u + \chi f + \alpha\right]\right.$$

$$\left. -\theta\ell_t|b_4 u + \chi g|^2\right\}dxdt \qquad (4.49)$$

$$\leq C\mathbb{E}\int_Q \Big\{\theta^2\big[b_1 u_t + b_2\cdot\nabla u + b_3 u + \chi f + \alpha\big]^2 + \lambda\theta^2|b_4 u + \chi g|^2\Big\}dxdt$$

$$+\mathbb{E}\int_Q\Big(-2\ell_t v_t + \sum_{j,k=1}^n b^{jk}\ell_{x_j}v_{x_k} + \Psi v\Big)^2 dxdt$$

$$\leq C\Big\{\mathbb{E}\int_Q \theta^2(f^2+\alpha^2+\lambda|g|^2)dxdt + |b_1|^2_{L^\infty_{\mathbb{F}}(0,T;L^\infty(G))}\mathbb{E}\int_Q \theta^2 u_t^2 dxdt$$

$$+\mathbb{E}\int_Q \theta^2 b_3^2 u^2 dxdt + |b_2|^2_{L^\infty_{\mathbb{F}}(0,T;L^\infty(G,\mathbb{R}^n))}\mathbb{E}\int_Q \theta^2|\nabla u|^2 dxdt$$

$$+\lambda|b_4|^2_{L^\infty_{\mathbb{F}}(0,T;L^\infty(G))}\mathbb{E}\int_Q \theta^2 u^2 dxdt\Big\}$$

$$+\mathbb{E}\int_Q\Big(-2\ell_t v_t + \sum_{j,k=1}^n b^{jk}\ell_{x_j}v_{x_k} + \Psi v\Big)^2 dxdt.$$

Recalling the definition of r_2 in (4.11), and using successively Hölder's and Sobolev's inequalities, we get

$$|b_3\theta u|^2_{L^2_{\mathbb{F}}(0,T;L^2(G))} \leq r_2|\theta u|^2_{L^2_{\mathbb{F}}(0,T;L^s(G))}$$

$$\leq r_2|\theta u|^2_{L^2_{\mathbb{F}}(0,T;H^{n/p})}\quad\text{for }\frac{1}{p}+\frac{1}{s}=\frac{1}{2}. \tag{4.50}$$

For any $\rho\in L^2_{\mathcal{F}_T}(\Omega;H^1(\mathbb{R}^n))$, by Hölder's inequality, one has

$$|\rho|^2_{L^2_{\mathcal{F}_T}(\Omega;H^{n/p}(\mathbb{R}^n))} = \mathbb{E}\int_{\mathbb{R}^n}(1+|\xi|^2)^{n/p}|\hat{\rho}(\xi)|^{2n/p}|\hat{\rho}(\xi)|^{2(1-n/p)}d\xi$$

$$\leq |\rho|^{2n/p}_{L^2_{\mathcal{F}_T}(\Omega;H^1(\mathbb{R}^n))}|\rho|^{2(1-n/p)}_{L^2_{\mathcal{F}_T}(\Omega;L^2(\mathbb{R}^n))}.$$

Hence, we know that there is a constant $C>0$ such that for any $\tilde{\rho}\in L^2_{\mathcal{F}_T}(\Omega;H^1_0(G))$, we have

$$|\tilde{\rho}|^2_{L^2_{\mathcal{F}_T}(\Omega;H^{n/p}(G))} \leq C|\tilde{\rho}|^{2n/p}_{L^2_{\mathcal{F}_T}(\Omega;H^1_0(G))}|\tilde{\rho}|^{2(1-n/p)}_{L^2_{\mathcal{F}_T}(\Omega;L^2(G))}.$$

Therefore, there is a constant $C>0$ such that for any $\bar{\rho}\in L^2_{\mathbb{F}}(0,T;H^1_0(G))$, it holds that

$$|\bar{\rho}|^2_{L^2_{\mathbb{F}}(0,T;H^{n/p}(G))} \leq C|\bar{\rho}|^{2n/p}_{L^2_{\mathbb{F}}(0,T;H^1_0(G))}|\bar{\rho}|^{2(1-n/p)}_{L^2_{\mathbb{F}}(0,T;L^2(G))}.$$

This, together with the inequality (4.50), implies that

$$|b_3\theta u|^2_{L^2_{\mathbb{F}}(0,T;L^2(G))}$$

$$\leq C|b_3\theta u|^{2n/p}_{L^2_{\mathbb{F}}(0,T;H^1_0(G))}|b_3\theta u|^{2(1-n/p)}_{L^2_{\mathbb{F}}(0,T;L^2(G))} \tag{4.51}$$

$$\leq \varepsilon\lambda|b_3\theta u|^2_{L^2_{\mathbb{F}}(0,T;H^1_0(G))} + C(\varepsilon)r_2^{2p/(p-n)}\lambda^{-n/(p-n)}|b_3\theta u|^2_{L^2_{\mathbb{F}}(0,T;L^2(G))},$$

where ε is small enough and $\mathcal{C}(\varepsilon)$ depends on ε.

Taking

$$\lambda_2 = \max\left\{\lambda_1, \mathcal{C}\left(r_1^2 + r_2^{\frac{1}{3/2-n/p}} + 1\right)\right\},$$

combining (4.48), (4.49) and (4.51), for any $\lambda \geq \lambda_2$, we have that

$$\mathcal{C}\lambda\mathbb{E}\int_{\Sigma_0}\theta^2\left(\sum_{j,k=1}^n b^{jk}\nu_j\nu_k\right)\left(\sum_{j',k'=1}^n b^{j'k'}\psi_{x_j},\nu_{k'}\right)\left|\frac{\partial u}{\partial\nu}\right|^2 d\Sigma$$

$$+\mathcal{C}\mathbb{E}\int_Q \theta^2(f^2 + \alpha^2 + \lambda g^2)dxdt \tag{4.52}$$

$$\geq \mathbb{E}\int_Q \theta^2\left(\lambda u_t^2 + \lambda|\nabla u|^2 + \lambda^3 u^2\right)dxdt.$$

Recalling the property of χ (see (4.37)) and $u = \chi z$, from (4.52), we find that

$$\mathcal{C}(r_1 + 1)\left[\mathbb{E}\int_0^{\frac{T}{2}-\varepsilon_1 T}\int_G \theta^2\left(z_t^2 + |\nabla z|^2 + z^2\right)dxdt\right.$$

$$\left.+\mathbb{E}\int_{\frac{T}{2}+\varepsilon_1 T}^T\int_G \theta^2\left(z_t^2 + |\nabla z|^2 + z^2\right)dxdt\right]$$

$$+\mathcal{C}\lambda\mathbb{E}\int_{\Sigma_0}\left(\sum_{j,k=1}^n b^{jk}\nu_j\nu_k\right)\left(\sum_{j',k'=1}^n b^{j'k'}\psi_{x_j},\nu_{k'}\right)\theta^2\left|\frac{\partial z}{\partial\nu}\right|^2 d\Sigma \tag{4.53}$$

$$+\mathcal{C}\mathbb{E}\int_Q \theta^2(f^2 + \lambda g^2)dxdt$$

$$\geq \mathbb{E}\int_{\frac{T}{2}-\varepsilon_0 T}^{\frac{T}{2}+\varepsilon_0 T}\int_G \theta^2\left(\lambda z_t^2 + \lambda|\nabla z|^2 + \lambda^3 z^2\right)dxdt.$$

Combining (4.15) and (4.53), we know that there is a

$$\lambda_3 = \mathcal{C}\left(r_1^2 + r_2^{\frac{1}{3/2-n/p}} + 1\right) \geq \lambda_2$$

such that for all $\lambda \geq \lambda_3$, it holds that

$$\mathcal{C}\lambda\mathbb{E}\int_{\Sigma_0}\theta^2\left(\sum_{j,k=1}^n b^{jk}\nu_j\nu_k\right)\left(\sum_{j',k'=1}^n b^{j'k'}\psi_{x_j},\nu_{k'}\right)\left|\frac{\partial z}{\partial\nu}\right|^2 d\Sigma$$

$$+\mathcal{C}\mathbb{E}\int_Q \theta^2(f^2 + \lambda g^2)dxdt \tag{4.54}$$

$$\geq e^{-\frac{T^2}{4}\lambda_3}\mathbb{E}\int_G\left(z_1^2 + |\nabla z_0|^2\right)dxdt.$$

Taking $\lambda = \lambda_3$, we obtain that

$$Ce^{\lambda_3 R_1^2} \left[\mathbb{E} \int_{\Sigma_0} \left| \frac{\partial z}{\partial \nu} \right|^2 d\Sigma + \mathbb{E} \int_Q (f^2 + g^2) dx dt \right]$$

$$\geq e^{-\frac{T^2}{4} \lambda_3} \mathbb{E} \int_G (z_1^2 + |\nabla z_0|^2) dx dt. \tag{4.55}$$

This leads to the inequality (4.34) immediately.   □

4.3.1. *Solution to the state observation problem for unknown initial data*

This subsection is devoted to a proof of Theorem 8.

Proof of Theorem 8. Set

$$y = \hat{y} - \tilde{y}.$$

Then y is the solution to (4.2) with

$$
\begin{cases}
b_1 = \displaystyle\int_0^1 \partial_\varrho F(\hat{y}, \tilde{y}_t + s(\hat{y}_t - \tilde{y}_t), \nabla\tilde{y}) ds, \\[2mm]
b_2 = \displaystyle\int_0^1 \partial_\zeta F(\hat{y}, \hat{y}_t, \nabla\tilde{y} + s(\nabla\hat{y} - \nabla\tilde{y})) ds, \\[2mm]
b_3 = \displaystyle\int_0^1 \partial_\eta F(\tilde{y} + s(\hat{y} - \tilde{y}), \hat{y}_t, \nabla y) ds, \\[2mm]
b_4 = \displaystyle\int_0^1 \partial_\eta K(\tilde{y} + s(\hat{y} - \tilde{y})) ds.
\end{cases}
$$

The inequality (4.3) follows from Theorem 10.   □

Next, we give a brief idea on the reconstruction of the state y. Denote by φ the observation on $(0, T) \times \Gamma_0$ and by

$$\mathcal{W} \overset{\triangle}{=} \{ y \in \mathbb{H}_T \,|\, y \text{ solves (4.2) for some initial data}$$
$$(y_0, y_1) \in L_{\mathcal{F}_0}^2 (\Omega; H_0^1(G) \times L^2(G)) \}.$$

Put

$$\mathcal{J}_1(y) = \mathbb{E} \int_0^T \int_{\Gamma_0} \left| \frac{\partial y}{\partial \nu} - \varphi \right|^2 d\Gamma dt, \quad \text{for } y \in \mathcal{W}.$$

Let $\tilde{y}$ be the state of (4.2) corresponding to the observation φ. Then, it is clear that the state $\tilde{y}$ satisfies that

$$\mathcal{J}_1(\tilde{y}) = \min_{y \in \mathcal{W}} \mathcal{J}_1(y) = 0.$$

Hence, the construction of $\tilde{y}$ can be transformed to the following optimization problem:

(**P3**) *Find a* $y \in \mathcal{W}$ *which minimizes* $\mathcal{J}_1(\cdot)$.

Remark 11. In the above formulation, we reduce the reconstruction problems in (**P1**) to the optimization problem (**P3**). However, this reduction is based on the assumption that there is no noise in the observation, which is unrealistic. Once the noise is considered, the Tikhonov functional method should be employed. In this case, one may use the quasi-reversibility method in the form of the Tikhonov functional to handle noise. Such kind of method was formulated for the state observation problem of deterministic hyperbolic equations (e.g., [42]). As far as we know, there is no published works for stochastic PDEs.

4.4. *State observation for three unknowns*

Since the (random) source $g(\cdot)$ appears in the right-hand side of (4.34), one cannot apply Theorem 10 to solve problem (**P2**). A new Carleman estimate for (4.4) that the source term g can be bounded above by the observation should be established. We should avoid employing the usual *energy estimate* because, when applying this sort of estimate to (4.4), the source term g would appear as a "bad" term. Meanwhile, since we also expect to identity the initial data, we need to bound above the initial data by the observation, too. In a word, we should obtain the estimate on the initial data and the source term in the Carleman inequality simultaneously.

4.4.1. *Carleman estimate for stochastic hyperbolic equations*

As mentioned before, we will prove Theorem 9 by establishing a new Carleman estimate for the equation (4.4).

Let us choose

$$\theta = e^{\ell}, \qquad \ell = \lambda\big[\psi(x) - c_1(t - T)^2\big],$$

where $\lambda > 0$ is a parameter, $\psi(\cdot)$ is the function given in Condition 4.1, and c_1 is the constant in Condition 4.3.

Our global Carleman estimate for (4.4) is as follows.

Theorem 11. *Let Conditions* 4.1 *and* 4.3 *hold. Then, there exists a constant* $\tilde{\lambda} > 0$ *such that for any* $\lambda \geq \tilde{\lambda}$ *and any solution* $z \in \mathbb{H}_T$ *of the equation* (4.4) *satisfying* $z(T) = 0$ *in* G, $\mathbb{P}$-*a.s., it holds that*

$$\mathbb{E} \int_G \theta^2 \left(\lambda |z_1|^2 + \lambda |\nabla z_0|^2 + \lambda^3 |z_0|^2 \right) dx + \lambda \mathbb{E} \int_Q (T-t) \theta^2 g^2 \, dx dt$$

$$\leq C \lambda \mathbb{E} \int_0^T \int_{\Gamma_0} \theta^2 \left| \frac{\partial z}{\partial \nu} \right|^2 d\Gamma dt. \tag{4.56}$$

Once Theorem 11 is established, similar to (4.55), one can prove Theorem 9 immediately.

Proof of Theorem 11. In what follows, we shall apply Lemma 1 to the equation (4.4) with

$$\phi = 1, \quad u = z, \quad p^{jk} = b^{jk}, \quad \Psi = \ell_{tt} + \sum_{j,k=1}^{n} (b^{jk} \ell_{x_j})_{x_k} - \lambda c_0$$

(recall Condition 4.3 for the constant c_0), and then estimate the terms in (4.24) one by one.

The proof is divided into three steps.

Step 1. In this step, we analyze the terms which stand for the "energy" of the solution to (4.4).

First, it is clear that the coefficient of $|v_t|^2$ reads:

$$\ell_{tt} + \sum_{j,k=1}^{n} \left(b^{jk} \ell_{x_j} \right)_{x_k} - \Psi = c_0 \lambda. \tag{4.57}$$

Noting that b^{jk} $(1 \leq j, k \leq n)$ are independent of t and $\ell_{tx_k} = \ell_{x_k t} = 0$, we find that

$$\sum_{j,k=1}^{n} \left[\left(b^{jk} \ell_{x_k} \right)_t + b^{jk} \ell_{tx_k} \right] v_{x_j} v_t = 0. \tag{4.58}$$

By Condition 4.1, we see that

$$\sum_{j,k=1}^{n} \left\{ (b^{jk} \ell_t)_t + \Psi b^{jk} + \sum_{j',k'=1}^{n} \left[2 b^{jk'} \left(b^{j'k} \ell_{x_{j'}} \right)_{x_{k'}} - \left(b^{jk} b^{j'k'} \ell_{x_{j'}} \right)_{x_{k'}} \right] \right\} v_{x_j} v_{x_k}$$

$$\geq \lambda (\mu_0 - 4c_1 - c_0) \sum_{j,k=1}^{n} b^{jk} v_{x_j} v_{x_k}. \tag{4.59}$$

Recalling (4.25), we find that

$$\mathcal{A} = \lambda^2 \left[4 c_1^2 (t-T)^2 - \sum_{j,k=1}^{n} b^{jk} \psi_{x_j} \psi_{x_k} \right] + O(\lambda). \tag{4.60}$$

Hence, by the definition of $\mathcal{B}$ (in (4.25)), we conclude that

$$\mathcal{B} = (4c_1 + c_0)\lambda^3 \sum_{j,k=1}^{n} b^{jk} \psi_{x_j} \psi_{x_k} + \lambda^3 \sum_{j,k=1}^{n} \sum_{j',k'=1}^{n} b^{jk} \psi_{x_j} \left(b^{j'k'} \psi_{x_{j'}} \psi_{x_{k'}} \right)_{x_k}$$
$$- 4\left(8c_1^3 + c_0 c_1^2\right)(t-T)^2 \lambda^3 + O(\lambda^2).$$

Similar to (4.44), one has

$$\mu_0 \sum_{j,k=1}^{n} b^{jk} \psi_{x_j} \psi_{x_k} \leq \sum_{j,k=1}^{n} \sum_{j',k'=1}^{n} b^{jk} \psi_{x_j} \left(b^{j'k'} \psi_{x_{j'}} \psi_{x_{k'}} \right)_{x_k}. \tag{4.61}$$

Therefore, by Condition 4.3, we obtain that

$$\mathcal{B} \geq (4c_1 + c_0)\lambda^3 \sum_{j,k=1}^{n} b^{jk} \psi_{x_j} \psi_{x_k} + \mu_0 \lambda^3 \sum_{j,k=1}^{n} b^{jk} \psi_{x_j} \psi_{x_k}$$
$$- 4(8c_1 + c_0)c_1^2 (t-T)^2 \lambda^3 + O(\lambda^2) \tag{4.62}$$
$$= (4c_1 + c_0)\lambda^3 \sum_{j,k=1}^{n} b^{jk} \psi_{x_j} \psi_{x_k} + O(\lambda^2).$$

Hence, there exists a $\tilde{\lambda}_0 > 0$ such that for any $\lambda \geq \tilde{\lambda}_0$, it holds that

$$\mathcal{B}v^2 \geq \mathcal{C}\lambda^3 v^2. \tag{4.63}$$

Step 2. In this step, we analyze the terms corresponding to $t = 0$ and $t = T$. For the time $t = 0$, we have that

$$\sum_{j,k=1}^{n} b^{jk} \ell_t v_{x_j} v_{x_k} - 2 \sum_{j,k=1}^{n} b^{jk} \ell_{x_j} v_{x_k} v_t + \ell_t v_t^2 - \Psi v_t v + \left(\mathcal{A}\ell_t + \frac{1}{2}\Psi_t \right)v^2$$
$$= 2c_1 T\lambda \sum_{j,k=1}^{n} b^{jk} v_{x_j} v_{x_k} - 2\lambda \sum_{j,k=1}^{n} b^{jk} \psi_{x_j} v_{x_k} v_t$$
$$- \lambda \left[-2c_1 + \sum_{j,k=1}^{n} (b^{jk} \psi_{x_j})_{x_k} - c_0 \right] v_t v + 2c_1 T\lambda v_t^2$$
$$+ \left[2c_1 T\left(4c_1^2 T^2 - \sum_{j,k=1}^{n} b^{jk} \psi_{x_j} \psi_{x_k} \right)\lambda^3 + O(\lambda^2) \right]v^2 \tag{4.64}$$
$$\geq 2c_1 T\lambda \sum_{j,k=1}^{n} b^{jk} v_{x_j} v_{x_k} - \lambda \left(\sum_{j,k=1}^{n} b^{jk} \psi_{x_j} \psi_{x_k} \right)^{\frac{1}{2}} \sum_{j,k=1}^{n} b^{jk} v_{x_j} v_{x_k}$$
$$- \lambda \left(\sum_{j,k=1}^{n} b^{jk} \psi_{x_j} \psi_{x_k} \right)^{\frac{1}{2}} v_t^2 + 2c_1 T\lambda v_t^2 - v_t^2$$

$$+\Big[2c_1T\Big(4c_1^2T^2 - \sum_{j,k=1}^{n} b^{jk}\psi_{x_j}\psi_{x_k}\Big)\lambda^3 + O(\lambda^2)\Big]v^2.$$

It follows from Condition 4.3 that

$$4c_1^2T^2 - \sum_{j,k=1}^{n} b^{jk}\psi_{x_j}\psi_{x_k} > 0$$

and that

$$2c_1T - \Big(\sum_{j,k=1}^{n} b^{jk}\psi_{x_j}\psi_{x_k}\Big)^{\frac{1}{2}} > 0.$$

Hence there exists a $\tilde{\lambda}_1 > 0$ such that for any $\tilde{\lambda} \geq \tilde{\lambda}_1$ and when $t = 0$, it holds that

$$\sum_{j,k=1}^{n} b^{jk}\ell_t v_{x_j} v_{x_k} - 2\sum_{j,k=1}^{n} b^{jk}\ell_{x_j} v_{x_k} v_t + \ell_t v_t^2 - \Psi v_t v + \Big(A\ell_t + \frac{1}{2}\Psi_t\Big)v^2$$

$$\geq C\Big[\lambda(v_t^2 + |\nabla v|^2) + \lambda^3 v^2\Big]. \tag{4.65}$$

On the other hand, since $\ell_t(T) = 0$, for $t = T$, it holds that

$$\sum_{j,k=1}^{n} b^{jk}\ell_t v_{x_j} v_{x_k} - 2\sum_{j,k=1}^{n} b^{jk}\ell_{x_j} v_{x_k} v_t + \ell_t v_t^2 - \Psi v_t v + \Big(A\ell_t + \frac{1}{2}\Psi_t\Big)v^2$$

$$= -2\sum_{j,k=1}^{n} b^{jk}\ell_{x_j}v_{x_k}v_t - \Psi v_t v. \tag{4.66}$$

Noting that $z(T) = 0$ in G, $\mathbb{P}$-a.s., we have $v(T) = 0$ and $v_{x_k}(T) = 0$ in G $(j = 1, 2, \cdots, n)$, $\mathbb{P}$-a.s. Thus, from the equality (4.66), we end up with

$$\Big[\sum_{j,k=1}^{n} b^{jk}\ell_t v_{x_j} v_{x_k} - 2\sum_{j,k=1}^{n} b^{jk}\ell_{x_j} v_{x_k} v_t + \ell_t v_t^2 - \Psi v_t v$$

$$+\Big(A\ell_t + \frac{1}{2}\Psi_t\Big)v^2\Big]\Big|_{t=T} = 0, \quad \mathbb{P}\text{-a.s.} \tag{4.67}$$

Step 3. Integrating (4.24) in Q, taking expectation in Ω and by the argument above, for $\tilde{\lambda} \geq \max\{\tilde{\lambda}_0, \tilde{\lambda}_1\}$, we obtain that

$$\mathbb{E}\int_Q \theta\Big\{\Big(-2\ell_t v_t + 2\sum_{j,k=1}^{n} b^{jk}\ell_{x_j}v_{x_k} + \Psi v\Big)\Big[dz_t - \sum_{j,k=1}^{n}(b^{jk}z_{x_j})_{x_k}dt\Big]\Big\}dx$$

$$+\lambda\mathbb{E}\int_\Sigma \sum_{j,k=1}^{n}\sum_{j',k'=1}^{n}\Big(2b^{jk}b^{j'k'}\psi_{x_{j'}}v_{x_j}v_{x_{k'}} - b^{jk}b^{j'k'}\psi_{x_j}v_{x_{j'}}v_{x_{k'}}\Big)\nu_k d\Sigma$$

$$\geq C\Big\{\mathbb{E}\int_Q\Big[\theta^2\Big(\lambda z_t^2+\lambda|\nabla z|^2+\lambda^3 z^2\Big) \tag{4.68}$$

$$+\Big(-2\ell_t v_t+2\sum_{j,k=1}^n b^{jk}\ell_{x_j}v_{x_k}+\Psi v\Big)^2\Big]dxdt$$

$$+\mathbb{E}\int_G\theta^2\Big[\lambda(|\nabla z_0|^2+|z_1|^2)+\lambda^3|z_0|^2\Big]dx+\mathbb{E}\int_Q\theta^2\ell_t|dz_t|^2\Big\}.$$

For the boundary term, noting that $z=0$ on Σ, similar to (4.47), we have that

$$\mathbb{E}\int_\Sigma\sum_{j,k=1}^n\sum_{j',k'=1}^n\Big(2b^{jk}b^{j'k'}\psi_{x_{j'}}v_{x_j}v_{x_{k'}}-b^{jk}b^{j'k'}\psi_{x_{j'}}v_{x_j}v_{x_{k'}}\Big)\nu_k d\Sigma$$

$$=\mathbb{E}\int_\Sigma\Big(\sum_{j,k=1}^n b^{jk}\nu_j\nu_k\Big)\Big(\sum_{j',k'=1}^n b^{j'k'}\psi_{x_{j'}}\nu_{j'}\Big)\Big|\frac{\partial v}{\partial\nu}\Big|^2 d\Sigma. \tag{4.69}$$

It follows from (4.68) and (4.69) that

$$\mathbb{E}\int_Q\theta\Big\{\Big(-2\ell_t v_t+2\sum_{j,k=1}^n b^{jk}\ell_{x_j}v_{x_k}+\Psi v\Big)\times\Big[dz_t-\sum_{j,k=1}^n(b^{jk}z_{x_j})_{x_k}dt\Big]\Big\}dx$$

$$+\lambda\mathbb{E}\int_\Sigma\Big(\sum_{j,k=1}^n b^{jk}\nu_j\nu_k\Big)\Big(\sum_{j',k'=1}^n b^{j'k'}\psi_{x_{j'}}\nu_{k'}\Big)\Big|\frac{\partial v}{\partial\nu}\Big|^2 d\Sigma$$

$$\geq C\Big\{\mathbb{E}\int_Q\Big[\theta^2\Big(\lambda z_t^2+\lambda|\nabla z|^2+\lambda^3 z^2\Big) \tag{4.70}$$

$$+\Big(-2\ell_t v_t+2\sum_{j,k=1}^n b^{jk}\ell_{x_j}v_{x_k}+\Psi v\Big)^2\Big]dxdt$$

$$+\mathbb{E}\int_G\theta^2\Big[\lambda(|\nabla z_0|^2+|z_1|^2)+\lambda^3|z_0|^2\Big]dx$$

$$+\lambda\mathbb{E}\int_Q(T-t)\theta^2|b_4 z+g|^2 dxdt\Big\}.$$

Since

$$|b_4 z+g|^2\geq\frac{1}{2}|g|^2-2|b_4 z|^2,$$

we get that

$$\lambda\mathbb{E}\int_Q(T-t)\theta^2|b_4 z+g|^2 dxdt$$

$$\geq\frac{1}{2}\lambda\mathbb{E}\int_Q(T-t)\theta^2 g^2 dxdt-2\lambda T\mathbb{E}\int_Q\theta^2 b_4^2 z^2 dxdt. \tag{4.71}$$

On the other hand, by the equation (4.4), it is clear that

$$
\mathbb{E} \int_Q \theta \Big\{ \Big(-2\ell_t v_t + 2 \sum_{j,k=1}^n b^{jk} \ell_{x_j} v_{x_k} + \Psi v \Big) \times \Big[dz_t - \sum_{j,k=1}^n \big(b^{jk} z_{x_j} \big)_{x_k} dt \Big] \Big\} dx
$$

$$
\leq \mathbb{E} \int_Q \Big(-2\ell_t v_t + \sum_{j,k=1}^n b^{jk} \ell_{x_j} v_{x_k} + \Psi v \Big)^2 dx dt \qquad (4.72)
$$

$$
+ \mathcal{C} \Big[|b_1|_{L_{\mathbb{F}}^\infty(0,T;L^\infty(G))}^2 \mathbb{E} \int_Q \theta^2 z_t^2 dx dt
$$

$$
+ \Big(|b_2|_{L_{\mathbb{F}}^\infty(0,T;L^\infty(G,\mathbb{R}^n))}^2 + |b_3|_{L_{\mathbb{F}}^\infty(0,T;L^n(G))}^2 \Big) \mathbb{E} \int_Q \theta^2 |\nabla z|^2 dx dt
$$

$$
+ \lambda^2 |b_3|_{L_{\mathbb{F}}^\infty(0,T;L^n(G))}^2 \mathbb{E} \int_Q \theta^2 z^2 dx dt \Big].
$$

Finally, taking $\tilde{\lambda} = \max \big\{ \mathcal{C}(r_1^2 + r_2^2), \tilde{\lambda}_0, \tilde{\lambda}_1 \big\}$, combining (4.9), (4.70), (4.71) and (4.72), for any $\lambda \geq \tilde{\lambda}$, we have (4.56).   □

5. Local state observation for stochastic hyperbolic equations

In Section 4, we consider some global observation problems, that is, we are to determine the whole unknown state/nonhomogeneous terms by means of a suitable observation. This section is devoted to the study of a local state observation problem for stochastic hyperbolic equations, that is, we consider the problem of what can we get from a given observation. The main content of this section is taken from [26].

5.1. *Formulation of the problem*

Consider the following stochastic hyperbolic equation:

$$
\sigma dz_t - \Delta z dt = (b_1 z_t + b_2 \cdot \nabla z + b_3 z) dt + b_4 z dW(t) \qquad \text{in } Q, \qquad (5.1)
$$

where, $\sigma \in C^1(\overline{Q})$ is positive,

$$
b_1 \in L_{\mathbb{F}}^\infty(0,T;L_{\mathrm{loc}}^\infty(G)), \quad b_2 \in L_{\mathbb{F}}^\infty(0,T;L_{\mathrm{loc}}^\infty(G;\mathbb{R}^n)),
$$

$$
b_3 \in L_{\mathbb{F}}^\infty(0,T;L_{\mathrm{loc}}^n(G)), \quad b_4 \in L_{\mathbb{F}}^\infty(0,T;L_{\mathrm{loc}}^\infty(G)).
$$

Recall that for $p \in [1,\infty]$,

$$
L_{\mathrm{loc}}^p(G) \overset{\triangle}{=} \{ f \mid \forall\, G' \subset\subset G, \chi_{G'} f \in L^p(G) \}
$$

and

$$
H_{\mathrm{loc}}^1(G) \overset{\triangle}{=} \{ f \mid \forall\, G' \subset\subset G, \chi_{G'} f \in H^1(G) \}.
$$

Since no boundary condition and initial condition are given, and the lower order terms do not satisfy (2.12), the equation (5.1) is not a special case of the equation (2.11). Hence, we should first introduce the solution to (5.1). To this end, put

$$\mathbb{H}_{T,\text{loc}} \triangleq L^2_{\mathbb{F}}(\Omega; C([0,T]; H^1_{\text{loc}}(G))) \cap L^2_{\mathbb{F}}(\Omega; C^1([0,T]; L^2_{\text{loc}}(G))). \tag{5.2}$$

Definition 19. We call $z \in \mathbb{H}_{T,\text{loc}}$ a solution to the equation (5.1) if for each $t \in [0,T]$, $G' \subset\subset G$ and $\eta \in H^1_0(G')$, it holds that

$$\int_{G'} z_t(t,x)\eta(x)dx - \int_{G'} z_t(0,x)\eta(x)dx - \int_0^t \int_{G'} \sigma_t(s,x)z_t(s,x)\eta(x)dxds$$

$$= \int_0^t \int_{G'} \Big[-\nabla z(s,x) \cdot \nabla\eta(x) + \big(b_1 z_t + b_2 \cdot \nabla z + b_3 z\big)\eta(x)\Big]dxds \tag{5.3}$$

$$+ \int_0^t \int_{G'} b_4 z\eta(x)dxdW(s), \quad \mathbb{P}\text{-a.s.}$$

Let $S \subset\subset G$ be a C^2-hypersurface. Let $x_0 \in S \setminus \partial G$ and suppose that S divides the ball $B_\rho(x_0) \subset G$, centered at x_0 and with radius ρ, into two parts $\mathcal{D}_\rho^+$ and $\mathcal{D}_\rho^-$. Denote as usual by $\nu(x)$ the unit normal vector to S at x inward to $\mathcal{D}_\rho^+$.

Let y be a solution to the equation (5.1). Let $\varepsilon > 0$. Consider the following local state observation problem:

(P4) *Can z in $\mathcal{D}_\rho^+ \times (\varepsilon, T - \varepsilon)$ be uniquely determined by the values of z in $\mathcal{D}_\rho^- \times (0,T)$?*

In other words, Problem **(P4)** concerns that whether the state in one side of S uniquely determines the state in the other side. Clearly, it is equivalent to the following unique continuation problem:

(P5) *Can we conclude that $z = 0$ in $\mathcal{D}_\rho^+ \times (\varepsilon, T - \varepsilon)$, provided that $z = 0$ in $\mathcal{D}_\rho^- \times (0,T)$?*

To present the main result of this section, let us first introduce the following notion.

Definition 20. Let $x_0 \in S$ and $K \geq 0$. S is said to satisfy the outer paraboloid condition with K at x_0 if there exists a neighborhood $\mathcal{V}$ of x_0 and a paraboloid $\mathcal{P}$ tangential to S at x_0 and $\mathcal{P} \cap \mathcal{V} \subset \mathcal{D}_\rho^-$ with $\mathcal{P}$ congruent to $x_1 = K \sum_{j=2}^n x_j^2$.

The main result in this section is the following one.

Theorem 12. *Let $x_0 \in S \setminus \partial S$ such that $\frac{\partial \sigma(x_0, T/2)}{\partial \nu} < 0$, and let S satisfy the outer paraboloid condition with*

$$K < \frac{-\frac{\partial \sigma}{\partial \nu}(x_0, T/2)}{4(|\sigma|_{L^\infty(B_\rho(x_0, T/2))} + 1)}. \tag{5.4}$$

Let $z \in \mathbb{H}_{T,\text{loc}}$ be a solution to the equation (5.1) satisfying that

$$z = \frac{\partial z}{\partial \nu} = 0 \quad \text{on } (0, T) \times S, \ \mathbb{P}\text{-a.s.} \tag{5.5}$$

Then, there is a neighborhood $\mathcal{V}$ of x_0 and $\varepsilon \in (0, T/2)$ such that

$$z = 0 \quad \text{in } (\mathcal{V} \cap \mathcal{D}_\rho^+) \times (\varepsilon, T - \varepsilon), \ \mathbb{P}\text{-a.s.} \tag{5.6}$$

Remark 12. In Theorem 12, we assume that $\frac{\partial \sigma(x_0, T/2)}{\partial \nu} < 0$. It is related to the propagation of the wave. This is a reasonable assumption since the unique continuation property may not hold if it is not fulfilled (e.g., [43]). It can be regarded as a kind of pseudoconvex condition (e.g., [11, Chapter XXVII]).

Remark 13. If S is a hyperplane, then Condition 5.4 always satisfies since we can take $K = 0$.

As an immediate corollary of Theorem 12, we have the following unique continuation property.

Corollary 3. *Let $x_0 \in S \setminus \partial S$ such that $\frac{\partial \sigma(x_0, T/2)}{\partial \nu} < 0$, and let S satisfy (5.4). Then for any $z \in \mathbb{H}_{T,\text{loc}}$ which solves equation (5.1) satisfying that*

$$z = 0 \quad \text{on } \mathcal{D}_\rho^- \times (0, T), \ \mathbb{P}\text{-a.s.}, \tag{5.7}$$

there is a neighborhood $\mathcal{V}$ of x_0 and $\varepsilon \in (0, T/2)$ such that

$$z = 0 \quad \text{in } (\mathcal{V} \cap \mathcal{D}_\rho^+) \times (\varepsilon, T - \varepsilon), \ \mathbb{P}\text{-a.s.} \tag{5.8}$$

Corollary 3 concludes that the value of z on $(\mathcal{V} \cap \mathcal{D}_\rho^+) \times (\varepsilon, T - \varepsilon)$ can be determined by the observation on $\mathcal{D}_\rho^- \times (0, T)$, $\mathbb{P}$-a.s., which answers the state observation problem (**P5**).

The rest of this section is organized as follows. In Subsection 5.2, we explain the choice of weight function in the Carleman estimate. Subsection 5.3 is devoted to the proof of a Carleman estimate while Section 5.4 is addressed to the proof of the main result.

5.2. *Choice of the weight function*

In this subsection, we explain the choice of the weight function which will be used to establish our global Carleman estimate. That function is already used in [44] to establish Carleman estimate for the deterministic wave equation.

Without loss of generality, we assume that $0 = (0, \cdots, 0) \in S \setminus \partial S$ and $\nu(0) = (1, \cdots, 0)$. Since S is C^2, for some $r > 0$, we can parameterize S in the neighborhood of the origin by

$$x_1 = \gamma(x_2, \cdots, x_n), \quad |x_2|^2 + \cdots + |x_n|^2 < r. \tag{5.9}$$

In the rest of this section, we set

$$\begin{cases} B_r\left(0, \dfrac{T}{2}\right) = \left\{(x, t) \in \mathbb{R}^{n+1} \,\Big|\, |x|^2 + \left(t - \dfrac{T}{2}\right)^2 < r^2\right\}, \\ B_r(0) = \{x \in \mathbb{R}^n \,|\, |x| < r\}. \end{cases} \tag{5.10}$$

By (5.4), we have that

$$\begin{cases} -\alpha_0 = \dfrac{\partial \sigma}{\partial \nu}\left(0, \dfrac{T}{2}\right) < 0, \\ K < \dfrac{\alpha_0}{4(|\sigma|_{L^\infty(B_r(0,T/2))} + 1)}, \\ -K\displaystyle\sum_{j=2}^{n} x_j^2 < \gamma(x_2, \cdots, x_n), \quad \text{if } \displaystyle\sum_{j=2}^{n} x_j^2 < r. \end{cases} \tag{5.11}$$

Let

$$M_1 = \max\left\{|\sigma|_{C^1(B_r(0,0))}, 1\right\}. \tag{5.12}$$

Denote

$$\mathcal{D}_r^- = \left\{x \in B_r(0) \,\big|\, x_1 < \gamma(x_2, \cdots, x_n)\right\}, \quad \mathcal{D}_r^+ = B_r(0) \setminus \overline{\mathcal{D}_r^-}.$$

For any $\alpha \in (0, \alpha_0)$, from the continuity of $\frac{\partial \sigma(x,t)}{\partial \nu}$ and the first inequality in (5.11), we know that there exists a $\delta_0 > 0$, small enough, such that $0 < \delta_0 < \min\{1, r^2\}$, which would be specified later, and

$$\frac{\partial \sigma(x,t)}{\partial \nu} < -\alpha \quad \text{if } |x|^2 + \left(t - \frac{T}{2}\right)^2 \leq \delta_0. \tag{5.13}$$

Letting $M_0 = |\sigma|_{L^\infty(B_r(0,T/2))}$, by the second inequality in (5.11), we can always choose $K > 0$ and $h > 0$ such that

$$K < \frac{1}{2h} < \frac{\alpha}{4(M_0 + 1)}. \tag{5.14}$$

Following immediately from (5.14), we have that

$$1 - 2hK > 0, \quad h\alpha - 2(M_0 + 1) > 0. \tag{5.15}$$

For K and h such chosen, take $\tau \in (0,1)$ such that

$$\left| \max\left\{ \frac{K}{1-2hK}, \frac{1}{2h} \right\} \right|^2 \tau^2 + \frac{2\tau}{1-2hK} \le \delta_0. \tag{5.16}$$

For convenience of notations, by denoting $\mu_0(\tau)$ the term in the left-hand side of (5.16) and letting $\sigma_0 = \min\{\sigma, 1\}$, we further assume that

$$\begin{cases} h^2\sigma_0 > 2hM_1\sqrt{\mu_0(\tau)} + 2M_1\mu_0(\tau), \\ \alpha h > 2(M_1^2 + M_1)\sqrt{\mu_0(\tau)} - (M_0^2 + nM_0) - (n-1). \end{cases} \tag{5.17}$$

Let

$$\varphi(x,t) = hx_1 + \frac{1}{2}\sum_{j=2}^{n} x_j^2 + \frac{1}{2}\left(t - \frac{T}{2}\right)^2 + \frac{1}{2}\tau. \tag{5.18}$$

For any positive number μ with $2\mu > \tau$, let

$$Q_\mu = \left\{ (x,t) \in \mathbb{R}^{n+1} \,\Big|\, x_1 > \gamma(x_2, x_3, \cdots, x_n), \ \sum_{j=2}^{N} x_j^2 < \delta_0, \ \varphi(x,t) < \mu \right\}. \tag{5.19}$$

By the third inequality in (5.11), we know that

$$\gamma(x_2, x_3, \cdots, x_n) > -K\sum_{j=2}^{n} x_j^2.$$

This, together with the first inequality in (5.15), implies that

$$\begin{aligned} \varphi(x,t) &\ge -hK\sum_{j=2}^{n} x_j^2 + \frac{1}{2}\sum_{j=2}^{n} x_j^2 + \frac{1}{2}\left(t - \frac{T}{2}\right)^2 + \frac{1}{2}\tau \\ &= \left(\frac{1}{2} - Kh\right)\sum_{j=2}^{n} x_j^2 + \frac{1}{2}\left(t - \frac{T}{2}\right)^2 + \frac{1}{2}\tau \\ &> \frac{1}{2}\tau. \end{aligned}$$

Noting that $(x,t) \in Q_\mu$ implies $\varphi(x,t) < \mu$, together with $2\mu > \tau$, we see by definition that $Q_\mu \neq \emptyset$.

In what follows, we will show that how to determine the number δ_0 in (5.16). Let $(x,t) \in \overline{Q}_\tau$. From the definition of Q_τ (see (5.19)) and noting that

$$\gamma(x_2, x_3, \cdots, x_n) > -K\sum_{j=2}^{n} x_j^2,$$

we find that

$$x_1 \le -\frac{\tau}{2h}\sum_{j=2}^{n}x_j^2 - \frac{1}{2h}\left(t-\frac{T}{2}\right)^2 + \frac{\tau}{2h} \le \frac{\tau}{2h}. \tag{5.20}$$

On the other hand, by $-K\sum_{j=2}^{n}x_j^2 \le x_1$, we obtain that

$$-Kh\sum_{j=2}^{n}x_j^2 + \frac{1}{2}\sum_{j=2}^{n}x_j^2 + \frac{1}{2}\left(t-\frac{T}{2}\right)^2 + \frac{1}{2}\tau \le \tau.$$

Thus

$$\sum_{j=2}^{n}x_j^2 < \frac{\tau}{1-2Kh}.$$

Then we get that

$$-x_1 \le K\sum_{j=2}^{n}x_j^2 < \frac{K\tau}{1-2Kh}. \tag{5.21}$$

Combining (5.20) and (5.21), we arrive at

$$|x_1| \le \max\left\{\frac{K}{1-2hK}, \frac{1}{2h}\right\}\tau, \tag{5.22}$$

which implies that

$$\tau > \varphi(x,t) = hx_1 + \frac{1}{2}\sum_{j=2}^{n}x_j^2 + \frac{1}{2}\left(t-\frac{T}{2}\right)^2 + \frac{\tau}{2}$$
$$> -\frac{Kh\tau}{1-2Kh} + \frac{1}{2}\sum_{j=2}^{n}x_j^2 + \frac{1}{2}\left(t-\frac{T}{2}\right)^2 + \frac{\tau}{2}. \tag{5.23}$$

This gives that

$$\left(t-\frac{T}{2}\right)^2 < \frac{2Kh\tau}{1-2Kh} + \tau = \frac{\tau}{1-2Kh}. \tag{5.24}$$

Correspondingly, we have that

$$|x|^2 + \left(t-\frac{T}{2}\right)^2 = x_1^2 + \sum_{j=2}^{n}x_j^2 + \left(t-\frac{T}{2}\right)^2$$
$$\le \left|\max\left\{\frac{K}{1-2Kh}, \frac{1}{2h}\right\}\right|^2\tau^2 + \frac{2\tau}{1-2Kh}.$$

Returning back to (5.13), by (5.21), (5.22) and (5.24), we choose the δ_0 satisfying

$$\delta_0 > \mu_0(\tau) = \left|\max\left\{\frac{K}{1-2Kh}, \frac{1}{2h}\right\}\right|^2\tau^2 + \frac{2\tau}{1-2Kh}. \tag{5.25}$$

5.3. *A global Carleman estimate*

Let $\ell = s\varphi^{-\lambda}$ with φ the weight function given by (5.18), and let $\theta = e^\ell$. We have the following global Carleman estimate.

Theorem 13. *Let u be an $H^2_{\mathrm{loc}}(G)$-valued $\mathbf{F}$-adapted process such that u_t is an $L^2_{\mathrm{loc}}(G)$-valued Itô process. If u is supported in Q_τ, then there exist a constant $\mathcal{C} > 0$ and an $s_0 > 0$ depending on σ, τ such that for all $s \geq s_0$, we have that[b]*

$$\mathbb{E}\int_{Q_\tau} \theta\big(-2\sigma\ell_t v_t + 2\nabla\ell\cdot\nabla v\big)\big(\sigma du_t - \Delta u dt\big)dx$$

$$\geq \mathcal{C}\mathbb{E}\int_{Q_\tau}\Big[s\lambda^2\varphi^{-\lambda-2}(|\nabla v|^2 + v_t^2) + s^3\lambda^4\varphi^{-3\lambda-4}v^2\Big]dxdt \qquad (5.26)$$

$$+\mathbb{E}\int_{Q_\tau}\big(-2\sigma\ell_t v_t + 2\nabla\ell\cdot\nabla v\big)^2 dxdt + \mathcal{C}\mathbb{E}\int_{Q_\tau}\sigma^2\theta^2\ell_t(du_t)^2 dx.$$

Proof. Let $(p^{jk})_{1\leq j,k\leq n}$ be the unit matrix of n-th order and let $\Psi = 0$ in (4.24). Then we have that

$$\theta\big(-2\sigma\ell_t v_t + 2\nabla\ell\cdot\nabla v\big)\big(\sigma du_t - \Delta u dt\big)$$
$$+\nabla\cdot\big[2(\nabla v\cdot\nabla\ell)\nabla v - |\nabla v|^2\nabla\ell - 2\ell_t v_t\nabla v + \sigma v_t^2\nabla\ell - \mathcal{A}\nabla\ell v^2\big]dt$$
$$+d\big(\sigma\ell_t|\nabla v|^2 - 2\sigma\nabla\ell\cdot\nabla v v_t + \sigma^2\ell_t v_t^2 + \sigma\mathcal{A}\ell_t v^2\big)$$
$$= \Big\{\big[(\sigma^2\ell_t)_t + \nabla\cdot(\sigma\nabla\ell)\big]v_t^2 - 2\big[(\sigma\nabla\ell)_t + \nabla(\sigma\ell_t)\big]\cdot\nabla v v_t \qquad (5.27)$$
$$+\big[(\sigma\ell_t)_t + \Delta\ell\big]|\nabla v|^2 + \mathcal{B}v^2 + \big(-2\sigma\ell_t v_t + 2\nabla\ell\cdot\nabla v\big)^2\Big\}dt + \sigma^2\theta^2\ell_t(du_t)^2,$$

where

$$\begin{cases} \mathcal{A} = \sigma(\ell_t^2 - \ell_{tt}) - (|\nabla\ell|^2 - \Delta\ell), \\ \mathcal{B} = (\sigma\mathcal{A}\ell_t)_t - \nabla\cdot(\mathcal{A}\nabla\ell). \end{cases} \qquad (5.28)$$

Some simple computations show that

$$\begin{cases} \ell_t = -s\lambda\varphi_t\varphi^{-\lambda-1} = -s\lambda\Big(t - \dfrac{T}{2}\Big)\varphi^{-\lambda-1}, \\[2mm] \ell_{tt} = s\lambda(\lambda+1)\Big(t - \dfrac{T}{2}\Big)^2\varphi^{-\lambda-2} - s\lambda\varphi^{-\lambda-1}, \\[2mm] \nabla\ell = -s\lambda\varphi^{-\lambda-1}\nabla\varphi, \\[2mm] \Delta\ell = s\lambda(\lambda+1)\varphi^{-\lambda-2}|\nabla\varphi|^2 - s\lambda\varphi^{-\lambda-1}\Delta\varphi, \\[2mm] \nabla\ell_t = s\lambda(\lambda+1)\varphi^{-\lambda-2}\Big(t - \dfrac{T}{2}\Big)\nabla\varphi. \end{cases} \qquad (5.29)$$

---

[b]See Remark 2 for the notation $(du_t)^2$.

Let us analyze the terms in the right-hand side of (5.27) one by one. The first one reads

$$
\begin{aligned}
&[(\sigma^2 \ell_t)_t + \nabla \cdot (\sigma \nabla \ell)] v_t^2 \\
&= (2\sigma \sigma_t \ell_t + \sigma^2 \ell_{tt} + \nabla \sigma \cdot \nabla \ell + \sigma \Delta \ell) v_t^2 \\
&= \Big[2\sigma \sigma_t \ell_t + \sigma^2 \ell_{tt} - s\lambda(\nabla \sigma \cdot \nabla \varphi + \sigma \Delta \varphi)\varphi^{-\lambda-1} + s\lambda(\lambda+1)\sigma|\nabla \varphi|^2 \varphi^{-\lambda-2} \Big] v_t^2 \\
&= -s\lambda \varphi^{-\lambda-1} \Big[2\sigma \sigma_t \Big(t - \frac{T}{2}\Big) + \sigma^2 + (\nabla \sigma \cdot \nabla \varphi + \sigma \Delta \varphi) \Big] v_t^2 \\
&\quad + s\lambda(\lambda+1)\varphi^{-\lambda-2} \Big[\sigma^2 \Big(t - \frac{T}{2}\Big)^2 + \sigma|\nabla \varphi|^2 \Big] v_t^2 \qquad (5.30) \\
&\geq -s\lambda \varphi^{-\lambda-1} \Big\{ h\alpha + 2(M_1^2 + M_1)\sqrt{\mu_0(\tau)} + [M_0^2 + (n-1)M_0] \Big\} v_t^2 \\
&\quad + s\lambda(\lambda+1)h^2 \sigma \varphi^{-\lambda-2} v_t^2 \\
&\geq s\lambda \varphi^{-\lambda-1} \Big\{ h\alpha - 2(M_1^2 + M_1)\sqrt{\mu_0(\tau)} - [M_0^2 + (n-1)M_0] \Big\} v_t^2 \\
&\quad + h^2 \sigma s\lambda(\lambda+1)\varphi^{-\lambda-2} v_t^2.
\end{aligned}
$$

Likewise, the second term in the right-hand side of (5.27) reads

$$
\begin{aligned}
&-2\Big[(\sigma \nabla \ell)_t + \nabla(\sigma \ell_t)\Big] \cdot \nabla v\, v_t \\
&= -2\big[\sigma_t \nabla \ell + \sigma \nabla \ell_t + \ell_t \nabla \sigma + \sigma \nabla \ell_t\big] \cdot \nabla v\, v_t \\
&= \Big\{ 2s\lambda \varphi^{-\lambda-1}\Big[\sigma_t \nabla \varphi + \Big(t - \frac{T}{2}\Big)\nabla \sigma\Big] - 2s\lambda(\lambda+1)\varphi^{-\lambda-2}\sigma\Big(t - \frac{T}{2}\Big)\nabla \varphi \Big\} \cdot \nabla v\, v_t \\
&= 2s\lambda \varphi^{-\lambda-2}\Big\{ \Big[\sigma_t \nabla \varphi + \Big(t - \frac{T}{2}\Big)\nabla \sigma\Big]\varphi - (\lambda+1)\sigma\Big(t - \frac{T}{2}\Big)\nabla \varphi \Big\} \cdot \nabla v\, v_t \\
&\geq -s\lambda \varphi^{-\lambda-2}\Big(M_1 h + 2M_1\sqrt{\mu_0(\tau)} \Big)\tau\Big(|\nabla v|^2 + v_t^2\Big) \qquad (5.31) \\
&\quad - s\lambda(\lambda+1)\varphi^{-\lambda-2}\Big(hM_1\sqrt{\mu_0(\tau)} + M_1\mu_0(\tau) \Big)\Big(|\nabla v|^2 + v_t^2\Big).
\end{aligned}
$$

Thus, there exists $\lambda_0 > 0$ such that for $\lambda > \lambda_0$, it holds that

$$
\begin{aligned}
&-2\Big[(\sigma \nabla \ell)_t + \nabla(\sigma \ell_t)\Big] \cdot \nabla v\, v_t \\
&\geq -2s\lambda(\lambda+1)\varphi^{-\lambda-2}\Big(hM_1\sqrt{\mu_0(\tau)} + M_1\mu_0(\tau) \Big)\Big(|\nabla v|^2 + v_t^2\Big). \quad (5.32)
\end{aligned}
$$

For the third term in the right-hand side of (5.27), we have that

$$
\begin{aligned}
[(\sigma \ell_t)_t + \Delta \ell]|\nabla v|^2 &= (\sigma_t \ell_t + \sigma \ell_{tt} + \Delta \ell)|\nabla v|^2 \\
&= -s\lambda \varphi^{-\lambda-1}\Big[\sigma_t\Big(t - \frac{T}{2}\Big) + \sigma + \Delta \varphi\Big]|\nabla v|^2 \\
&\quad + s\lambda(\lambda+1)\varphi^{-\lambda-2}\Big[\sigma\Big(t - \frac{T}{2}\Big)^2 + |\nabla \varphi|^2\Big]|\nabla v|^2
\end{aligned}
$$

$$\geq -s\lambda\varphi^{-\lambda-1}\Big[M_1\sqrt{\mu_0(\tau)} + M_0 + (n-1)\Big]|\nabla v|^2$$
$$+h^2 s\lambda(\lambda+1)\varphi^{-\lambda-2}|\nabla v|^2. \tag{5.33}$$

Following (5.30), (5.32), (5.33) and noting (5.17), we find that for all $\lambda > \lambda_0$,

$$[(\sigma^2\ell_t)_t + \nabla\cdot(\sigma\nabla\ell)]v_t^2 - 2[(\sigma\nabla\ell)_t + \nabla(\sigma\ell_t)]\cdot\nabla v\, v_t + [(\sigma\ell_t)_t + \Delta\ell]|\nabla v|^2$$
$$\geq Cs\lambda^2\varphi^{-\lambda-2}(|\nabla v|^2 + v_t^2). \tag{5.34}$$

Next, from (5.28), we get that

$$\mathcal{A} = s^2\lambda^2\varphi^{-2\lambda-2}\Big[\sigma\Big(t - \frac{T}{2}\Big)^2 - |\nabla\varphi|^2\Big]$$
$$+s\lambda(\lambda+1)\varphi^{-\lambda-2}\Big[|\nabla\varphi|^2 - \sigma\Big(t - \frac{T}{2}\Big)^2\Big] \tag{5.35}$$
$$+s\lambda\varphi^{-\lambda-1}[\sigma - (n-1)]$$

and that

$$\mathcal{B} = (\sigma\mathcal{A}\ell_t)_t - \nabla\cdot(\mathcal{A}\nabla\ell)$$
$$= \sigma_t\mathcal{A}\ell_t + \sigma\mathcal{A}_t\ell_t + \sigma\mathcal{A}\ell_{tt} - \nabla\mathcal{A}\cdot\nabla\ell - \mathcal{A}\Delta\ell$$
$$= 3s^3\lambda^2(\lambda+1)^2\Big(t - \frac{T}{2}\Big)^2\Big[\Big(t - \frac{T}{2}\Big)^2 - |\nabla\varphi|^2\Big]\varphi^{-3\lambda-4}$$
$$+3s^3\lambda^2(\lambda+1)^2|\nabla\varphi|^2\Big[|\nabla\varphi|^2 - \Big(t - \frac{T}{2}\Big)^2\Big]\varphi^{-3\lambda-4} \tag{5.36}$$
$$+s^3O(\lambda^3\varphi^{-3\lambda-3}) + s^2O(\lambda^4\varphi^{-3\lambda-4}).$$

Clearly, there exist $\lambda_1 > 0$ and $s_0 > 0$ such that for all $\lambda \geq \lambda_1$ and $s \geq s_0$,

$$\mathcal{B}v^2 \geq Cs^3\lambda^4\varphi^{-3\lambda-4}v^2. \tag{5.37}$$

Integrating (5.27) over Q_τ and taking mathematical expectation, we obtain that

$$\mathbb{E}\int_{Q_\tau}\theta\big(-2\sigma\ell_t v_t + 2\nabla\ell\cdot\nabla v\big)(\sigma du_t - \Delta u dt)dx$$
$$\geq C\mathbb{E}\int_{Q_\tau}\Big[s\lambda^2\varphi^{-\lambda-2}(|\nabla v|^2 + v_t^2) + s^3\lambda^4\varphi^{-3\lambda-4}v^2\Big]dxdt \tag{5.38}$$
$$+\mathbb{E}\int_{Q_\tau}\big(-2\sigma\ell_t v_t + 2\nabla\ell\cdot\nabla v\big)^2 dxdt + C\mathbb{E}\int_{Q_\tau}\sigma^2\theta^2\ell_t(du_t)^2 dx.$$

This completes the proof. $\qquad\square$

5.4. *Proof of the local observation result*

This section is dedicated to the proof of Theorem 12.

Proof. Without loss of generality, we assume that

$$x_0 = (0, 0, \cdots, 0), \quad \nu(x_0) = (1, 0, \cdots, 0)$$

and S is parameterized as in Section 5.2 near 0. Also, K, δ_0, h and τ are all given as in Subsection 5.2. By the definition of $\varphi(x, t)$ and Q_μ, for any $\mu \in (0, \tau]$, the boundary Γ_μ of Q_μ is composed of the following three parts:

$$\begin{cases} \Gamma_\mu^1 = \left\{ (x, t) \in \mathbb{R}^{n+1} \middle| x_1 = \gamma(x_2, x_3, \cdots, x_n), \sum_{j=2}^{n} x_j^2 < \delta_0, \varphi(x, t) < \mu \right\}, \\ \Gamma_\mu^2 = \left\{ (x, t) \in \mathbb{R}^{n+1} \middle| x_1 > \gamma(x_2, x_3, \cdots, x_n), \sum_{j=2}^{n} x_j^2 < \delta_0, \varphi(x, t) = \mu \right\}, \\ \Gamma_\mu^3 = \left\{ (x, t) \in \mathbb{R}^{n+1} \middle| x_1 > \gamma(x_2, x_3, \cdots, x_n), \sum_{j=2}^{n} x_j^2 = \delta_0, \varphi(x, t) < \mu \right\}. \end{cases}$$

$$(5.39)$$

Next, we show that $\Gamma_\mu^3 = \emptyset$. By $\gamma(x_2, x_3, \cdots, x_n) > -K \sum_{j=2}^{n} x_j^2$ and the definition of φ, we have that

$$(1 - 2Kh) \sum_{j=2}^{n} x_j^2 + \left(t - \frac{T}{2} \right)^2 < 2 \left[h x_1 + \sum_{j=2}^{n} x_j^2 + \left(t - \frac{T}{2} \right)^2 \right]$$

$$= 2\varphi - \tau < 2\mu - \tau < \tau. \qquad (5.40)$$

Also, note that Γ_μ^3 is subordinated to $\sum_{j=2}^{n} x_j^2 = \delta_0$. From (5.40), it follows that $\delta_0 < \dfrac{\tau}{1 - 2Kh}$, a contradiction to $\delta_0 > \dfrac{\tau}{1 - 2Kh}$ introduced in (5.25). Consequently, $\Gamma_\mu = \Gamma_\mu^1 \cup \Gamma_\mu^2$.

It is clear that

$$\Gamma_\mu^1 \cup \Gamma_\mu^2 \subset \overline{Q_\tau}.$$

Let

$$t_0 = \sqrt{\frac{\tau}{1 - 2Kh}}.$$

By (5.24), it follows

$$\begin{cases} \Gamma_\mu^1 \subset \{x \mid x_1 = \gamma(x_2, x_3, \cdots, x_n)\} \times \{t \mid |t - T/2| \le t_0\}, \\ \Gamma_\mu^2 \subset \{x \mid \varphi(x, t) = \mu\}, \quad \mu \in (0, \tau]. \end{cases} \qquad (5.41)$$

It is clear that

$$\Gamma_\mu^j \subset \Gamma_\tau^j, \quad j = 1, 2.$$

For convenience in the later statement, denote $Q_\tau = Q_1$. Fixing an arbitrarily small number $\tilde{\tau} \in (0, \frac{\tau}{8})$, let

$$Q_{k+1} = \left\{ (t, x) \middle| \varphi(x, t) < \tau - k\tilde{\tau}, k = 1, 2, 3 \right\}.$$

Hence, $Q_4 \subset Q_3 \subset Q_2 \subset Q_1$.

Introduce a truncation function $\chi \in C_0^\infty(Q_2)$ as

$$\chi \in [0, 1] \quad \text{and} \quad \chi = 1 \quad \text{in} \quad Q_3.$$

Let z be the solution of (5.1). Let $u = \chi z$. Then

$$\begin{cases} \sigma du - \Delta u dt = (b_1 u_t + b_2 \cdot \nabla u + b_3 u + f) dt + b_4 u dW(t), & \text{in } Q_\tau, \\ u = 0, \ \dfrac{\partial u}{\partial \nu} = 0, & \text{on } \Gamma_\tau. \end{cases} \tag{5.42}$$

Here

$$f = \sigma \chi_{tt} z + 2\sigma \chi_t z_t - 2\nabla\chi \cdot \nabla z - z\Delta\chi - b_1 \chi_t z - b_2 \cdot z\nabla\chi.$$

From the definition of χ, f is clearly supported in $Q_2 \setminus \overline{Q}_3$.

From (5.26), we have that

$$\mathbb{E} \int_{Q_\varepsilon} \theta \left(-2\sigma \ell_t v_t + 2\nabla\ell \cdot \nabla v \right) \left(\sigma du_t - \Delta u dt \right) dx$$

$$\geq C\mathbb{E} \int_{Q_\varepsilon} \left[s\lambda^2 \varphi^{-\lambda-2} (|\nabla v|^2 + v_t^2) + s^3 \lambda^4 \varphi^{-3\lambda-4} v^2 \right] dx dt \tag{5.43}$$

$$+ \mathbb{E} \int_{Q_\tau} \left(-2\sigma \ell_t v_t + 2\nabla\ell \cdot \nabla v \right)^2 dx dt + C\mathbb{E} \int_{Q_\tau} \sigma^2 \theta^2 \ell_t (du_t)^2 dx.$$

Due to the elementary property of Itô integral, it is clear that

$$\mathbb{E} \int_{Q_\tau} \theta \left(-2\sigma \ell_t v_t + 2\nabla\ell \cdot \nabla v \right) \left(\sigma du_t - \Delta u dt \right) dx$$

$$= \mathbb{E} \int_{Q_\tau} \theta \left(-2\sigma \ell_t v_t + 2\nabla\ell \cdot \nabla v \right) \left(b_1 u_t + b_2 \cdot \nabla u + b_3 u + f \right) dx dt$$

$$+ \mathbb{E} \int_{Q_\tau} \theta \left(-2\sigma \ell_t v_t + 2\nabla\ell \cdot \nabla v \right) b_4 u dW(t) dx$$

$$\leq \mathbb{E} \int_{Q_\tau} \theta^2 \left(b_1 u_t + b_2 \cdot \nabla u + b_3 u + f \right)^2 dx dt$$

$$+ \mathbb{E} \int_{Q_\varepsilon} \left(-2\sigma \ell_t v_t + 2\nabla\ell \cdot \nabla v \right)^2 dx dt.$$

Thus,

$$\mathbb{E}\int_{Q_\tau}\left[s\lambda^2\varphi^{-\lambda-2}(|\nabla v|^2+v_t^2)+s^3\lambda^4\varphi^{-3\lambda-4}v^2\right]dxdt+\mathbb{E}\int_{Q_\tau}\sigma^2\theta^2\ell_t(du_t)^2dx$$
$$\leq C\mathbb{E}\int_{Q_\tau}\theta^2\big(b_1u_t+b_2\cdot\nabla u+b_3u+f\big)^2dxdt.$$

Let us now do some estimates for the right hand side of the above inequality.

$$\mathbb{E}\int_{Q_\tau}\theta^2\big(b_1u_t+b_2\cdot\nabla u+b_3u+f\big)^2dxdt$$
$$\leq 2\mathbb{E}\int_{Q_\tau}\theta^2\big(b_1u_t+b_2\cdot\nabla u+b_3u\big)^2dxdt+2\mathbb{E}\int_{Q_\tau}\theta^2|f|^2dxdt. \tag{5.44}$$

Noting that f is supported in $Q_2\setminus\overline{Q}_3$, we have that

$$\mathbb{E}\int_{Q_\tau}\theta^2|f|^2dxdt$$
$$=\mathbb{E}\int_{Q_\tau}\theta^2|\sigma\chi_{tt}z+2\sigma\chi_t z_t-2\nabla\chi\cdot\nabla z-z\Delta\chi-b_1\chi_t z-b_2\cdot z\nabla\chi|^2dxdt$$
$$\leq C\mathbb{E}\int_{Q_2\setminus\overline{Q}_3}\theta^2\big(z_t^2+|\nabla z|^2+z^2\big)dxdt.$$

Thus, we achieve that

$$\mathbb{E}\int_{Q_\tau}\theta^2\big(b_1u_t+b_2\cdot\nabla u+b_3u+f\big)^2dxdt$$
$$\leq C\mathbb{E}\int_{Q_1}\theta^2\big(u_t^2+|\nabla u|^2+u^2\big)dxdt+C\mathbb{E}\int_{Q_2\setminus\overline{Q}_3}\theta^2\big(z_t^2+|\nabla z|^2+z^2\big)dxdt. \tag{5.45}$$

And then

$$\mathbb{E}\int_{Q_\tau}\left[s\lambda^2\varphi^{-\lambda-2}(|\nabla v|^2+v_t^2)+s^3\lambda^4\varphi^{-3\lambda-4}v^2\right]dxdt+\mathbb{E}\int_{Q_\epsilon}\sigma^2\theta^2\ell_t(du_t)^2dx$$
$$\leq C\mathbb{E}\int_{Q_1}\theta^2\big(u_t^2+|\nabla u|^2+u^2\big)dxdt+C\mathbb{E}\int_{Q_2\setminus\overline{Q}_3}\theta^2\big(z_t^2+|\nabla z|^2+z^2\big)dxdt. \tag{5.46}$$

Therefore

$$\mathbb{E}\int_{Q_\tau}\left[s\lambda^2\varphi^{-\lambda-2}(|\nabla v|^2+v_t^2)+s^3\lambda^4\varphi^{-3\lambda-4}v^2\right]dxdt$$
$$\leq C\mathbb{E}\int_{Q_1}\theta^2\big(u_t^2+|\nabla u|^2+u^2\big)dxdt+\mathbb{E}\int_{Q_\tau}\theta^2\sigma^2 s\lambda t\varphi^{-\lambda-1}b_4^2u^2dxdt$$

$$+ C\mathbb{E} \int_{Q_2 \setminus \overline{Q}_3} \theta^2 \big(z_t^2 + |\nabla z|^2 + z^2\big) dx dt. \tag{5.47}$$

Then for s and λ large enough, it follows that

$$\mathbb{E} \int_{Q_\tau} \big[s\lambda^2 \varphi^{-\lambda-2}(|\nabla v|^2 + v_t^2) + s^3\lambda^4 \varphi^{-3\lambda-4}v^2\big] dx dt$$

$$\leq C\mathbb{E} \int_{Q_\tau} \theta^2 \big(u_t^2 + |\nabla u|^2 + u^2\big) dx dt + C\mathbb{E} \int_{Q_2 \setminus \overline{Q}_3} \theta^2 \big(z_t^2 + |\nabla z|^2 + z^2\big) dx dt. \tag{5.48}$$

Since $v = \theta u$, we have that

$$|\nabla v|^2 + v_t^2 \geq C\theta^2 \big(s^2\lambda^2 \varphi^{-2\lambda-2}u^2 + |\nabla u|^2 + u_t^2\big).$$

Thus for large s and λ, we have that

$$\mathbb{E} \int_{Q_\tau} \theta^2 \big[s\lambda^2 \varphi^{-\lambda-2}(|\nabla u|^2 + u_t^2) + s^3\lambda^4 \varphi^{-3\lambda-4}u^2\big] dx dt$$

$$\leq C\mathbb{E} \int_{Q_2 \setminus \overline{Q}_3} \theta^2 \big(z_t^2 + |\nabla z|^2 + z^2\big) dx dt. \tag{5.49}$$

Recall that and $u = z$ in $Q_3 \subset Q_\tau$. It follows from (5.49) that

$$\mathbb{E} \int_{Q_3} \theta^2 \big[s\lambda^2 \varphi^{-\lambda-2}\big(|\nabla z|^2 + z_t^2\big) + s^3\lambda^4 \varphi^{-3\lambda-4}z^2\big] dx dt$$

$$\leq C\mathbb{E} \int_{Q_2 \setminus \overline{Q}_3} \theta^2 \big(z_t^2 + |\nabla z|^2 + z^2\big) dx dt. \tag{5.50}$$

Note that in Q_4, $\varphi(x,t) < \tau - 3\tilde{\tau}$, then

$$\theta = e^{s\varphi^{-\lambda}} > e^{s(\tau - 3\tilde{\tau})^{-\lambda}}.$$

Moreover, in $Q_2 \setminus \overline{Q}_3$,

$$\tau - 2\tilde{\tau} < \varphi(x,t) < \tau - \tilde{\tau},$$

then

$$e^{s(\tau - \tilde{\tau})^{-\lambda}} < \theta < e^{s(\tau - 2\tilde{\tau})^{-\lambda}}.$$

Therefore

$$\mathbb{E} \int_{Q_4} \big[s\lambda^2 \varphi^{-\lambda-2}(|\nabla z|^2 + z_t^2) + s^3\lambda^4 \varphi^{-3\lambda-4}z^2\big] dx dt$$

$$\leq C e^{2[s(\tau - 2\tilde{\tau})^{-\lambda} - s(\tau - 3\tilde{\tau})^{-\lambda}]}\mathbb{E} \int_{Q_2 \setminus \overline{Q}_3} \big(z_t^2 + |\nabla z|^2 + z^2\big) dx dt \tag{5.51}$$

$$\leq C e^{2[s(\tau - 2\tilde{\tau})^{-\lambda} - s(\tau - 3\tilde{\tau})^{-\lambda}]}\mathbb{E} \int_{Q_\tau} \big(z_t^2 + |\nabla z|^2 + z^2\big) dx dt.$$

Consequently,

$$
\begin{aligned}
\mathbb{E} \int_{Q_4} &(|\nabla z|^2 + z_t^2 + z^2)dxdt \\
&\leq Ce^{2s[(\tau - 2\widetilde{\tau})^{-\lambda} - (\tau - 3\widetilde{\tau})^{-\lambda}]}\mathbb{E}\int_{Q_\tau}\left(z_t^2 + |\nabla z|^2 + z^2\right)dxdt.
\end{aligned}
\tag{5.52}
$$

By letting $s \to +\infty$, we find $z = 0$ in Q_4. Taking Q_4 the desired region, we complete the proof. $\qquad\square$

6. State observation problem for stochastic parabolic equations

This section is devoted to a global state observation problem for stochastic parabolic equations. The main content is taken from [45]. We do some minor modifications.

6.1. *Formulation of the problem*

Let $T > 0$, and $G \in \mathbb{R}^n$ ($n \in \mathbb{N}$) be a given bounded domain with the C^4 boundary Γ. Let G_0 be a nonempty subset of G. Put

$$
Q \stackrel{\Delta}{=} (0,T) \times G, \quad \Sigma \stackrel{\Delta}{=} (0,T) \times \Gamma, \quad Q_0 \stackrel{\Delta}{=} (0,T) \times G_0.
$$

Recall that $(b^{jk})_{1 \leq j,k \leq n} \in C^1(G; \mathbb{R}^{n \times n})$ satisfying that $b^{jk} = b^{kj}$ for $j,k = 1, 2, \cdots, n$ and

$$
\sum_{j,k=1}^n b^{jk}\xi_j\xi_k \geq s_0|\xi|^2, \quad \forall (x,\xi) \stackrel{\Delta}{=} (x,\xi_1,\cdots,\xi_n) \in G \times \mathbb{R}^n,
\tag{6.1}
$$

for some constant $s_0 > 0$.
    Let

$$
F(\eta,\zeta) : \mathbb{R} \times \mathbb{R}^n \to \mathbb{R}
$$

and

$$
K(\eta) : \mathbb{R} \to \mathbb{R}
$$

be two nonlinear functions satisfying that

$$
|F(\eta_1,\zeta_1) - F(\eta_2,\zeta_2)| \leq L(|\eta_1 - \eta_2| + |\zeta_1 - \zeta_2|_{\mathbb{R}^n}),
$$
$$
\forall (\eta_i,\zeta_i) \in \mathbb{R} \times \mathbb{R}^n, \ i = 1,2,
$$
$$
|F(0,\zeta)| \leq L|\zeta|_{\mathbb{R}^n}, \ \forall \zeta \in \mathbb{R}^n,
$$
$$
|K(0)| \in L^2_{\mathbb{F}}(0,T), \ |K(\eta_1) - K(\eta_2)| \leq L|\eta_1 - \eta_2|, \ \forall \eta_1,\eta_2 \in \mathbb{R},
$$

for some constant $L > 0$.

Consider the following semilinear stochastic parabolic equation:

$$\begin{cases} dy - \displaystyle\sum_{j,k=1}^{n} (b^{jk} y_{x_j})_{x_k} dt = F(y, \nabla y)dt + K(y)dW(t), & \text{in } Q, \\ y = 0, & \text{on } \Sigma, \\ y(0) = y_0, & \text{in } G, \end{cases} \tag{6.2}$$

where the initial data $y_0 \in L^2_{\mathcal{F}_0}(\Omega; L^2(G))$.

By Theorem 5, the equation (6.2) admits a unique (mild) solution

$$y \in \mathcal{H}_T \triangleq L^2_{\mathbb{F}}(\Omega; C([0,T]; L^2(G))) \cap L^2_{\mathbb{F}}(0,T; H^1_0(G)).$$

Define a map $\mathcal{M}_2$ as follows:

$$\begin{cases} \mathcal{M}_2 : L^2_{\mathcal{F}_0}(\Omega; L^2(G)) \to L^2_{\mathbb{F}}(0,T; L^2(G_0)), \\ \mathcal{M}_2(y_0) = y|_{(0,T) \times G_0}, \end{cases}$$

where y solves the equation (6.2) with the initial datum y_0.

Consider the following state observation problem:

(P6) *Does there exist a constant $C > 0$ such that for any initial data $y_0, \hat{y}_0 \in L^2_{\mathcal{F}_0}(\Omega; L^2(G))$ and $t \in (0,T]$,*

$$|y(t) - \hat{y}(t)|_{L^2_{\mathcal{F}_t}(\Omega; L^2(G))} \leq C|\mathcal{M}_2(y_0) - \mathcal{M}_2(\hat{y}_0)|_{L^2_{\mathbb{F}}(0,T; L^2(G_0))}. \tag{6.3}$$

Here y and $\hat{y}$ are the solution to (6.2) with the initial data y_0 and $\hat{y}_0$, respectively.

Remark 14. Due to the decay of the solution to (6.2), one cannot expect that the left hand side of (6.3) to be $|y_0 - \hat{y}_0|_{L^2_{\mathcal{F}_0}(\Omega; L^2(G))}$.

We have the following result.

Theorem 14. *There exists a constant $C > 0$ such that for any initial data $y_0, \hat{y}_0 \in L^2_{\mathcal{F}_0}(\Omega; L^2(G))$, the inequality (6.3) holds.*

Remark 15. Theorem 14 indicates that the state $y(t)$ of (6.2) (for $t \in (0,T]$) can be uniquely determined from the observation $y|_{(0,T) \times G_0}$, $\mathbb{P}$-a.s., and continuously depends on it. Therefore, it provides positive answers to the first and second questions for the state observation problem of the system (6.2).

6.2. *A weighted identity and Carleman estimate for a stochastic parabolic-like operator*

Throughout this section, we assume that

$$p^{jk} = p^{kj} \in L^2_{\mathbb{F}}(\Omega; C^1([0,T]; W^{2,\infty}(G))), \quad j,k = 1,2,\cdots,n, \qquad (6.4)$$

$\ell \in C^{1,3}((0,T) \times G)$ and $\Psi \in C^{1,2}((0,T) \times G)$.

We have the following identity.

Lemma 2. [45, *Theorem 3.1*] *Let u be an $H^2(G)$-valued continuous Itô process. Set $\theta = e^\ell$ and $w = \theta u$. Then, for any $t \in [0,T]$ and a.e. $(x,\omega) \in G \times \Omega$,*

$$2\theta\Big[-\sum_{j,k=1}^n \big(p^{jk}w_{x_j}\big)_{x_k} + \mathcal{A}w\Big]\Big[du - \sum_{j,k=1}^n \big(p^{jk}u_{x_j}\big)_{x_k} dt\Big]$$

$$+2\sum_{j,k=1}^n \big(p^{jk}w_{x_j}dw\big)_{x_k} + 2\sum_{j,k=1}^n \Big[\sum_{j',k'=1}^n \big(2p^{jk}p^{j'k'}\ell_{x_{j'}}w_{x_j}w_{x_{k'}}\big)$$

$$-p^{jk}p^{j'k'}\ell_{x_j}w_{x_{j'}}w_{x_{k'}}\big) + \Psi p^{jk}w_{x_j}w - p^{jk}\Big(\mathcal{A}\ell_{x_j} + \frac{\Psi_{x_j}}{2}\Big)w^2\Big]_{x_k} dt$$

$$= 2\sum_{j,k=1}^n c^{jk}w_{x_j}w_{x_k}dt + \mathcal{B}w^2 dt + d\Big(\sum_{j,k=1}^n p^{jk}w_{x_j}w_{x_k} + \mathcal{A}w^2\Big) \qquad (6.5)$$

$$+2\Big[-\sum_{j,k=1}^n \big(p^{jk}w_{x_j}\big)_{x_k} + \mathcal{A}w\Big]^2 dt$$

$$-\theta^2 \sum_{j,k=1}^n p^{jk}\big(du_{x_j} + \ell_{x_j}du\big)\big(du_{x_k} + \ell_{x_k}du\big) - \theta^2 \mathcal{A}(du)^2,$$

where

$$\begin{cases} \mathcal{A} \triangleq -\sum_{j,k=1}^n \big(p^{jk}\ell_{x_j}\ell_{x_k} - p^{jk}_{x_k}\ell_{x_j} - p^{jk}\ell_{x_jx_k}\big) - \Psi - \ell_t, \\[2mm] \mathcal{B} \triangleq 2\Big[\mathcal{A}\Psi - \sum_{j,k=1}^n \big(\mathcal{A}p^{jk}\ell_{x_j}\big)_{x_k}\Big] - \mathcal{A}_t - \sum_{j,k=1}^n \big(p^{jk}\Psi_{x_k}\big)_{x_j}, \\[2mm] c^{jk} \triangleq \sum_{j',k'=1}^n \Big[2p^{jk'}\big(p^{j'k}\ell_{x_{j'}}\big)_{x_{k'}} - \big(p^{jk}p^{j'k'}\ell_{x_{j'}}\big)_{x_{k'}}\Big] - \frac{p^{jk}_t}{2} + \Psi p^{jk}. \end{cases} \qquad (6.6)$$

Proof. The proof is divided into four steps.

Step 1. Recalling $\theta = e^\ell$ and $w = \theta u$, one has

$$du = \theta^{-1}(dw - \ell_t w dt)$$

and

$$u_{x_j} = \theta^{-1}(w_{x_j} - \ell_{x_j} w) \text{ for } j = 1, 2, \cdots, n.$$

By (6.4), we have

$$\sum_{j,k=1}^{n} p^{jk}\left(\ell_{x_j} w_{x_k} + \ell_{x_k} w_{x_j}\right) = 2\sum_{j,k=1}^{n} p^{jk}\ell_{x_j} w_{x_k}.$$

Hence,

$$\theta \sum_{j,k=1}^{n} \left(p^{jk} u_{x_j}\right)_{x_k}$$

$$= \theta \sum_{j,k=1}^{n} \left[\theta^{-1} p^{jk}\left(w_{x_j} - \ell_{x_j} w\right)\right]_{x_k}$$

$$= \sum_{j,k=1}^{n} \left[p^{jk}\left(w_{x_j} - \ell_{x_j} w\right)\right]_{x_k} - \sum_{j,k=1}^{n} p^{jk}\left(w_{x_j} - \ell_{x_j} w\right)\ell_{x_k}$$

$$= \sum_{j,k=1}^{n} \left[\left(p^{jk} w_{x_j}\right)_{x_k} - p^{jk}\left(\ell_{x_j} w_{x_k} + \ell_{x_k} w_{x_j}\right)\right. \tag{6.7}$$

$$+ \left.\left(p^{jk}\ell_{x_j}\ell_{x_k} - p^{jk}_{x_k}\ell_{x_j} - p^{jk}\ell_{x_j x_k}\right) w\right]$$

$$= \sum_{j,k=1}^{n} \left[\left(p^{jk} w_{x_j}\right)_{x_k} - 2p^{jk}\ell_{x_j} w_{x_k} + \left(p^{jk}\ell_{x_j}\ell_{x_k} - p^{jk}_{x_k}\ell_{x_j} - p^{jk}\ell_{x_j x_k}\right) w\right].$$

Put

$$\begin{cases} I \stackrel{\triangle}{=} -\sum_{j,k=1}^{n} \left(p^{jk} w_{x_j}\right)_{x_k} + \mathcal{A}w, \\[2mm] I_1 \stackrel{\triangle}{=} \left[-\sum_{j,k=1}^{n} \left(p^{jk} w_{x_j}\right)_{x_k} + \mathcal{A}w\right]dt, \\[2mm] I_2 \stackrel{\triangle}{=} dw + 2\sum_{j,k=1}^{n} p^{jk}\ell_{x_j} w_{x_k} dt, \\[2mm] I_3 \stackrel{\triangle}{=} \Psi w dt. \end{cases} \tag{6.8}$$

By (6.7) and (6.8), we see that

$$\theta\left[du - \sum_{j,k=1}^{n} \left(p^{jk} u_{x_j}\right)_{x_k} dt\right] = I_1 + I_2 + I_3.$$

Consequently,

$$2\theta\left[-\sum_{j,k=1}^{n}\left(p^{jk}w_{x_j}\right)_{x_k}+\mathcal{A}w\right]\left[du-\sum_{j,k=1}^{n}\left(p^{jk}u_{x_j}\right)_{x_k}dt\right]=2I\big(I_1+I_2+I_3\big).$$

(6.9)

Step 2. Let us compute $2II_2$. Utilizing (6.4) again, and noting that

$$\sum_{j,k,j',k'=1}^{n}\left(p^{jk}p^{j'k'}\ell_{x_{j'}}w_{x_j}w_{x_k}\right)_{x_{k'}}=\sum_{j,k,j',k'=1}^{n}\left(p^{jk}p^{j'k'}\ell_{x_j}w_{x_{j'}}w_{x_{k'}}\right)_{x_k},$$

we get

$$2\sum_{j,k,j',k'=1}^{n}p^{jk}p^{j'k'}\ell_{x_{j'}}w_{x_j}w_{x_kx_{k'}}$$

$$=\sum_{j,k,j',k'=1}^{n}p^{jk}p^{j'k'}\ell_{x_{j'}}\left(w_{x_j}w_{x_kx_{k'}}+w_{x_k}w_{x_jx_{k'}}\right)$$

$$=\sum_{j,k,j',k'=1}^{n}p^{jk}p^{j'k'}\ell_{x_{j'}}\left(w_{x_j}w_{x_k}\right)_{x_{k'}}$$

(6.10)

$$=\sum_{j,k,j',k'=1}^{n}\left(p^{jk}p^{j'k'}\ell_{x_{j'}}w_{x_j}w_{x_k}\right)_{x_{k'}}-\sum_{j,k,j',k'=1}^{n}\left(p^{jk}p^{j'k'}\ell_{x_{j'}}\right)_{x_{k'}}w_{x_j}w_{x_k}$$

$$=\sum_{j,k,j',k'=1}^{n}\left(p^{jk}p^{j'k'}\ell_{x_j}w_{x_{j'}}w_{x_{k'}}\right)_{x_k}-\sum_{j,k,j',k'=1}^{n}\left(p^{jk}p^{j'k'}\ell_{x_{j'}}\right)_{x_{k'}}w_{x_j}w_{x_k}.$$

Hence, by (6.10), and noting that

$$\sum_{j,k,j',k'=1}^{n}p^{jk}\left(p^{j'k'}\ell_{x_{j'}}\right)_{x_k}w_{x_j}w_{x_{k'}}=\sum_{j,k,j',k'=1}^{n}p^{jk'}\left(p^{j'k}\ell_{x_{j'}}\right)_{x_{k'}}w_{x_j}w_{x_k},$$

we obtain that

$$4\left[-\sum_{j,k=1}^{n}\left(p^{jk}w_{x_j}\right)_{x_k}+\mathcal{A}w\right]\sum_{j,k=1}^{n}p^{jk}\ell_{x_j}w_{x_k}$$

$$=-4\sum_{j,k,j',k'=1}^{n}\left(p^{jk}p^{j'k'}\ell_{x_{j'}}w_{x_j}w_{x_{k'}}\right)_{x_k}+4\sum_{j,k,j',k'=1}^{n}p^{jk}\left(p^{j'k'}\ell_{x_{j'}}\right)_{x_k}w_{x_j}w_{x_{k'}}$$

$$+4\sum_{j,k,j',k'=1}^{n}p^{jk}p^{j'k'}\ell_{x_{j'}}w_{x_j}w_{x_kx_{k'}}+2\mathcal{A}\sum_{j,k=1}^{n}p^{jk}\ell_{x_j}(w^2)_{x_k}$$

(6.11)

$$=-2\sum_{j,k=1}^{n}\left[\sum_{j',k'=1}^{n}\left(2p^{jk}p^{j'k'}\ell_{x_{j'}}w_{x_j}w_{x_{k'}}-p^{jk}p^{j'k'}\ell_{x_j}w_{x_{j'}}w_{x_{k'}}\right)\right.$$

$$-\mathcal{A}p^{jk}\ell_{x_j}w^2\Big]_{x_k} + 2\sum_{j,k,j',k'=1}^{n}\Big[2p^{jk'}\big(p^{j'k}\ell_{x_{j'}}\big)_{x_{k'}} - \big(p^{jk}p^{j'k'}\ell_{x_{j'}}\big)_{x_{k'}}\Big]w_{x_j}w_{x_k}$$

$$-2\sum_{j,k=1}^{n}\big(\mathcal{A}p^{jk}\ell_{x_j}\big)_{x_k}w^2.$$

Using Itô's formula, we have

$$2\Big[-\sum_{j,k=1}^{n}\big(p^{jk}w_{x_j}\big)_{x_k} + \mathcal{A}w\Big]dw$$

$$= -2\sum_{j,k=1}^{n}\big(p^{jk}w_{x_j}dw\big)_{x_k} + 2\sum_{j,k=1}^{n}p^{jk}w_{x_j}dw_{x_k} + 2\mathcal{A}wdw \qquad (6.12)$$

$$= -2\sum_{j,k=1}^{n}\big(p^{jk}w_{x_j}dw\big)_{x_k} + d\Big(\sum_{j,k=1}^{n}p^{jk}w_{x_j}w_{x_k} + \mathcal{A}w^2\Big)$$

$$- \sum_{j,k=1}^{n}p_t^{jk}w_{x_j}w_{x_k}dt - \mathcal{A}_t w^2 dt - \sum_{j,k=1}^{n}p^{jk}dw_{x_j}dw_{x_k} - \mathcal{A}(dw)^2.$$

Now, from (6.8), (6.11) and (6.12), we get that

$$2II_2$$

$$= -2\sum_{j,k=1}^{n}\Big[\sum_{j',k'=1}^{n}\big(2p^{jk}p^{j'k'}\ell_{x_j},w_{x_j}w_{x_{k'}} - p^{jk}p^{j'k'}\ell_{x_j}w_{x_{j'}},w_{x_{k'}}\big) \qquad (6.13)$$

$$-\mathcal{A}p^{jk}\ell_{x_j}w^2\Big]_{x_k}dt - 2\sum_{j,k=1}^{n}(p^{jk}w_{x_j}dw)_{x_k} + d\Big(\sum_{j,k=1}^{n}p^{jk}w_{x_j}w_{x_k} + \mathcal{A}w^2\Big)$$

$$+2\sum_{j,k=1}^{n}\Big\{\sum_{j',k'=1}^{n}\Big[2p^{jk'}\big(p^{j'k}\ell_{x_{j'}}\big)_{x_{k'}} - \big(p^{jk}p^{j'k'}\ell_{x_{j'}}\big)_{x_{k'}}\Big] - \frac{p_t^{jk}}{2}\Big\}w_{x_j}w_{x_k}dt$$

$$-\Big[\mathcal{A}_t + 2\sum_{j,k=1}^{n}\big(\mathcal{A}p^{jk}\ell_{x_j}\big)_{x_k}\Big]w^2 dt - \sum_{j,k=1}^{n}p^{jk}dw_{x_j}dw_{x_k} - \mathcal{A}(dw)^2.$$

Step 3. Let us compute $2II_3$. By (6.8), we get

$$2II_3 = 2\Big[-\sum_{j,k=1}^{n}\big(p^{jk}w_{x_j}\big)_{x_k} + \mathcal{A}w\Big]\Psi w dt$$

$$= \Big[-2\sum_{j,k=1}^{n}\big(\Psi p^{jk}w_{x_j}w\big)_{x_k} + 2\Psi\sum_{j,k=1}^{n}p^{jk}w_{x_j}w_{x_k}$$

$$+ \sum_{j,k=1}^{n}p^{jk}\Psi_{x_k}(w^2)_{x_j} + 2\mathcal{A}\Psi w^2\Big]dt \qquad (6.14)$$

$$= \left\{ -\sum_{j,k=1}^{n} \left(2\Psi p^{jk} w_{x_j} w - p^{jk} \Psi_{x_j} w^2 \right)_{x_k} + 2\Psi \sum_{j,k=1}^{n} p^{jk} w_{x_j} w_{x_k} \right.$$

$$\left. + \left[-\sum_{j,k=1}^{n} \left(p^{jk} \Psi_{x_k} \right)_{x_j} + 2\mathcal{A}\Psi \right] w^2 \right\} dt.$$

Step 4. Finally, combining the equalities (6.9), (6.13) and (6.14), and noting that

$$\sum_{j,k=1}^{n} p^{jk} dw_{x_j} dw_{x_k} + \mathcal{A}(dw)^2$$

$$= \theta^2 \sum_{j,k=1}^{n} p^{jk} \left(du_{x_j} + \ell_{x_j} du \right) \left(du_{x_k} + \ell_{x_k} du \right) + \theta^2 \mathcal{A}(du)^2,$$

we obtain the desired equality (6.5). □

6.3. *Global Carleman estimate for stochastic parabolic equations*

In this subsection, as a preliminary to prove Theorem 14, we shall derive a global Carleman estimate for the following stochastic parabolic equation:

$$\begin{cases} dz - \displaystyle\sum_{j,k=1}^{n} (b^{jk} z_{x_j})_{x_k} dt = f dt + g dW(t), & \text{in } Q, \\ z = 0, & \text{on } \Sigma, \\ z(0) = z_0, & \text{in } G, \end{cases} \tag{6.15}$$

where $z_0 \in L^2_{\mathcal{F}_0}(\Omega; L^2(G))$, $f \in L^2_{\mathbb{F}}(0,T; L^2(G))$ and $g \in L^2_{\mathbb{F}}(0,T; H^1(G))$.

We begin with the following known technical result (see [15, p. 4, Lemma 1.1] for its proof), which shows the existence of a nonnegative function with an arbitrary given critical point location in G.

Lemma 3. *For any nonempty open subset G_1 of G, there is a $\psi \in C^4(\overline{G})$ such that $\psi > 0$ in G, $\psi = 0$ on Γ, and $|\nabla \psi(x)| > 0$ for all $x \in \overline{G} \setminus G_1$.*

For any (large) parameters $\lambda > 1$ and $\mu > 1$, we choose

$$\theta = e^\ell, \quad \ell = \lambda \alpha, \quad \alpha(t,x) = \frac{e^{\mu\psi(x)} - e^{2\mu|\psi|_{C(\overline{G})}}}{t(T-t)}, \quad \varphi(t,x) = \frac{e^{\mu\psi(x)}}{t(T-t)}, \tag{6.16}$$

and

$$\Psi = 2 \sum_{j,k=1}^{n} b^{jk} \ell_{x_j x_k}. \tag{6.17}$$

Recall that for a positive integer r, we write $O(\mu^r)$ for a function of order μ^r for large μ (which is independent of λ). Meanwhile, we denote by $O_\mu(\lambda^r)$ a function of order λ^r for fixed μ and for large λ. For $j, k = 1, 2, \cdots, n$, it is easy to check that

$$\ell_t = \lambda\alpha_t, \quad \ell_{x_j} = \lambda\mu\varphi\psi_{x_j}, \quad \ell_{x_j x_k} = \lambda\mu^2\varphi\psi_{x_j}\psi_{x_k} + \lambda\mu\varphi\psi_{x_j x_k} \quad (6.18)$$

and that

$$\alpha_t = \varphi^2 O\big(e^{2\mu|\psi|_{C(\overline{G})}}\big), \qquad \varphi_t = \varphi^2 O\big(e^{\mu|\psi|_{C(\overline{G})}}\big). \quad (6.19)$$

The desired Carleman estimate for (6.15) is stated as follows:

Theorem 15. *There is a constant* $\mu_0 = \mu_0(G, G_0, (b^{jk})_{n\times n}) > 0$ *such that for all* $\mu \geq \mu_0$, *one can find two constants* $C = C(\mu) > 0$ *and* $\lambda_0 = \lambda_0(\mu) > 0$ *such that for any* $\lambda \geq \lambda_0$, $z_0 \in L^2_{\mathcal{F}_0}(\Omega; L^2(G))$, $f \in L^2_{\mathbb{F}}(0, T; L^2(G))$ *and* $g \in L^2_{\mathbb{F}}(0, T; H^1(G))$, *the corresponding solution* z *to* (6.15) *satisfies*

$$\lambda^3\mu^4\mathbb{E}\int_Q \varphi^3\theta^2 z^2\,dxdt + \lambda\mu^2\mathbb{E}\int_Q \varphi\theta^2|\nabla z|^2\,dxdt$$

$$\leq C\mathbb{E}\Big[\int_Q \theta^2\big(f^2 + |\nabla g|^2 + \lambda^2\mu^2\varphi^2 g^2\big)dxdt + \lambda^3\mu^4\int_{Q_0}\varphi^3\theta^2 z^2\,dxdt\Big]. \quad (6.20)$$

Proof. We divide the proof into three steps.

Step 1. Let $b^{jk} = p^{jk}$ for $j, k = 1, \cdots, n$ in (6.5). Noting (6.17)–(6.18), from (6.6), we have $\ell_{x_j x_k} = \lambda\mu^2\varphi\psi_{x_j}\psi_{x_k} + \lambda\varphi O(\mu)$ and that

$$\sum_{j,k=1}^n c^{jk}w_{x_j}w_{x_k}$$

$$= \sum_{j,k=1}^n \Big\{ \sum_{j',k'=1}^n \Big[2b^{jk'}b^{j'k}\ell_{x_{j'}x_{k'}} + b^{jk}b^{j'k'}\ell_{x'_j x'_k} + 2b^{jk'}b^{j'k}_{x_{k'}}\ell_{x_{j'}}$$

$$- \big(b^{jk}b^{j'k'}\big)_{x_{k'}}\ell_{x_{j'}}\Big] - \frac{b^{jk}_t}{2}\Big\}w_{x_j}w_{x_k}$$

$$= \sum_{j,k=1}^n \Big\{ \sum_{j',k'=1}^n \Big[2\lambda\mu^2\varphi b^{jk'}b^{j'k}\psi_{x_{j'}}\psi_{x_{k'}} + \lambda\mu^2\varphi b^{jk}b^{j'k'}\psi_{x_j}\psi_{x_{k'}}$$

$$+ \lambda\varphi O(\mu)\Big]\Big\}w_{x_j}w_{x_k} \quad (6.21)$$

$$= 2\lambda\mu^2\varphi\Big(\sum_{j,k=1}^n b^{jk}\psi_{x_j}w_{x_k}\Big)^2 + \lambda\mu^2\varphi\Big(\sum_{j,k=1}^n b^{jk}\psi_{x_j}\psi_{x_k}\Big)\Big(\sum_{j,k=1}^n b^{jk}w_{x_j}w_{x_k}\Big)$$

$$+ \lambda\varphi|\nabla w|^2 O(\mu)$$

$$\geq \big(s_0^2\lambda\mu^2\varphi|\nabla\psi|^2 + \lambda\varphi O(\mu)\big)|\nabla w|^2.$$

Similarly, by the definition of $\mathcal{A}$ in (6.6), and noting (6.19), we see that

$$\mathcal{A} = -\sum_{j,k=1}^{n} \left(b^{jk}\ell_{x_j}\ell_{x_k} - b^{jk}_{x_k}\ell_{x_j} + b^{jk}\ell_{x_j x_k} \right) - \ell_t$$

$$= -\lambda\mu \sum_{j,k=1}^{n} \left[b^{jk}\lambda\mu\varphi^2\psi_{x_j}\psi_{x_k} - b^{jk}_{x_k}\varphi\psi_{x_j} + b^{jk}\left(\mu\varphi\psi_{x_j}\psi_{x_k} + \varphi\psi_{x_j x_k}\right) \right]$$

$$+ \lambda\varphi^2 O\left(e^{2\mu|\psi|_{C(\overline{G})}}\right) \tag{6.22}$$

$$= -\lambda^2\mu^2\varphi^2 \sum_{j,k=1}^{n} b^{jk}\psi_{x_j}\psi_{x_k} + \lambda\varphi^2 O\left(e^{2\mu|\psi|_{C(\overline{G})}}\right).$$

Now, let us estimate $\mathcal{B}$ (recall (6.6) for the definition of $\mathcal{B}$). For this, by (6.18), and recalling the definition of Ψ (in (6.17)), we see that

$$\Psi = 2\lambda\mu \sum_{j,k=1}^{n} b^{jk}\left(\mu\varphi\psi_{x_j}\psi_{x_k} + \varphi\psi_{x_j x_k}\right)$$

$$= 2\lambda\mu^2\varphi \sum_{j,k=1}^{n} b^{jk}\psi_{x_j}\psi_{x_k} + \lambda\varphi O(\mu),$$

$$\ell_{x_{j'}x_{k'}x_k} = \lambda\mu^3\varphi\psi_{x_{j'}}\psi_{x_{k'}}\psi_{x_k} + \lambda\varphi O(\mu^2),$$

$$\ell_{x_{j'}x_{k'}x_j x_k} = \lambda\mu^4\varphi\psi_{x_{j'}}\psi_{x_{k'}}\psi_{x_j}\psi_{x_k} + \lambda\varphi O(\mu^3),$$

$$\Psi_{x_k} = 2\sum_{j',k'=1}^{n} \left(b^{j'k'}\ell_{x_{j'}x_{k'}} \right)_{x_k} = 2\sum_{j',k'=1}^{n} \left(b^{j'k'}_{x_k}\ell_{x_{j'}x_{k'}} + b^{j'k'}\ell_{x_{j'}x_{k'}x_k} \right)$$

$$= 2\lambda\mu^3\varphi \sum_{j',k'=1}^{n} b^{j'k'}\psi_{x_{j'}}\psi_{x_{k'}}\psi_{x_k} + \lambda\varphi O(\mu^2),$$

$$\Psi_{x_j x_k} = 2\sum_{j',k'=1}^{n} \left(b^{j'k'}_{x_j x_k}\ell_{x_{j'}x_{k'}} + b^{j'k'}\ell_{x_{j'}x_{k'}x_j x_k} + 2b^{j'k'}_{x_k}\ell_{x_{j'}x_{k'}x_j} \right)$$

$$= 2\lambda\mu^4\varphi \sum_{j',k'=1}^{n} b^{j'k'}\psi_{x_{j'}}\psi_{x_{k'}}\psi_{x_j}\psi_{x_k} + \lambda\varphi O(\mu^3),$$

$$-\sum_{j,k=1}^{n} \left(b^{jk}\Psi_{x_k} \right)_{x_j} = -\sum_{j,k=1}^{n} \left(b^{jk}_{x_j}\Psi_{x_k} + b^{jk}\Psi_{x_j x_k} \right)$$

$$= -2\lambda\mu^4\varphi \left(\sum_{j,k=1}^{n} b^{jk}\psi_{x_j}\psi_{x_k} \right)^2 + \lambda\varphi O(\mu^3).$$

228   *Qi Lü*

Hence, recalling the definition of $\mathcal{A}$ (in (6.6)), and using (6.18) and (6.19), we have that

$$\mathcal{A}\Psi = -2\lambda^3\mu^4\varphi^3\Big(\sum_{j,k=1}^{n} b^{jk}\psi_{x_j}\psi_{x_k}\Big)^2 + \lambda^3\varphi^3 O(\mu^3) + \lambda^2\varphi^3 O(\mu^2 e^{2\mu|\psi|_{C(\overline{G})}}),$$

$$\mathcal{A}_{x_k} = -\sum_{j',k'=1}^{n}\big(b^{j'k'}_{x_k}\ell_{x_{j'}}\ell_{x_{k'}} + 2b^{j'k'}\ell_{x_j}\ell_{x'_k x_k} - b^{j'k'}_{x'_k x_k}\ell_{x_{j'}} - b^{j'k'}_{x_k}\ell_{x_{j'}x_k}$$

$$+ b^{j'k'}_{x_k}\ell_{x_{j'}x_{k'}} + b^{j'k'}\ell_{x_{j'}x_{k'}x_k}\big) - \ell_{tx_k}$$

$$= -\sum_{j',k'=1}^{n}\big(2b^{j'k'}\ell_{x_{j'}}\ell_{x_{k'}x_k} + b^{j'k'}\ell_{x_{j'}x_{k'}x_k}\big) - \ell_{tx_k} + \big(\lambda\varphi+\lambda^2\varphi^2\big)O(\mu^2)$$

$$= -2\lambda^2\mu^3\varphi^2\sum_{j',k'=1}^{n}b^{j'k'}\psi_{x_{j'}}\psi_{x_{k'}}\psi_{x_k} + \lambda^2\varphi^2 O(\mu^2) + \lambda\varphi^2 O(\mu e^{2\mu|\psi|_{C(\overline{G})}}),$$

$$\sum_{j,k=1}^{n}\mathcal{A}_{x_k}b^{jk}\ell_{x_j} = -2\lambda^3\mu^4\varphi^3\Big(\sum_{j,k=1}^{n}b^{jk}\psi_{x_j}\psi_{x_k}\Big)^2 + \lambda^3\varphi^3 O(\mu^3)$$

$$+ \lambda^2\varphi^3 O(\mu^2 e^{2\mu|\psi|_{C(\overline{G})}}),$$

$$\sum_{j,k=1}^{n}\big(\mathcal{A}b^{jk}\ell_{x_j}\big)_{x_k} = \sum_{j,k=1}^{n}\mathcal{A}_{x_k}b^{jk}\ell_{x_j} + \mathcal{A}\sum_{j,k=1}^{n}\big(b^{jk}_{x_k}\ell_{x_j} + b^{jk}\ell_{x_j x_k}\big)$$

$$= -3\lambda^3\mu^4\varphi^3\Big(\sum_{j,k=1}^{n}b^{jk}\psi_{x_j}\psi_{x_k}\Big)^2 + \lambda^3\varphi^3 O(\mu^3)$$

$$+ \lambda^2\varphi^3 O(\mu^2 e^{2\mu|\psi|_{C(\overline{G})}}),$$

and that

$$\mathcal{A}_t = -\sum_{j,k=1}^{n}\big(b^{jk}\ell_{x_j}\ell_{x_k} - b^{jk}_{x_k}\ell_{x_j} + b^{jk}\ell_{x_j x_k} - \ell_t\big)_t$$

$$= -\sum_{j,k=1}^{n}\big[b^{jk}\big(\ell_{x_j}\ell_{x_k}\big)_t - b^{jk}_{x_k}\ell_{x_j t} + b^{jk}\ell_{x_j x_k t}\big]$$

$$+ \lambda^2\varphi^2 O(\mu^2) + \lambda\varphi^3 O(e^{2\mu|\psi|_{C(\overline{G})}})$$

$$= \lambda^2\varphi^3 O(\mu^2 e^{2\mu|\psi|_{C(\overline{G})}}) + \lambda\varphi^3 O(e^{2\mu|\psi|_{C(\overline{G})}}).$$

From the definition of $\mathcal{B}$ (see (6.6)), we have that

$$\mathcal{B} = -4\lambda^3\mu^4\varphi^3\Big(\sum_{j,k=1}^{n}b^{jk}\psi_{x_j}\psi_{x_k}\Big)^2 + \lambda^3\varphi^3 O(\mu^3) + \lambda^2\varphi^3 O(\mu^2 e^{2\mu|\psi|_{C(\overline{G})}})$$

$$+6\lambda^3\mu^4\varphi^3\Big(\sum_{j,k=1}^{n}b^{jk}\psi_{x_j}\psi_{x_k}\Big)^2+\lambda^3\varphi^3O(\mu^3)+\lambda^2\varphi^3O\big(\mu^2e^{2\mu|\psi|_{C(\overline{G})}}\big)$$

$$+\lambda^2\varphi^3O\big(\mu^2e^{2\mu|\psi|_{C(\overline{G})}}\big)+\lambda\varphi^3O\big(e^{2\mu|\psi|_{C(\overline{G})}}\big)$$

$$-2\lambda\mu^4\varphi\Big(\sum_{j,k=1}^{n}b^{jk}\psi_{x_j}\psi_{x_k}\Big)^2+\lambda\varphi O(\mu^3)$$

$$=2\lambda^3\mu^4\varphi^3\Big(\sum_{j,k=1}^{n}b^{jk}\psi_{x_j}\psi_{x_k}\Big)^2+\lambda^3\varphi^3O(\mu^3)$$

$$+\lambda^2\varphi^3O\big(\mu^2e^{2\mu|\psi|_{C(\overline{G})}}\big)+\lambda\varphi^3O\big(e^{2\mu|\psi|_{C(\overline{G})}}\big),$$

which leads to

$$\mathcal{B}\geq 2s_0^2\lambda^3\mu^4\varphi^3|\nabla\psi|^4+\lambda^3\varphi^3O(\mu^3)+\lambda^2\varphi^3O\big(\mu^2e^{2\mu|\psi|_{C(\overline{G})}}\big)$$
$$+\lambda\varphi^3O\big(e^{2\mu|\psi|_{C(\overline{G})}}\big). \tag{6.23}$$

Step 2. Integrating the equality (6.5) on Q, taking mathematical expectation in both sides, and noting (6.21)–(6.23), we conclude that there is a constant $c_0 > 0$ such that

$$2\mathbb{E}\int_Q\theta\Big[-\sum_{j,k=1}^{n}(b^{jk}w_{x_j})_{x_k}+\mathcal{A}w\Big]\Big[dz-\sum_{j,k=1}^{n}(b^{jk}z_{x_j})_{x_k}dt\Big]dx$$

$$+2\mathbb{E}\int_Q\sum_{j,k=1}^{n}(b^{jk}w_{x_j}dw)_{x_k}dx$$

$$+2\mathbb{E}\int_Q\sum_{j,k=1}^{n}\Big[\sum_{j',k'=1}^{n}\Big(2b^{jk}b^{j'k'}\ell_{x_{j'}}w_{x_j}w_{x_{k'}}-b^{jk}b^{j'k'}\ell_{x_j}w_{x_{j'}}w_{x_{k'}}\Big)$$

$$+\Psi b^{jk}w_{x_j}w-b^{jk}\Big(\mathcal{A}\ell_{x_j}+\frac{\Psi_{x_j}}{2}\Big)w^2\Big]_{x_k}dxdt \tag{6.24}$$

$$\geq 2c_0\mathbb{E}\int_Q\Big[\varphi\big(\lambda\mu^2|\nabla\psi|^2+\lambda O(\mu)\big)|\nabla w|^2+\varphi^3\big(\lambda^3\mu^4|\nabla\psi|^4+\lambda^3O(\mu^3)$$

$$+\lambda^2O\big(\mu^2e^{2\mu|\psi|_{C(\overline{G})}}\big)+\lambda O\big(e^{2\mu|\psi|_{C(\overline{G})}}\big)\big)w^2\Big]dxdt$$

$$+2\mathbb{E}\int_Q\Big|-\sum_{j,k=1}^{n}(b^{jk}w_{x_j})_{x_k}+\mathcal{A}w\Big|^2dxdt$$

$$-\mathbb{E}\int_Q\theta^2\sum_{j,k=1}^{n}b^{jk}\big(dz_{x_j}+\ell_{x_j}dz\big)\big(dz_{x_k}+\ell_{x_k}dz\big)dx-\mathbb{E}\int_Q\theta^2\mathcal{A}(dz)^2dx.$$

By (6.15), we find that

$$2\mathbb{E}\int_Q \theta\Big[-\sum_{j,k=1}^n \big(b^{jk}w_{x_j}\big)_{x_k} + \mathcal{A}w\Big]\Big[dz - \sum_{j,k=1}^n \big(b^{jk}z_{x_j}\big)_{x_k} dt\Big]dx$$

$$= 2\mathbb{E}\int_Q \theta\Big[-\sum_{j,k=1}^n \big(b^{jk}w_{x_j}\big)_{x_k} + \mathcal{A}w\Big]\big(fdt + gdW(t)\big)dx$$

$$= 2\mathbb{E}\int_Q \theta\Big[-\sum_{j,k=1}^n \big(b^{jk}w_{x_j}\big)_{x_k} + \mathcal{A}w\Big]fdtdx \tag{6.25}$$

$$\leq \mathbb{E}\int_Q \Big|-\sum_{j,k=1}^n \big(b^{jk}w_{x_j}\big)_{x_k} + \mathcal{A}w\Big|^2 dtdx + \mathbb{E}\int_Q \theta^2 f^2 dtdx.$$

Since $z|_\Sigma = 0$, we have

$$w|_\Sigma = 0 \quad \text{and} \quad w_{x_j}|_\Sigma = \theta z_{x_j}|_\Sigma = \theta \frac{\partial z}{\partial \nu}\nu_j\Big|_\Sigma, \quad \text{for } j = 1,\cdots,n,$$

where $\nu(x) = (\nu_1(x),\cdots,\nu_n(x))$ is the unit outward normal vector of G at $x \in \Gamma$. Similarly, by Lemma 3, we have

$$\ell_{x_j} = \lambda\mu\varphi\psi_{x_j} = \lambda\mu\varphi\frac{\partial\psi}{\partial\nu}\nu_j \quad \text{for } j = 1,\cdots,n, \quad \text{and} \quad \frac{\partial\psi}{\partial\nu} < 0 \quad \text{on } \Sigma.$$

Therefore, utilizing integration by parts and noting $w|_\Sigma = 0$, we obtain that

$$2\mathbb{E}\int_Q \sum_{j,k=1}^n \big(b^{jk}w_{x_j}dw\big)_{x_k}dx = 2\mathbb{E}\int_\Sigma \sum_{j,k=1}^n b^{jk}w_{x_j}\nu_j dwdx = 0, \tag{6.26}$$

and that

$$2\mathbb{E}\int_Q \sum_{j,k=1}^n\Big[\sum_{j',k'=1}^n\Big(2b^{jk}b^{j'k'}\ell_{x_{j'}}w_{x_j}w_{x_{k'}} - b^{jk}b^{j'k'}\ell_{x_j}w_{x_{j'}}w_{x_{k'}}\Big)$$

$$+\Psi b^{jk}w_{x_j}w - b^{jk}\Big(\mathcal{A}\ell_{x_j} + \frac{\Psi_{x_j}}{2}\Big)w^2\Big]_{x_k}dtdx$$

$$= 2\mathbb{E}\int_\Sigma \sum_{j,k=1}^n\Big[\sum_{j',k'=1}^n\Big(2b^{jk}b^{j'k'}\ell_{x_{j'}}w_{x_j}w_{x_{k'}} - b^{jk}b^{j'k'}\ell_{x_j}w_{x_{j'}}w_{x_{k'}}\Big)$$

$$+\Psi b^{jk}w_{x_j}w - b^{jk}\Big(\mathcal{A}\ell_{x_j} + \frac{\Psi_{x_j}}{2}\Big)w^2\Big]\nu_k d\Gamma dt \tag{6.27}$$

$$= 2\mathbb{E}\int_\Sigma \sum_{j,k=1}^n\Big[\sum_{j',k'=1}^n\Big(2b^{jk}b^{j'k'}\ell_{x_{j'}}w_{x_j}w_{x_{k'}} - b^{jk}b^{j'k'}\ell_{x_j}w_{x_{j'}}w_{x_{k'}}\Big)\Big]\nu_k d\Gamma dt$$

$$= 2\lambda\mu\mathbb{E}\int_{\Sigma}\varphi\frac{\partial\psi}{\partial\nu}\Big(\frac{\partial z}{\partial\nu}\Big)^2\sum_{j,k=1}^{n}\Big[\sum_{j',k'=1}^{n}\Big(2b^{jk}b^{j'k'}\nu_{j'}\nu_j\nu_{k'}$$

$$-b^{jk}b^{j'k'}\nu_j\nu_{j'}\nu_{k'}\Big)\Big]\nu_k\,d\Gamma dt$$

$$= 2\lambda\mu\mathbb{E}\int_{\Sigma}\varphi\frac{\partial\psi}{\partial\nu}\Big(\frac{\partial z}{\partial\nu}\Big)^2\Big(\sum_{j,k=1}^{n}b^{jk}\nu_j\nu_k\Big)^2 d\Gamma dt \le 0.$$

By (6.15), we see that

$$\mathbb{E}\int_Q\theta^2\sum_{j,k=1}^{n}b^{jk}\big(dz_{x_j}+\ell_{x_j}dz\big)\big(dz_{x_k}+\ell_{x_k}dz\big)dx + \mathbb{E}\int_Q\theta^2\mathcal{A}(dz)^2 dx$$

$$= \mathbb{E}\int_Q\theta^2\sum_{j,k=1}^{n}b^{jk}\big(g_{x_j}+\ell_{x_j}g\big)\big(g_{x_k}+\ell_{x_k}g\big)dxdt + \mathbb{E}\int_Q\theta^2\mathcal{A}g^2 dxdt.$$

$$(6.28)$$

Combining (6.24)–(6.28), we arrive at

$$2c_0\mathbb{E}\int_Q\Big[\varphi\big(\lambda\mu^2|\nabla\psi|^2+\lambda O(\mu)\big)|\nabla w|^2+\varphi^3\big(\lambda^3\mu^4|\nabla\psi|^4+\lambda^3 O(\mu^3)$$

$$+\lambda^2 O\big(\mu^2 e^{2\mu|\psi|_{C(\overline{G})}}\big)+\lambda O\big(e^{2\mu|\psi|_{C(\overline{G})}}\big)\big)w^2\Big]dxdt \qquad (6.29)$$

$$\le \mathbb{E}\int_Q\theta^2\Big[f^2+\sum_{j,k=1}^{n}b^{jk}\big(g_{x_j}+\ell_{x_j}g\big)\big(g_{x_k}+\ell_{x_k}g\big)+\mathcal{A}g^2\Big]dxdt.$$

By Lemma 3, $\min_{x\in G\backslash G_1}|\nabla\psi|>0$. Hence, there is a $\mu_0>0$ such that for all $\mu\ge\mu_0$, one can find a constant $\lambda_0=\lambda_0(\mu)$ so that for any $\lambda\ge\lambda_0$, it holds that

$$2\mathbb{E}\int_0^T\int_{G\backslash G_1}\Big[\lambda\varphi\big(\mu^2\min_{x\in G\backslash G_1}|\nabla\psi|^2+O(\mu)\big)|\nabla w|^2$$

$$+\varphi^3\big(\lambda^3\mu^4\min_{x\in G\backslash G_1}|\nabla\psi|^4+\lambda^3 O(\mu^3) \qquad (6.30)$$

$$+\lambda^2 O\big(\mu^2 e^{2\mu|\psi|_{C(\overline{G})}}\big)+\lambda O\big(e^{2\mu|\psi|_{C(\overline{G})}}\big)\big)w^2\Big]dxdt$$

$$\ge c_1\lambda\mu^2\mathbb{E}\int_0^T\int_{G\backslash G_1}\varphi\big(|\nabla w|^2+\lambda^2\mu^2\varphi^2 w^2\big)dxdt,$$

where

$$c_1\overset{\triangle}{=}\min\big(\min_{x\in G\backslash G_1}|\nabla\psi|^2,\ \min_{x\in G\backslash G_1}|\nabla\psi|^4\big).$$

By

$$z_{x_j}=\theta^{-1}\big(w_{x_j}-\ell_{x_j}w\big)=\theta^{-1}\big(w_{x_j}-\lambda\mu\varphi\psi_{x_j}w\big)$$

and
$$w_{x_j} = \theta\big(z_{x_j} + \ell_{x_j}z\big) = \theta\big(z_{x_j} + \lambda\mu\varphi\psi_{x_j}z\big),$$

we obtain that
$$\frac{1}{C}\theta^2\big(|\nabla z|^2 + \lambda^2\mu^2\varphi^2 z^2\big) \le |\nabla w|^2 + \lambda^2\mu^2\varphi^2 w^2 \le C\theta^2\big(|\nabla z|^2 + \lambda^2\mu^2\varphi^2 z^2\big).$$

Therefore, it follows from (6.30) that

$$\lambda\mu^2\mathbb{E}\int_Q \varphi\theta^2\big(|\nabla z|^2 + \lambda^2\mu^2\varphi^2 z^2\big)dxdt$$

$$= \lambda\mu^2\mathbb{E}\int_0^T \Big(\int_{G\backslash G_1} + \int_{G_1}\Big)\varphi\theta^2\big(|\nabla z|^2 + \lambda^2\mu^2\varphi^2 z^2\big)dxdt \qquad (6.31)$$

$$\le C\Big\{\mathbb{E}\int_Q \Big[\varphi\big(\lambda\mu^2|\nabla\psi|^2 + \lambda O(\mu)\big)|\nabla w|^2 + \varphi^3\big(\lambda^3\mu^4|\nabla\psi|^4 + \lambda^3 O(\mu^3)\big)$$

$$+ \lambda^2 O\big(\mu^2 e^{2\mu|\psi|_{C(\overline{G})}}\big) + \lambda O\big(e^{2\mu|\psi|_{C(\overline{G})}}\big)\big)w^2\Big]dxdt$$

$$+ \lambda\mu^2\mathbb{E}\int_0^T\int_{G_1}\varphi\theta^2\big(|\nabla z|^2 + \lambda^2\mu^2\varphi^2 z^2\big)dxdt\Big\}.$$

Combining (6.29) and (6.31), we end up with

$$\lambda\mu^2\mathbb{E}\int_Q \varphi\theta^2\big(|\nabla z|^2 + \lambda^2\varphi^2\mu^2 z^2\big)dtdx$$

$$\le C\Big[\mathbb{E}\int_Q \theta^2\big(f^2 + |\nabla g|^2 + \lambda^2\mu^2\varphi^2 g^2\big)dxdt \qquad (6.32)$$

$$+ \lambda\mu^2\mathbb{E}\int_0^T\int_{G_1}\varphi\theta^2\big(|\nabla z|^2 + \lambda^2\mu^2\varphi^2 z^2\big)dxdt\Big].$$

Step 3. Choose a cut-off function $\zeta \in C_0^\infty(G_0; [0,1])$ so that $\zeta \equiv 1$ on G_1. By

$$d(\varphi\theta^2 z^2) = z^2(\varphi\theta^2)_t dt + 2\varphi\theta^2 z dz + \varphi\theta^2(dz)^2,$$

recalling $\lim\limits_{t\to 0^+}\varphi(t,\cdot) = \lim\limits_{t\to T^-}\varphi(t,\cdot) \equiv 0$ and using (6.15), we find that

$$0 = \mathbb{E}\int_{Q_0}\zeta^2\big[z^2(\varphi\theta^2)_t dt + 2\varphi\theta^2 z dz + \varphi\theta^2(dz)^2\big]dx$$

$$= \mathbb{E}\int_{Q_0}\zeta^2\theta^2\Big\{z^2\big(\varphi_t + 2\lambda\varphi\alpha_t\big) + 2\varphi z\Big[\sum_{j,k=1}^n \big(b^{jk}z_{x_j}\big)_{x_k} + f\Big] + \varphi g^2\Big\}dxdt$$

$$= \mathbb{E}\int_{Q_0}\theta^2\Big[\zeta^2 z^2\big(\varphi_t + 2\lambda\varphi\alpha_t\big) - 2\zeta^2\varphi\sum_{j,k=1}^n b^{jk}z_{x_j}z_{x_k} \qquad (6.33)$$

$$-2\mu\zeta^2\varphi z \sum_{j,k=1}^{n} b^{jk} z_{x_j} \psi_{x_k} - 4\lambda\mu\zeta^2\varphi^2 z \sum_{j,k=1}^{n} b^{jk} z_{x_j} \psi_{x_k}$$

$$-4\zeta\varphi z \sum_{j,k=1}^{n} b^{jk} z_{x_j} \zeta_{x_k} + 2\zeta^2\varphi f z + \zeta^2\varphi g^2 \Big] dxdt.$$

Hence, for any $\varepsilon > 0$, one has

$$2\mathbb{E} \int_{Q_0} \zeta^2\varphi\theta^2 \sum_{j,k=1}^{n} b^{jk} z_{x_j} z_{x_k} dxdt$$

$$= \mathbb{E} \int_{Q_0} \theta^2 \Big[\zeta^2 z^2 (\varphi_t + 2\lambda\varphi\alpha_t) - 2\mu\zeta^2\varphi z \sum_{j,k=1}^{n} b^{jk} z_{x_j} \psi_{x_k}$$

$$-4\lambda\mu\zeta^2\varphi^2 z \sum_{j,k=1}^{n} b^{jk} z_{x_j} \psi_{x_k} - 4\zeta\varphi z \sum_{j,k=1}^{n} b^{jk} z_{x_j} \zeta_{x_k}$$

$$+2\zeta^2\varphi f z + \zeta^2\varphi g^2 \Big] dxdt \qquad (6.34)$$

$$\leq \varepsilon\mathbb{E} \int_{Q_0} \zeta^2\varphi\theta^2 |\nabla z|^2 dxdt + \frac{\mathcal{C}}{\varepsilon}\mathbb{E} \int_{Q_0} \theta^2 \Big(\frac{1}{\lambda^2\mu^2} f^2 + \lambda^2\mu^2\varphi^3 z^2 \Big) dxdt$$

$$+\mathbb{E} \int_{Q_0} \varphi\theta^2 g^2 dxdt.$$

From (6.34), we conclude that

$$\mathbb{E} \int_0^T \int_{G_1} \varphi\theta^2 |\nabla z|^2 dxdt$$

$$\leq \mathcal{C}\Big[\mathbb{E} \int_Q \theta^2 \Big(\frac{1}{\lambda^2\mu^2} f^2 + \varphi g^2 \Big) dxdt + \lambda^2\mu^2 \mathbb{E} \int_{Q_0} \varphi^3\theta^2 z^2 dxdt\Big]. \qquad (6.35)$$

Finally, combining (6.32) and (6.35), we obtain (6.20). $\qquad \square$

Remark 16. In Theorem 15, it is assumed that $g \in L^2_{\mathbb{F}}(0,T;H^1(G))$, which can be relaxed (e.g., [46]).

6.4. *Solution to the state observation problem*

Proof of Theorem 14. Set

$$z = y - \hat{y}.$$

Then, it is easy to see that z satisfies

$$\begin{cases} dz - \sum_{j,k=1}^{n} (b^{jk} z_{x_j})_{x_k} dt = \big[F(\nabla y, y) - F(\nabla\hat{y}, \hat{y})\big] dt \\ \qquad\qquad\qquad\qquad + \big[K(y) - K(\hat{y})\big] dW(t), & \text{in } Q, \\ z = 0, & \text{on } \Sigma, \\ z(0) = y_0 - \hat{y}_0, & \text{in } G. \end{cases}$$

Clearly,

$$F(\nabla y, y) - F(\nabla \hat{y}, \hat{y}) \in L^2_{\mathbb{F}}(0, T; L^2(G))$$

and

$$K(y) - K(\hat{y}) \in L^2_{\mathbb{F}}(0, T; H^1(G)).$$

Hence, we know that z solves the equation (6.15) with

$$f = F(\nabla y, y) - F(\nabla \hat{y}, \hat{y}), \quad g = K(y) - K(\hat{y}).$$

By Theorem 15, there exist a $\lambda_1 > 0$ and a $\mu_1 > 0$ so that for all $\lambda \geq \lambda_1$ and $\mu \geq \mu_1$, it holds that

$$\lambda^3 \mu^4 \mathbb{E} \int_Q \varphi^3 \theta^2 z^2 dx dt + \lambda \mu^2 \mathbb{E} \int_Q \varphi \theta^2 |\nabla z|^2 dx dt$$

$$\leq C\mathbb{E}\left[\int_Q \theta^2 (f^2 + |\nabla g|^2 + \lambda^2 \mu^2 \varphi^2 g^2) dx dt + \lambda^3 \mu^4 \int_{Q_0} \varphi^3 \theta^2 z^2 dx dt\right].$$

By the choice of f, we see that

$$\mathbb{E} \int_Q \theta^2 |f|^2 dx dt \leq \mathbb{E} \int_Q \theta^2 |F(\nabla y, y) - F(\nabla \hat{y}, \hat{y})|^2 dx dt$$

$$\leq L^2 \mathbb{E} \int_Q \theta^2 (|\nabla y - \nabla \hat{y}|^2 + |y - \hat{y}|^2) dx dt$$

$$\leq L^2 \mathbb{E} \int_Q \theta^2 (|\nabla z|^2 + |z|^2) dx dt.$$

Similarly,

$$\mathbb{E} \int_Q \theta^2 (|\nabla g|^2 + \lambda^2 \mu^2 \varphi^2 |g|^2) dx dt \leq L^2 \mathbb{E} \int_Q \theta^2 (|\nabla z|^2 + \lambda^2 \mu^2 \varphi^2 |z|^2) dx dt.$$

Hence, we obtain that

$$\mathbb{E} \int_Q \theta^2 (\lambda^3 \mu^4 \varphi^3 |z|^2 + \lambda \mu^2 \varphi |\nabla z|^2) dx dt$$

$$\leq C\left[L^2 \mathbb{E} \int_Q \theta^2 (|\nabla z|^2 + \lambda^2 \mu^2 \varphi^2 |z|^2) dx dt + \lambda^3 \mu^4 \int_{Q_0} \varphi^3 \theta^2 z^2 dx dt\right].$$

Thus, there is a $\mu_2 \geq \max\{\mu_1, CL^2\}$ such that for all $\lambda \geq \lambda_1$ and $\mu \geq \mu_2$, it holds that

$$\mathbb{E} \int_Q \theta^2 (\lambda^3 \mu^4 \varphi^3 |z|^2 + \lambda \mu \varphi |\nabla z|^2) dx dt \leq C\lambda^3 \mu^4 \int_{Q_0} \varphi^3 \theta^2 z^2 dx dt. \quad (6.36)$$

By the definition of ℓ and θ (see (6.16)), we find that

$$
\mathbb{E} \int_Q \theta^2 \left(\varphi^3 |z|^2 + \varphi |\nabla z|^2 \right) dx dt
$$
$$
\geq \min_{x \in \overline{G}} \left(\varphi\left(\frac{T}{2}, x\right) \theta^2\left(\frac{T}{4}, x\right) \right) \mathbb{E} \int_{\frac{T}{4}}^{\frac{3T}{4}} \int_G \left(|z|^2 + |\nabla z|^2 \right) dx dt \tag{6.37}
$$

and that

$$
\mathbb{E} \int_{Q_0} \varphi^3 \theta^2 z^2 dx dt \leq \max_{(x,t) \in \overline{Q}} \left(\varphi(t,x) \theta^2(t,x) \right) \mathbb{E} \int_{Q_0} z^2 dx dt. \tag{6.38}
$$

Further, by a standard energy estimate (similar to the proof of Proposition 5), we can obtain that for any $0 \leq t \leq s \leq T$, it holds

$$
\mathbb{E} |y(t)|^2_{L^2(G)} \leq e^{CL^2} \mathbb{E} |y(s)|^2_{L^2(G)}. \tag{6.39}
$$

Combining (6.36)–(6.38) and (6.39), we obtain the inequality (6.3). □

7. Local state observation problem for stochastic parabolic equations

In this section, we consider a local state observation problem for stochastic parabolic equations. The main content of this paper is taken from [47].

7.1. *Formulation of the problem*

Let $T > 0$, $G \subset \mathbb{R}^n (n \in \mathbb{N})$ be a given bounded domain. Let $G_0 \subset\subset G$ be a given subdomain of G. Put $Q_0 = (0,T) \times G_0$.

Assume that $b^{jk} \in W^{1,\infty}(0,T; W^{2,\infty}_{\mathrm{loc}}(G)), j, k = 1, 2, \cdots, n$ and for any subset G_1, there exists a constant $s_0 = s_0(G_1) > 0$ such that

$$
\sum_{j,k=1}^n b^{jk} \xi_j \xi_k \geq s_0 |\xi|^2, \tag{7.1}
$$
$$
\forall (t,x,\xi) \equiv (t,x,\xi^1,\xi^2,\cdots,\xi^n) \in (0,T) \times G_1 \times \mathbb{R}^n.
$$

Consider the following stochastic parabolic equation:

$$
dy - \sum_{j,k=1}^n (b^{jk} y_{x_j})_{x_k} dt = a_1 \cdot \nabla y dt + a_2 y dt + a_3 y dW(t) \quad \text{in } Q. \tag{7.2}
$$

Here, $a_1 \in L^\infty_{\mathbb{F}}(0,T; L^\infty_{\mathrm{loc}}(G; \mathbb{R}^n))$, $a_2 \in L^\infty_{\mathbb{F}}(0,T; L^\infty_{\mathrm{loc}}(G))$ and $a_3 \in L^\infty_{\mathbb{F}}(0,T; H^1_{\mathrm{loc}}(G))$.

We call $y \in L_{\mathbb{F}}^2(\Omega; C([0,T]; L_{\text{loc}}^2(G))) \cap L_{\mathbb{F}}^2(0,T; H_{\text{loc}}^1(G))$ a solution to the equation (7.2) if for any $t \in [0,T]$ and any $\rho \in H_0^1(G')$, it holds that

$$
\int_{G'} y(t,x)\rho(x)dx - \int_{G'} y(0,x)\rho(x)dx
$$

$$
= \int_0^t \int_{G'} \left[-\sum_{j,k=1}^n b^{jk} y_{x_j}(s,x)\rho_{x_k}(x) + a_1 \cdot \nabla y(s,x)\rho(x) \right. \tag{7.3}
$$

$$
\left. + a_2 y(s,x)\rho(x) \right] dxds + \int_0^t \int_{G'} a_3 y(s,x)\rho(x)dxdW(s), \quad \mathbb{P}\text{-a.s.}
$$

Our goal is to prove that the value of the solution for (7.2) can be determined by its observation on any arbitrary open subset of G, which is implied by the following local Carleman estimate for (7.2).

Theorem 16. *For any $G' \subset\subset G$ with $G_0 \subset\subset G'$ and $\kappa \in (0, \frac{1}{2})$ arbitrarily fixed, there exists a subdomain $\widetilde{G} \subset\subset G$ with $G' \subset\subset \widetilde{G}$, such that for any solution y of the equation (7.2), it holds that*

$$
\mathbb{E} \int_{\frac{T}{2}-\kappa T}^{\frac{T}{2}+\kappa T} \int_{G'} (|\nabla y|^2 + y^2)dxdt
$$

$$
\leq C \left[\exp\left(\frac{C}{\varepsilon}\right) \mathbb{E} \int_{Q_0} (y^2 + |\nabla y|^2)dxdt + \varepsilon \mathbb{E} \int_0^T \int_{\widetilde{G}} (y^2 + |\nabla y|^2)dxdt \right] \tag{7.4}
$$

for some constant C.

7.2. *Proof of the local Carleman estimate for (7.2)*

Proof of Theorem 16. Without loss of generality, we can assume that the boundary ∂G of G is smooth. Otherwise, we can choose G'' such that $G' \subset\subset G'' \subset\subset G$ with the smooth boundary $\partial G''$ and study the local Carleman estimate on G''. Let $G_0 \subset\subset G$ be a region. By Lemma 3, there exists a function $\psi \in C^4(G)$ such that

$$
\begin{cases} \psi > 0, & \text{in } G, \\ \psi = 0, & \text{on } \partial G, \\ |\nabla \psi| > 0, & \text{in } G \backslash G_0. \end{cases}
$$

Since $G' \subset\subset G'' \subset\subset G$, it is clear that by the choice of ψ we can fix N large enough such that

$$
G' \subset \left\{ x \in G \,\middle|\, \psi(x) > \frac{4}{N} |\psi|_{L^\infty(G)} \right\}. \tag{7.5}
$$

For $\kappa \in (0, \frac{1}{2})$, take $\delta = \frac{1}{\sqrt{2}}(\frac{1}{2} - \kappa)T > 0$, then $\sqrt{2}\delta = (\frac{1}{2} - \kappa)T$. Choose a constant $c > 0$ satisfying

$$c\delta^2 < |\psi|_{L^\infty(G)} < 2c\delta^2. \tag{7.6}$$

Let $t_0 \in [\sqrt{2}\delta, T - \sqrt{2}\delta]$. Set

$$\phi(t, x) \overset{\Delta}{=} \psi(x) - c(t - t_0)^2, \quad \alpha(t, x) \overset{\Delta}{=} \exp(\mu\phi).$$

Put

$$\beta_k \overset{\Delta}{=} \exp\left(\mu\left(\frac{k}{N}|\psi|_{L^\infty(G)} - \frac{c}{N}\delta^2\right)\right), \quad k = 1, 2, 3, 4$$

and

$$Q_k \overset{\Delta}{=} \left\{(t, x) \in [0, T] \times \overline{G} \mid \alpha(t, x) > \beta_k\right\}.$$

It is obvious that $Q_{k+1} \subset Q_k, k = 1, 2, 3$. For any $(t, x) \in [0, T] \times \overline{G}$ with

$$|t - t_0| < \frac{\delta}{\sqrt{N}}, \quad x \in G', \tag{7.7}$$

it is clear from (7.5) that $\psi(x) > \frac{4}{N}|\psi|_{L^\infty(G)}$ and $-c(t - t_0)^2 > -\frac{c}{N}\delta^2$. Then

$$\psi(x) - c(t - t_0)^2 > \frac{4}{N}|\psi|_{L^\infty(G)} - \frac{c}{N}\delta^2.$$

Therefore, the pair (t, x) with the property (7.7) belongs to Q_4.

Next, if $(t, x) \in Q_1$, then

$$\psi(x) - c(t - t_0)^2 > \frac{1}{N}|\psi|_{L^\infty(G)} - \frac{c}{N}\delta^2$$

and then we get from (7.6) that

$$c(t - t_0)^2 < \psi(x) - \frac{1}{N}|\psi|_{L^\infty(G)} + \frac{c}{N}\delta^2 < \left(2 - \frac{1}{N}\right)c\delta^2.$$

Thus

$$|t - t_0| < \sqrt{2}\,\delta.$$

As a result, we conclude that

$$\left(t_0 - \frac{\delta}{\sqrt{N}}, t_0 + \frac{\delta}{\sqrt{N}}\right) \times G' \subset Q_4 \subset Q_1 \subset (t_0 - \sqrt{2}\delta, t_0 + \sqrt{2}\delta) \times \overline{G}.$$

Let $\eta \in C_0^\infty(Q_2)$ be such that

$$\eta \in [0, 1] \text{ and } \eta = 1 \quad \text{in } Q_3.$$

Let $z = \eta y$. Then

$$
\begin{cases}
dz - \displaystyle\sum_{j,k=1}^{n} \left(b^{jk} z_{x_j}\right)_{x_k} dt = a_1 \cdot \nabla z dt + a_2 z dt + f dt + a_3 z dW(t), & \text{in } Q_1 \\
z = 0, & \text{on } \partial Q_1,
\end{cases}
\tag{7.8}
$$

where

$$
f = \eta_t y - 2 \sum_{j,k=1}^{n} b^{jk} \eta_{x_j} y_{x_k} - y \sum_{j,k=1}^{n} \left(b^{jk}\eta_{x_j}\right)_{x_k} - y a_1 \cdot \nabla\eta.
$$

Clearly, f is supported in $Q_2 \setminus Q_3$.

Replace p^{jk} by b^{jk} and u by z in (6.5) and then integrate (6.5) over $G \times (0, T)$. Noting that z is supported in Q_1, after taking mathematical expectation from both sides, we have

$$
2\mathbb{E}\int_{Q_1} \theta \left[-\sum_{j,k=1}^{n} \left(b^{jk} w_{x_j}\right)_{x_k} + \mathcal{A}w \right]\left[dz - \sum_{j,k=1}^{n} \left(b^{jk} z_{x_j}\right)_{x_k} dt \right] dx
$$

$$
+ 2\mathbb{E}\int_{Q_1} \sum_{j,k=1}^{n} \left(b^{jk} w_{x_j} dw\right)_{x_k} dx
$$

$$
+ 2\mathbb{E}\int_{Q_1} \sum_{j,k=1}^{n} \left[\sum_{j',k'=1}^{n} \left(2b^{jk}b^{j'k'}\ell_{x_{j'}} w_{x_j} w_{x_{k'}} - b^{jk}a^{i'j'}\ell_{x_j} w_{x_{j'}} w_{x_{k'}}\right) \right.
$$

$$
\left. + \Psi b^{jk} w_{x_j} w - b^{jk}\left(\mathcal{A}\ell_{x_j} + \frac{1}{2}\Psi_{x_j}\right)w^2 \right]_{x_k} dt dx
$$

$$
\geq 2\mathbb{E}\int_{Q_1} \sum_{j,k=1}^{n} c^{jk} w_{x_j} w_{x_k} dt dx + \mathbb{E}\int_{Q_1} \mathcal{B}w^2 dt
\tag{7.9}
$$

$$
+ 2\mathbb{E}\int_{Q_1} \left[-\sum_{j,k=1}^{n} \left(b^{jk} w_{x_j}\right)_{x_k} + \mathcal{A}w \right]^2 dt dx - \mathbb{E}\int_{Q_1} \theta^2 \mathcal{A}(dz)^2 dx
$$

$$
- 2\mathbb{E}\int_{Q_1} \left[-\sum_{j,k=1}^{n} \left(b^{jk} w_{x_j}\right)_{x_k} + \mathcal{A}w \right] \ell_t w dt dx
$$

$$
- \mathbb{E}\int_{Q_1} \theta^2 \sum_{j,k=1}^{n} b^{jk} \left[(dz_{x_j} + \ell_{x_j} dz)(dz_{x_k} + \ell_{x_k} dz) \right] dt dx.
$$

The first term in the left hand side of (7.9) satisfies that

$$
2\mathbb{E}\int_{Q_1} \theta \left[-\sum_{j,k=1}^{n} \left(b^{jk} w_{x_j}\right)_{x_k} + \mathcal{A}w \right]\left[dz - \sum_{j,k=1}^{n} \left(b^{jk} z_{x_j}\right)_{x_k} dt \right] dx
$$

$$= 2\mathbb{E} \int_{Q_1} \theta \left[-\sum_{j,k=1}^{n} \left(b^{jk} w_{x_j} \right)_{x_k} + \mathcal{A}w \right] \left(a_1 \cdot \nabla z + a_2 z + f \right) dxdt$$

$$+ 2\mathbb{E} \int_{Q_1} \theta \left[-\sum_{j,k=1}^{n} \left(b^{jk} w_{x_j} \right)_{x_k} + \mathcal{A}w \right] a_3 z dW(t) dx$$

$$\leq \mathbb{E} \int_{Q_1} \left| -\sum_{j,k=1}^{n} \left(b^{jk} w_{x_j} \right)_{x_k} + \mathcal{A}w \right|^2 dxdt$$

$$+ \mathbb{E} \int_{Q_1} \theta^2 \left(a_1 \cdot \nabla z + a_2 z + f \right)^2 dxdt.$$

Next, we find that

$$-2\mathbb{E} \int_{Q_1} \left[-\sum_{j,k=1}^{n} \left(b^{jk} w_{x_j} \right)_{x_k} + \mathcal{A}w \right] \ell_t w dxdt$$

$$\leq \mathbb{E} \int_{Q_1} \left| -\sum_{j,k=1}^{n} \left(b^{jk} w_{x_j} \right)_{x_k} + \mathcal{A}w \right|^2 dxdt + \mathbb{E} \int_{Q_1} \ell_t^2 w^2 dxdt.$$

Finally, we get that

$$\mathbb{E} \int_{Q_1} \theta^2 \left(a_1 \cdot \nabla z + a_2 z + f \right)^2 dtdx$$

$$\geq 2\mathbb{E} \int_{Q_1} \sum_{j,k=1}^{n} c^{jk} w_{x_j} w_{x_k} dtdx + \mathbb{E} \int_{Q_1} \mathcal{B}w^2 dxdt$$

$$- \mathbb{E} \int_{Q_1} \theta^2 \mathcal{A}(dz)^2 dx - \mathbb{E} \int_{Q_1} \ell_t^2 w^2 dtdx$$

$$- \mathbb{E} \int_{Q_1} \theta^2 \sum_{j,k=1}^{n} b^{jk} \left[(dz_{x_j} + \ell_{x_j} dz)(dz_{x_k} + \ell_{x_k} dz) \right] dx.$$

(7.10)

We now deal with the left hand side of (7.10). First,

$$\mathbb{E} \int_{Q_1} \theta^2 f^2 dxdt = \mathbb{E} \int_{Q_1} \theta^2 \Big| \eta_t y - 2 \sum_{j,k=1}^{n} b^{jk} \eta_{x_j} y_{x_k}$$

$$- y \sum_{j,k=1}^{n} \left(b^{jk} \eta_{x_j} \right)_{x_k} - y a_1 \cdot \nabla \eta \Big|^2 dxdt$$

$$\leq C\mathbb{E} \int_{Q_2 \backslash Q_3} \theta^2 \left(|\nabla y|^2 + y^2 \right) dxdt.$$

Therefore

$$\mathbb{E}\int_{Q_1}\theta^2\big(a_1\cdot\nabla z+a_2z+f\big)^2dxdt\le 3|a_1|^2_{L^\infty_{\mathbb{F}}(0,T;L^\infty(G''))}\mathbb{E}\int_{Q_1}\theta^2|\nabla z|^2dxdt$$

$$+3|a_2|^2_{L^\infty_{\mathbb{F}}(0,T;L^\infty(G'';\mathbb{R}^n))}\mathbb{E}\int_{Q_1}\theta^2z^2dxdt$$

$$+C\mathbb{E}\int_{Q_2\setminus Q_3}\theta^2(|\nabla y|^2+y^2)dxdt.\quad(7.11)$$

In what follows, we hand the terms in the right hand side of inequality (7.10). Recalling the definitions of α and ψ, we have that

$$\ell_t=\lambda\mu\alpha\phi_t,\quad \ell_{x_j}=\lambda\mu\alpha\psi_{x_j},\quad \ell_{x_jx_k}=\lambda\mu^2\alpha\psi_{x_j}\psi_{x_k}+\lambda\mu\alpha\psi_{x_jx_k},$$

$$\ell_{x_jx_kx_{k'}}=\lambda\mu^3\alpha\psi_{x_j}\psi_{x_k}\psi_{x_{k'}}+\lambda\mu^2\alpha(\psi_{x_jx_{k'}}\psi_{x_k}+\psi_{x_j}\psi_{x_kx_{k'}}+\psi_{x_jx_k}\psi_{x_{k'}})$$

$$+\lambda\mu\alpha\psi_{x_jx_kx_{k'}}$$

and

$$\ell_{x_jx_kx_{j'}x_{k'}}$$
$$=\lambda\mu^4\alpha\psi_{x_j}\psi_{x_k}\psi_{x_{j'}}\psi_{x_{k'}}+\lambda\mu^3\alpha\big(\psi_{x_jx_{j'}}\psi_{x_k}\psi_{x_{k'}}+\psi_{x_kx_{k'}}\psi_{x_j}\psi_{x_{j'}}$$
$$+\psi_{x_j}\psi_{x_{k'}}\psi_{x_{j'}x_k}+\psi_{x_{j'}}\psi_{x_k}\psi_{x_jx_{k'}}+\psi_{x_j}\psi_{x_k}\psi_{x_{j'}x_{k'}}+\psi_{x_{j'}}\psi_{x_{k'}}\psi_{x_jx_k}\big)$$
$$+\lambda\mu^2\alpha\big(\psi_{x_jx_{j'}x_{k'}}\psi_{x_k}+\psi_{x_{j'}}\psi_{x_jx_kx_{k'}}+\psi_{x_jx_{k'}}\psi_{x_{j'}x_k}+\psi_{x_jx_{j'}}\psi_{x_kx_{k'}}$$
$$+\psi_{x_j}\psi_{x_kx_{j'}x_{k'}}+\psi_{x_k}\psi_{x_{j'}x_jx_{k'}}+\psi_{x_jx_k}\psi_{x_{j'}x_{k'}}\big)+\lambda\mu\alpha\psi_{x_jx_kx_{j'}x_{k'}}.$$

Noting that $\Psi=2\displaystyle\sum_{j,k=1}^n b^{jk}\ell_{x_jx_k}$, we have

$$\Psi=2\lambda\mu^2\alpha\sum_{j,k=1}^n b^{jk}\psi_{x_j}\psi_{x_k}+2\lambda\mu\alpha\sum_{j,k=1}^n b^{jk}\psi_{x_jx_k}.\quad(7.12)$$

From (6.6), we get that

$$\sum_{j,k=1}^n c^{jk}v_{x_j}v_{x_k}$$

$$=\sum_{j,k=1}^n\bigg\{\sum_{j',k'=1}^n\Big[2b^{jk'}(b^{j'k}\ell_{x_{j'}})_{x_{k'}}-(b^{jk}b^{j'k'}\ell_{x_{j'}})_{x_{k'}}\Big]-\frac{b^{jk}_t}{2}+\Psi b^{jk}\bigg\}v_{x_j}v_{x_k}$$

$$=\sum_{j,k=1}^n\bigg[\sum_{j',k'=1}^n\Big(2b^{jk'}b^{j'k}\ell_{x_{j'}x_{k'}}+b^{jk}b^{j'k'}\ell_{x_{j'}x_{k'}}+2b^{jk'}b^{j'k}_{x_{k'}}\ell_{x_{j'}}\Big)-\frac{b^{jk}_t}{2}\bigg]v_{x_j}v_{x_k}$$

$$=2\lambda\mu^2\alpha\bigg(\sum_{j,k=1}^n b^{jk}\psi_{x_j}v_{x_k}\bigg)^2+\lambda\mu^2\alpha\sum_{j,k=1}^n b^{jk}\psi_{x_j}\psi_{x_k}\sum_{j,k=1}^n b^{jk}v_{x_j}v_{x_k}$$

$$+\lambda\alpha O(\mu)|\nabla v|^2+O(1)|\nabla v|^2\quad(7.13)$$

and

$$
\begin{aligned}
\mathcal{A} &= \sum_{j,k=1}^{n} b^{jk} \ell_{x_j x_k} + \sum_{j,k=1}^{n} b^{jk}_{x_k} \ell_{x_j} - \sum_{j,k=1}^{n} b^{jk} \ell_{x_j} \ell_{x_k} - \Psi \\
&= - \sum_{j,k=1}^{n} b^{jk} \ell_{x_j x_k} + \sum_{j,k=1}^{n} b^{jk}_{x_k} \ell_{x_j} - \sum_{j,k=1}^{n} b^{jk} \ell_{x_j} \ell_{x_k} \\
&= -\lambda^2 \mu^2 \alpha^2 \sum_{j,k=1}^{n} b^{jk} \psi_{x_j} \psi_{x_k} + \lambda \alpha O(\mu).
\end{aligned}
\tag{7.14}
$$

By (7.14), we find that

$$
\begin{aligned}
\theta^2 \mathcal{A}(dz)^2 = \Big(&- \lambda^2 \mu^2 \alpha^2 a_3^2 \sum_{j,k=1}^{n} b^{jk} \psi_{x_j} \psi_{x_k} \\
&- \lambda \mu^2 \alpha a_3^2 \sum_{j,k=1}^{n} b^{jk} \psi_{x_j} \psi_{x_k} + \lambda \alpha a_3^2 O(\mu) \Big) v^2.
\end{aligned}
\tag{7.15}
$$

According to (7.12) and (7.14), we obtain that

$$
\begin{aligned}
2\mathcal{A}\Psi = &-4\lambda^3 \mu^4 \alpha^3 \Big(\sum_{j,k=1}^{n} b^{jk} \psi_{x_j} \psi_{x_k} \Big)^2 \\
&-4\lambda^3 \mu^3 \alpha^3 \sum_{j,k=1}^{n} b^{jk} \psi_{x_j} \psi_{x_k} \sum_{j,k=1}^{n} b^{jk} \psi_{x_j x_k} \\
&-4\lambda^2 \mu^4 \alpha^2 \Big(\sum_{j,k=1}^{n} b^{jk} \psi_{x_j} \psi_{x_k} \Big)^2 + \lambda^2 \alpha^2 O(\mu^2)
\end{aligned}
\tag{7.16}
$$

and that

$$
2 \sum_{j,k=1}^{n} (\mathcal{A}b^{jk} \ell_{x_j})_{x_k} = 2 \sum_{j,k=1}^{n} b^{jk} \ell_{x_j} \mathcal{A}_{x_k} + 2 \sum_{j,k=1}^{n} \mathcal{A} b^{jk}_{x_k} \ell_{x_j} + 2 \sum_{j,k=1}^{n} \mathcal{A} b^{jk} \ell_{x_j x_k}.
\tag{7.17}
$$

Now we turn to estimate the terms in the right hand side of (7.17) one by one. The first one reads

$$
\begin{aligned}
&2 \sum_{j,k=1}^{n} b^{jk} \ell_{x_j} \mathcal{A}_{x_k} \\
&= -2 \sum_{j,k=1}^{n} b^{jk} \ell_{x_j} \sum_{j',k'=1}^{n} \Big(b^{j'k'}_{x_k} \ell_{x_{j'}} \ell_{x_{k'}} + 2 b^{j'k'} \ell_{x_j} \ell_{x_{k'} x_k} - b^{j'k'}_{x_k x_k} \ell_{x_{j'}} \\
&\qquad - b^{j'k'}_{x_{k'}} \ell_{x_j} \ell_{x_k} + b^{j'k'}_{x_k} \ell_{x_{j'} x_{k'}} + b^{j'k'} \ell_{x_{j'} x_{k'} x_k} \Big)
\end{aligned}
\tag{7.18}
$$

$$= -4\lambda^3\mu^4\alpha^3 \Big(\sum_{j,k=1}^{n} b^{jk}\psi_{x_j}\psi_{x_k} \Big)^2 - 2\lambda^2\mu^4\alpha^2 \Big(\sum_{j,k=1}^{n} b^{jk}\psi_{x_j}\psi_{x_k} \Big)^2$$
$$+\lambda^3\alpha^3 O(\mu^3) + \lambda^2\alpha^2 O(\mu^3).$$

The second one is

$$2\sum_{j,k=1}^{n} \mathcal{A} b^{jk}_{x_k}\ell_{x_j}$$
$$= \sum_{j,k=1}^{n} b^{jk}_{x_k}\ell_{x_j} \sum_{j',k'=1}^{n} \Big(-b^{j'k'}\ell_{x_{j'}x_{k'}} + b^{j'k'}_{x_{k'}}\ell_{x_{j'}} - b^{j'k'}\ell_{x_{j'}}\ell_{x_{k'}} \Big) \quad (7.19)$$
$$= \lambda^2\alpha^2 O(\mu^3).$$

The third one reads

$$2\sum_{j,k=1}^{n} \mathcal{A} b^{jk}\ell_{x_j x_k}$$
$$= 2\sum_{j,k=1}^{n} b^{jk} \sum_{j',k'=1}^{n} \Big(-b^{j'k'}\ell_{x_{j'}x_{k'}} + b^{j'k'}_{x_{k'}} - b^{j'k'}\ell_{x_{j'}x_{k'}} \Big)$$
$$= -2\lambda^3\mu^4\alpha^3 \Big(\sum_{j,k=1}^{n} b^{jk}\psi_{x_j}\psi_{x_k} \Big)^2 - 2\lambda^3\mu^3\alpha^3 \Big(\sum_{j,k=1}^{n} b^{jk}\psi_{x_j}\psi_{x_k} \Big)^2 \quad (7.20)$$
$$-2\lambda^2\mu^4\alpha^2 \Big(\sum_{j,k=1}^{n} b^{jk}\psi_{x_j}\psi_{x_k} \Big)^2 + \lambda^2\alpha^2 O(\mu^3).$$

Due to (7.18)–(7.20), we obtain that

$$2\sum_{j,k=1}^{n} \big(\mathcal{A} b^{jk}\ell_{x_j} \big)_{x_k}$$
$$= -6\lambda^3\mu^4\alpha^3 \Big(\sum_{j,k=1}^{n} b^{jk}\psi_{x_j}\psi_{x_k} \Big)^2 - 4\lambda^2\mu^4\alpha^2 \Big(\sum_{j,k=1}^{n} b^{jk}\psi_{x_j}\psi_{x_k} \Big)^2 \quad (7.21)$$
$$+\lambda^3\alpha^3 O(\mu^3) + \lambda^2\alpha^2 O(\mu^2).$$

From (6.6) again, we obtain that

$$\mathcal{A}_t = -2\lambda^2\mu^3\alpha^2\phi_t \sum_{j,k=1}^{n} b^{jk}\psi_{x_j}\psi_{x_k} - \lambda^2\mu^2\alpha^2 \sum_{j,k=1}^{n} b^{jk}_t\psi_{x_j}\psi_{x_k}$$
$$-\lambda\mu^3\alpha\phi_t \sum_{j,k=1}^{n} b^{jk}\psi_{x_j}\psi_{x_k} - \lambda\mu^2\alpha \sum_{j,k=1}^{n} b^{jk}_t\psi_{x_j}\psi_{x_k} \quad (7.22)$$

$$+\lambda\mu^2\alpha\phi_t \sum_{j,k=1}^{n} \left(b_{x_k}^{jk}\psi_{x_j} - b^{jk}\psi_{x_jx_k}\right) + \lambda\mu\alpha \sum_{j,k=1}^{n} \left(b_{x_kt}^{jk}\psi_{x_j} - b_t^{jk}\psi_{x_jx_k}\right)$$
$$= \lambda^2\alpha^2 O(\mu^3) + \lambda\alpha O(\mu^3).$$

Further, we have that

$$\sum_{j,k=1}^{n} b_{x_j}^{jk}\Psi_{x_k} = \sum_{j,k=1}^{n} b_{x_j}^{jk} \sum_{j',k'=1}^{n} \left(b_{x_k}^{j'k'}\ell_{x_{j'}x_{k'}} + b^{j'k'}\ell_{x_{j'}x_{k'}x_j}\right)$$
$$= \lambda\alpha O(\mu^3)$$
(7.23)

and that

$$\sum_{j,k=1}^{n} b^{jk}\Psi_{x_jx_k} = \sum_{j,k=1}^{n} b^{jk} \sum_{j',k'=1}^{n} \left(b_{x_jx_k}^{j'k'}\ell_{x_{j'}x_{k'}} + b_{x_k}^{j'k'}\ell_{x_{j'}x_{k'}x_j}\right.$$
$$\left. +b_{x_j}^{j'k'}\ell_{x_{j'}x_{k'}x_k} + b^{j'k'}\ell_{x_{j'}x_{k'}x_jx_k}\right)$$
(7.24)
$$= 2\lambda\mu^4\alpha \sum_{j,k=1}^{n} \left(b^{jk}\psi_{x_j}\psi_{x_k}\right)^2 + \lambda\alpha O(\mu^3).$$

Combining (7.23) and (7.24), we get that

$$\sum_{j,k=1}^{n} (b^{jk}\Psi_{x_k})_{x_j} = \sum_{j,k=1}^{n} b_{x_j}^{jk}\Psi_{x_k} + \sum_{j,k=1}^{n} b^{jk}\Psi_{x_jx_k}$$
(7.25)
$$= 2\lambda\mu^4\alpha \left(\sum_{j,k=1}^{n} b^{jk}\psi_{x_j}\psi_{x_k}\right)^2 + \lambda\alpha O(\mu^3).$$

According to (7.16), (7.21), (7.22) and (7.25), we obtain that

$$\mathcal{B} = 2\lambda^3\mu^4\alpha^3 \left(\sum_{j,k=1}^{n} b^{jk}\psi_{x_j}\psi_{x_k}\right)^2 + 2\lambda^2\mu^4\alpha^2 \left(\sum_{j,k=1}^{n} b^{jk}\psi_{x_j}\psi_{x_k}\right)^2$$
(7.26)
$$-2\lambda\mu^4\alpha \left(\sum_{j,k=1}^{n} b^{jk}\psi_{x_j}\psi_{x_k}\right)^2 + \lambda^3\alpha^3 O(\mu^3).$$

Next, by Itô's formula, we see that

$$(dz_{x_j} + \ell_{x_j}dz)(dz_{x_k} + \ell_{x_k}dz)$$
$$= (a_3z)_{x_j}(a_3z)_{x_k}dt + \ell_{x_j}dz_{x_k}dz + \ell_{x_j}dz_{x_k}dz + \ell_{x_j}\ell_{x_k}a_3^2z^2dt.$$

Then

$$\mathbb{E}\int_Q \theta^2 \sum_{j,k=1}^{n} b^{jk}(dz_{x_j} + \ell_{x_j}dz)(dz_{x_k} + \ell_{x_k}dz)dx$$
$$\leq \mathcal{C}|a_3|_{L_{\mathbb{F}}^\infty(0,T;H^1(G''))}^2 \left[\mathbb{E}\int_Q \theta^2(z^2 + |\nabla z|^2)dxdt + \lambda^2\mu^2\mathbb{E}\int_Q \theta^2\alpha^2|z|^2dxdt\right].$$

Noting that $v = \theta z$ and $z = \theta^{-1} v$, we get that

$$\frac{1}{C} \theta^2 \left(|\nabla z|^2 + \lambda^2 \mu^2 \alpha^2 z^2 \right) \leq |\nabla v|^2 + \lambda^2 \mu^2 \alpha^2 |v|^2 \leq C \theta^2 \left(|\nabla z|^2 + \lambda^2 \mu^2 \alpha^2 z^2 \right).$$

$$(7.27)$$

Recalling the definitions of α, ϕ and ψ, it is easy to verify that if we fix a constant $\mu_0 > 1$, which may be chosen to be so large that those terms involving $\lambda^3 \mu^3 \alpha^3, \lambda^{k_1} \mu^{k_2} \alpha^{k_1}, k_1 \leq 2, k_2 \leq 4$, can all be absorbed by the term involving $\lambda^3 \mu^4 \alpha^3$. Also, for large λ being chosen, those terms with bounded coefficients can be absorbed. Consequently, it is clear that for such a constant $\mu_0 > 1$, for any $\mu > \mu_0$, there exists a constant λ_0 depending on μ, which we denote explicitly by $\lambda_0(\mu)$, such that for any $\lambda > \lambda_0(\mu)$, due to (7.10), (7.11), (7.13), (7.15), (7.26) and (7.27), it follows that

$$\mathbb{E} \int_{Q_2 \backslash Q_3} \theta^2 (|\nabla y|^2 + y^2) dx dt$$

$$\geq C \lambda^3 \mu^4 \mathbb{E} \int_{Q_1 \backslash Q_0} \alpha^3 \left(\sum_{j,k=1}^n b^{jk} \psi_{x_j} \psi_{x_k} \right)^2 v^2 dx dt \qquad (7.28)$$

$$+ C \lambda \mu^2 \mathbb{E} \int_{Q_1 \backslash Q_0} \alpha \left(\sum_{j,k=1}^n b^{jk} \psi_{x_j} \psi_{x_k} \right)^2 |\nabla v|^2 dx dt.$$

From (7.28), (7.1) and the properties of ψ, we obtain that

$$\lambda^3 \mu^4 \mathbb{E} \int_{Q_1} \alpha^3 \theta^2 z^2 dx dt + \lambda \mu^2 \mathbb{E} \int_{Q_1} \alpha \theta^2 |\nabla z|^2 dx dt$$

$$\leq C \lambda^3 \mu^4 \mathbb{E} \int_{Q_0} \alpha^3 \theta^2 y^2 dx dt + C \lambda \mu^2 \mathbb{E} \int_{Q_0} \alpha \theta^2 |\nabla y|^2 dx dt \qquad (7.29)$$

$$+ C \mathbb{E} \int_{Q_2 \backslash Q_3} \theta^2 (|\nabla y|^2 + y^2) dx dt.$$

Here C is a constant depending on s_0.

Noting that $z = y$ in Q_3 and $Q_3 \subset Q_1$, then (7.29) implies that

$$\lambda^3 \mu^4 \mathbb{E} \int_{Q_3} \alpha^3 \theta^2 y^2 dx dt + \lambda \mu^2 \mathbb{E} \int_{Q_3} \alpha \theta^2 |\nabla y|^2 dx dt$$

$$\leq C \lambda^3 \mu^4 \mathbb{E} \int_{Q_0} \alpha^3 \theta^2 y^2 dx dt + \lambda \mu^2 \mathbb{E} \int_{Q_0} \alpha \theta^2 |\nabla y|^2 dx dt$$

$$+ C \mathbb{E} \int_{Q_2 \backslash Q_3} \theta^2 (|\nabla y|^2 + y^2) dx dt.$$

From the definition of $Q_k, k = 1, 2, 3, 4$ and G', we have that

$$\lambda^3 \mu^4 \mathbb{E} \int_{Q_3} \alpha^3 \theta^2 y^2 dx dt + \lambda \mu^2 \mathbb{E} \int_{Q_3} \alpha \theta^2 |\nabla y|^2 dx dt$$

$$\geq C \exp(2\lambda\beta_4)\mathbb{E} \int_{t_0-\frac{\delta}{\sqrt{N}}}^{t_0+\frac{\delta}{\sqrt{N}}} \int_{G'} \left(\lambda^3 y^2 + \lambda|\nabla y|^2\right) dxdt.$$

From the relationship between Q_2 and Q_3, we know that

$$\mathbb{E} \int_{Q_2 \backslash Q_3} \theta^2(|\nabla y|^2 + y^2) dxdt \leq \exp(2\lambda\beta_3)\mathbb{E} \int_{Q_1} (y^2 + |\nabla y|^2) dxdt.$$

Therefore, we get that

$$\exp(2\lambda\beta_4)\mathbb{E} \int_{t_0-\frac{\delta}{\sqrt{N}}}^{t_0+\frac{\delta}{\sqrt{N}}} \int_{G'} \left(\lambda^3 y^2 + \lambda|\nabla y|^2\right) dxdt$$

$$\leq C\lambda^3\mu^4\mathbb{E} \int_{Q_0} \alpha^3\theta^2 y^2 dxdt + C\lambda\mu^2\mathbb{E} \int_{Q_0} \alpha\theta^2 |\nabla y|^2 dxdt$$

$$+ C\exp(2\lambda\beta_3)\mathbb{E} \int_{Q_1} (y^2 + |\nabla y|^2) dxdt.$$

Hence, we know that there exists a constant $\mu_1 > \mu_0$ such that for any $\mu > \mu_1$, there corresponds a constant λ_1 depending on μ, which we denote by $\lambda_1(\mu)$, such that for any $\lambda > \lambda_1(\mu)$,

$$\exp(2\lambda\beta_4)\mathbb{E} \int_{t_0-\frac{\delta}{\sqrt{N}}}^{t_0+\frac{\delta}{\sqrt{N}}} \int_{G'} \left(\lambda^3 y^2 + \lambda|\nabla y|^2\right) dxdt$$

$$\leq C\exp(C\lambda)\mathbb{E} \int_{Q_0} \left(\lambda^3 y^2 + \lambda|\nabla y|^2\right) dxdt$$

$$+ C\exp(2\lambda\beta_3)\mathbb{E} \int_{Q_1} (y^2 + |\nabla y|^2) dxdt.$$

Consequently, there is a constant $\lambda_2 > \lambda_1$, such that for $\lambda > \lambda_2$ it holds from the above inequality that

$$\mathbb{E} \int_{t_0-\frac{\delta}{\sqrt{N}}}^{t_0+\frac{\delta}{\sqrt{N}}} \int_{G'} (y^2 + |\nabla y|^2) dxdt$$

$$\leq C\exp(C\lambda)\mathbb{E} \int_{Q_0} \left(y^2 + |\nabla y|^2\right) dxdt$$

$$+ C\exp\left(-2\lambda(\beta_4 - \beta_3)\right)\mathbb{E} \int_{Q_1} (y^2 + |\nabla y|^2) dxdt.$$

Put $\varepsilon_0 = \exp\left(-2\lambda_1(\beta_4 - \beta_3)\right)$. Then one finds that for any $0 < \varepsilon < \varepsilon_0$, it holds that

$$\mathbb{E} \int_{t_0-\frac{\delta}{\sqrt{N}}}^{t_0+\frac{\delta}{\sqrt{N}}} \int_{G'} (y^2 + |\nabla y|^2) dxdt \tag{7.30}$$

$$\leq C\exp\left(\frac{C}{\varepsilon}\right)\mathbb{E} \int_{Q_0} (y^2 + |\nabla y|^2) dxdt + C\varepsilon\mathbb{E} \int_{Q_1} (y^2 + |\nabla y|^2) dxdt.$$

Clearly, (7.30) holds for $\varepsilon > \varepsilon_0$. Hence, it is true for all $\varepsilon > 0$.

Due to this result, it is of course that (7.30) holds for any $\varepsilon > 0$.

Next, we deal with the left hand side of (7.30) to end the proof. Noticing that $t_0 \in [\sqrt{2}\delta, T - \sqrt{2}\delta]$, take

$$t_0 = \sqrt{2}\delta + \frac{k\delta}{\sqrt{N}}, \quad k = 0, 1, 2, \cdots, m$$

such that

$$\sqrt{2}\,\delta + \frac{m\delta}{\sqrt{N}} \leq T - \sqrt{2}\delta \leq \sqrt{2}\,\delta + \frac{(m+1)\delta}{\sqrt{N}}.$$

Then, it follows that

$$\mathbb{E} \int_{\frac{T}{2}-\kappa}^{\frac{T}{2}+\kappa} \int_{G'} \left(|\nabla y|^2 + y^2 \right) dxdt$$

$$= \mathbb{E} \int_{\sqrt{2}\,\delta}^{T-\sqrt{2}\,\delta} \int_{G'} \left(|\nabla y|^2 + y^2 \right) dxdt$$

$$\leq \sum_{k=1}^{m} \mathbb{E} \int_{\sqrt{2}\,\delta + \frac{(k-1)\delta}{\sqrt{N}}}^{\sqrt{2}\,\delta + \frac{(k+1)\delta}{\sqrt{N}}} \int_{G'} \left(|\nabla y|^2 + y^2 \right) dxdt$$

$$\leq \mathcal{C} \exp\left(\frac{\mathcal{C}}{\varepsilon}\right) \mathbb{E} \int_{Q_0} \left(y^2 + |\nabla y|^2 \right) dxdt + C\varepsilon \mathbb{E} \int_{Q_1} \left(y^2 + |\nabla y|^2 \right) dxdt.$$

Taking $(0, T) \times \widetilde{G} = Q_1$ in the above inequality, we complete the proof of Theorem 16. □

8. Notes and comments

In this note, we only focus on some recent works on state observation problems for stochastic hyperbolic equations and stochastic parabolic equations. There are some other interesting results on state observation problems recently (e.g., [18, 19, 29–31]). Further, for other control problems for SPDEs, we refer the readers to [48, 49] for concise lecture notes and [28, 50, 51] for treatises.

There are plenty of open problems in the topic of this lecture note. Some of them are particularly relevant and could need new ideas and further developments.

- **Efficiency algorithm for the construction of the state from the observation**

    In this note, we mainly consider the first and the second questions in the state observation problem presented in Section 3. The third one is still open. In Section 4, we have shown that the state observation problems can be solved by studying the corresponding optimization problems. However, it is hard to solve these optimization problems. For example, if one want to use the gradient method to solve (**P3**), then one has to solve some backward stochastic hyperbolic equations numerically, which is very difficult to be done. Indeed, the numerics of backward stochastic partial differential equations are far from being well understood.

- **Observability estimate and unique continuation property with less restrictive conditions**

  We prove the inequality (4.34) under Conditions (4.1) and (4.2). It is well known that a sharp sufficient condition for establishing observability estimate for deterministic hyperbolic equations with time invariant lower order terms is that the triple (G, Γ_0, T) satisfies the geometric control condition introduced in [52]. It would be quite interesting and challenging to extend this result to the stochastic setting. However, there are lots of things to be done before solving this problem. For instance, the propagation of singularities for stochastic partial differential equations, at least, for stochastic hyperbolic equations, should be established.

- **The optimal observation method**

  For practical problems, observations are made by a limited number of expensive sensors. Numerous applications from engineering provide evidence that this choice of sensor location may have a significant bearing upon the accuracy achievable in state observation problems. The problem of where to locate the observation sites then has to be solved. Interesting works are done for PDEs (e.g., [53]). As far as we know, there are no published works addressing this problem for SPDEs.

- **What can we benefit from the uncertainty?**

  From Subsection 4.4, we see that the uncertainty in SPDEs places some advantage in state observation problems. What can we benefit from the uncertainty in SPDEs for state observation problems, or more generally, for control problems, is far from being understood. We believe the study for this problem will lead to a new insight in uncertainty.

References

[1] R. A. Carmona and B. Rozovskii, eds., *Stochastic partial differential equations: six perspectives*. Vol. 64, *Mathematical Surveys and Monographs*, American Mathematical Society, Providence, RI (1999).

[2] G. D. Prato and J. Zabczyk, *Stochastic Equations in Infinite Dimensions.* Cambridge University Press (2009).

[3] P. E. Greenwood and L. M. Ward, *Stochastic Neuron Models.* Springer, Cham (2016).

[4] H. Holden, B. Øksendal, J. Ubøe, and T. Zhang, *Stochastic Partial Differential Equations. A modeling, white noise functional approach.* Springer, New York (2010).

[5] P. Kotelenez, *Stochastic Ordinary and Stochastic Partial Differential Equations. Transition from microscopic to macroscopic equations.* Springer, New York (2008).

[6] A. J. Majda, I. Timofeyev, and E. Vanden Eijnden, A mathematical framework for stochastic climate models, *Comm. Pure Appl. Math.* **54**(8), 891–974 (2001).

[7] J.-L. Lions, Exact controllability, stabilization and perturbations for distributed systems, *SIAM Rev.* **30**(1), 1–68 (1988).

[8] R. M. Murray, ed., *Control in an Information Rich World.* SIAM, Philadelphia (2003).

[9] K. Law, A. Stuart, and K. Zygalakis, *Data assimilation. A mathematical introduction.* Springer, Cham (2015).

[10] T. Carleman, Sur un problème d'unicité pur les systèmes d'équations aux dérivées partielles à deux variables indépendantes, *Ark. Mat., Astr. Fys.* **26**(17), 9 (1939).

[11] L. Hörmander, *The Analysis of Linear Partial Differential Operators IV. Fourier Integral Operators.* Springer, Berlin, Heidelberg (2009).

[12] N. Lerner, *Carleman inequalities. An introduction and more.* Springer, Cham (2019).

[13] M. Bellassoued and M. Yamamoto, *Carleman estimates and applications to inverse problems for hyperbolic systems.* Springer, Tokyo (2017).

[14] M. V. Klibanov and A. A. Timonov, *Carleman Estimates for Coefficient Inverse Problems and Numerical Applications.* VSP, Utrecht (2004).

[15] A. V. Fursikov and O. Y. Imanuvilov, *Controllability of Evolution Equations.* Research Institute of Mathematics, Seoul National University, Seoul, Korea (1996).

[16] X. Fu, Q. Lü, and X. Zhang, *Carleman estimates for second order partial differential operators and applications. A unified approach.* Springer, Cham (2019).

[17] X. Fu and X. Liu, A weighted identity for stochastic partial differential operators and its applications, *J. Differential Equations.* **262**(6), 3551–3582 (2017).

[18] X. Fu and X. Liu, Controllability and observability of some stochastic complex Ginzburg-Landau equations, *SIAM J. Control Optim.* **55**(2), 1102–1127 (2017).

[19] P. Gao, M. Chen, and Y. Li, Observability estimates and null controllability for forward and backward linear stochastic Kuramoto-Sivashinsky equations, *SIAM J. Control Optim.* **53**(1), 475–500 (2015).

[20] H. Li and Q. Lü, A quantitative boundary unique continuation for stochastic parabolic equations, *J. Math. Anal. Appl.* **402**(2), 518–526 (2013).

[21] Q. Lü, Carleman estimate for stochastic parabolic equations and inverse stochastic parabolic problems, *Inverse Problems.* **28**(4), 045008, 18 (2012).

[22] Q. Lü, Exact controllability for stochastic Schrödinger equations, *J. Differential Equations.* **255**(8), 2484–2504 (2013).

[23] Q. Lü, Observability estimate and state observation problems for stochastic hyperbolic equations, *Inverse Problems.* **29**(9), 095011, 22 (2013).

[24] Q. Lü, Exact controllability for stochastic transport equations, *SIAM J. Control Optim.* **52**(1), 397–419 (2014).

[25] Q. Lü, Observability estimate for stochastic Schrödinger equations and its applications, *SIAM J. Control Optim.* **51**(1), 121–144 (2013).

[26] Q. Lü and Z. Yin, Local state observation for stochastic hyperbolic equations, *ESAIM Control Optim. Calc. Var.* **26**, Paper No. 79, 19 (2020).

[27] Q. Lü and X. Zhang, Control theory for stochastic distributed parameter systems, an engineering perspective, *Annu. Rev. Control.* **51**, 268–330 (2021).

[28] Q. Lü and X. Zhang, *Mathematical theory for stochastic distributed parameter control systems.* Springer Nature Switzerland AG. (2021).

[29] B. Wu, Q. Chen, and Z. Wang, Carleman estimates for a stochastic degenerate parabolic equation and applications to null controllability and an inverse random source problem, *Inverse Problems.* **36**(7), 075014, 38 (2020).

[30] G. Yuan, Determination of two kinds of sources simultaneously for a stochastic wave equation, *Inverse Problems.* **31**(8), 085003, 13 (2015).

[31] G. Yuan, Conditional stability in determination of initial data for stochastic parabolic equations, *Inverse Problems.* **33**(3), 035014, 26 (2017).

[32] G. Yuan, Determination of two unknowns simultaneously for stochastic Euler-Bernoulli beam equations, *J. Math. Anal. Appl.* **450**(1), 137–151 (2017).

[33] X. Zhang, Unique continuation for stochastic parabolic equations, *Differential Integral Equations.* **21**(1–2), 81–93 (2008).

[34] X. Zhang, Carleman and observability estimates for stochastic wave equations, *SIAM J. Math. Anal.* **40**(2), 851–868 (2008).

[35] X. Zhang, A unified controllability/observability theory for some stochastic and deterministic partial differential equations. In *Proceedings of the International Congress of Mathematicians 2010 (ICM 2010)*, New Delhi (2011).

[36] W. Bryc, *The Normal Distribution. Characterizations with Applications.* Lecture Notes in Statistics, Springer, New York (1995).

[37] Q. Lü and X. Zhang, Global uniqueness for an inverse stochastic hyperbolic problem with three unknowns, *Comm. Pure Appl. Math.* **68**(6), 948–963 (2015).

[38] L. Beilina and M. V. Klibanov, *Approximate Global Convergence and Adaptivity for Coefficient Inverse Problems.* Springer US (2012).

[39] Y. Liu, Some sufficient conditions for the controllability of wave equations with variable coefficients, *Acta Appl. Math.* **128**, 181–191 (2013).

[40] J.-L. Lions, *Contrôlabilité exacte, perturbations et stabilisation de systèmes distribués. Tome 1.* Masson, Paris (1988).

[41] X. Fu, X. Liu, Q. Lü, and X. Zhang, An internal observability estimate for stochastic hyperbolic equations, *ESAIM Control Optim. Calc. Var.* **22**(4), 1382–1411 (2016).

[42] M. V. Klibanov, Carleman estimates for global uniqueness, stability and numerical methods for coefficient inverse problems, *J. Inverse Ill-Posed Probl.* **21**(4), 477–560 (2013).

[43] S. Alinhac, Non-unicité du problème de Cauchy, *Ann. of Math.* (2). **117**(1), 77–108 (1983).

[44] A. Amirov, *Integral Geometry and Inverse Problems for Kinetic Equations.* PhD thesis, Novosivbirsk (1988).

[45] S. Tang and X. Zhang, Null controllability for forward and backward stochastic parabolic equations, *SIAM J. Control Optim.* **48**(4), 2191–2216 (2009).

[46] X. Liu, Global carleman estimate for stochastic parabolic equations, and its application, *ESAIM Control Optim. Calc. Var.* **20**, 823–839 (2014).

[47] Z. Yin, A quantitative internal unique continuation for stochastic parabolic equations, *Math. Control Relat. Fields.* **5**(1), 165–176 (2015).

[48] Q. Lü and X. Zhang, A mini-course on stochastic control. In *Control and inverse problems for partial differential equations*, vol. 22, *Ser. Contemp. Appl. Math. CAM*, pp. 171–254. Higher Ed. Press, Beijing (2019).

[49] Q. Lü and X. Zhang, A concise introduction to control theory for stochastic partial differential equations, *Math. Control Relat. Fields* (2021).

[50] G. Fabbri, F. Gozzi, and A. Świech, *Stochastic optimal control in infinite dimension.* Springer, Cham (2017).

[51] P. S. Knopov and O. N. Deriyeva, *Estimation and control problems for stochastic partial differential equations.* Springer, New York (2013).

[52] C. Bardos, G. Lebeau, and J. Rauch, Sharp sufficient conditions for the observation, control, and stabilization of waves from the boundary, *SIAM J. Control Optim.* **30**(5), 1024–1065 (1992).

[53] D. Uciński, *Optimal measurement methods for distributed parameter system identification.* Systems and Control Series, CRC Press, Boca Raton, FL (2005).

Uniqueness Theorems Under Kalman's Rank Condition and Asymptotic Synchronization for Second Order Dissipative Systems*

Tatsien Li†

*School of Mathematical Sciences, Fudan University,
Shanghai 200433, China
dqli@fudan.edu.cn*

Bopeng Rao

*Institut de Recherche Mathématique Avancée, Université de Strasbourg,
67084 Strasbourg, France
bopeng.rao@math.unistra.fr*

Abstract. We first present the frequency domain approach and LaSalle's invariance principe, which reduce the asymptotic stability of evolution systems to the uniqueness of the corresponding adjoint problem with incomplete observations. We observe that the classic Kalman's rank condition is necessary for the uniqueness of solution to the over-determined adjoint system, but only Kalman's rank condition is not sufficient. In order to obtain the uniqueness of solution to this complex system, our basic idea is to combine the uniform observability of a scalar equation and the algebraic structure of the coupling matrices, namely, Kalman's rank condition. Thus, *"under Kalman's rank condition, the observability of a scalar problem implies the uniqueness of solution to a complex system."* Using this method, we provide a direct and efficient approach to solve a seemingly difficult problem of uniqueness of a complex system. We next study the uniform synchronization by compact perturbation method. Various applications are given for the system of wave equations

*This work is supported by National Natural Sciences Foundation of China under Grant 11831011.
†Shanghai Key Laboratory for Contemporary Applied Mathematics, Nonlinear Mathematical Modeling and Methods Laboratory.

with boundary feedback or (and) locally distributed feedback, and for the system of Kirchhoff plate with distributed feedback. Some open questions are raised at the end of the lecture for future development.

1. Introduction

Synchronization is a widespread natural phenomenon. It was first observed by Huygens[14] in 1665. The research on synchronization from a mathematical point of view dates back to Wiener[47] in the 1950s.

The previous studies focused on systems described by ordinary differential equations (ODEs), such as

$$X_i' = f(t, X_i) + \sum_{j=1}^{N} A_{ij} X_j \quad (i = 1, \cdots, N), \tag{1}$$

where $X_i (i = 1, \cdots, N)$ are n-dimensional state vectors, " $'$ " stands for the time derivative, $A_{ij}(i, j = 1, \cdots, N)$ are $n \times n$ matrices, and $f(t, X)$ is an n-dimensional vector function independent of $i = 1, \cdots, N$.[36,43] The system shows that every $X_i(i = 1, \cdots, N)$ possesses two basic features: satisfying a fundamental governing equation and bearing a coupled relation among one another. If for any given initial data

$$t = 0: \quad X_i = X_i^{(0)} \ (i = 1, \cdots, N), \tag{2}$$

the solution $X = (X_1, \cdots, X_N)^T = X(t)$ to the system satisfies

$$X_i(t) - X_j(t) \to 0 \quad (i, j = 1, \cdots, N) \quad \text{as } t \to +\infty, \tag{3}$$

namely, all the states $X_i(t) \ (i = 1, \cdots, N)$ tend to coincide with each other as $t \to +\infty$, then we say that the system possesses the synchronization in the consensus sense, or, in particular, if the solution $X = X(t)$ satisfies

$$X_i(t) - a(t) \to 0 \quad (i = 1, \cdots, N) \quad \text{as } t \to +\infty, \tag{4}$$

where $a(t)$ is a state vector which is a priori unknown, then we say that the system possesses the synchronization in the pinning sense. Obviously, the synchronization in the pinning sense implies that in the consensus sense. These kinds of synchronizations are all called the asymptotic synchronization, which should be realized on the infinite time interval $[0, +\infty)$.

The first attempt to extend, in both concept and method, the universal phenomena of synchronization from finite dimensional dynamical systems of ordinary differential equations to infinite dimensional dynamical systems of partial differential equations (PDEs) was published in a special issue of

Chinese Annals of Mathematics in honour of the scientific heritage of J.-L. Lions in 2013 in [26], the results of which were announced 2012 in [25].

For fixing the idea, we consider the following coupled system of wave equations with Dirichlet boundary conditions:

$$\begin{cases} U'' - \Delta U + AU = 0, & \text{in } (0, +\infty) \times \Omega, \\ U = 0, & \text{on } (0, +\infty) \times \Gamma_0, \\ U = DH, & \text{on } (0, +\infty) \times \Gamma_1 \end{cases} \tag{5}$$

with the initial data

$$t = 0: \quad U = \widehat{U}_0, \ U' = \widehat{U}_1 \quad \text{in } \Omega, \tag{6}$$

where Ω is a bounded domain with smooth boundary $\Gamma = \Gamma_0 \cup \Gamma_1$ such that $\overline{\Gamma}_0 \cap \overline{\Gamma}_1 = \emptyset$ and mes $(\Gamma_1) > 0$, $U = (u^{(1)}, \cdots, u^{(N)})^T$ is the state variable, $A \in \mathbb{M}^N(\mathbb{R})$ is an $N \times N$ coupling matrix with constant elements, the boundary control matrix D is an $N \times M$ full column-rank matrix $(M \leqslant N)$ with constant elements, $H = (h^{(1)}, \cdots, h^{(M)})^T$ stands for the boundary control.

Then, we can give artificial intervention to the evolution of state variables through appropriate boundary controls, which combines synchronization with controllability, and introduces the study of synchronization to the field of control. This is a new perspective to the investigation of synchronization for systems of partial differential equations. On the other hand, precisely due to the artificial intervention of control, we can make a higher demand, i.e., to meet the requirement of synchronization within a limited time, instead of waiting until $t \to +\infty$. Thus we have introduced the concepts of the exact boundary synchronization and the approximate boundary synchronization in the monograph [29]. Consequently, this kind of study of synchronization becomes a part of research in control theory. The optimal control for the exact synchronization of parabolic system was recently investigated in [46].

For systems governed by PDEs, we can similarly consider the asymptotic synchronization on an infinite time interval as in the case of systems governed by ODEs, namely, we may ask the following questions: under what conditions do the system states with any given initial data possess the asymptotic synchronization in the consensus sense:

$$u^{(i)}(t, \cdot) - u^{(j)}(t, \cdot) \to 0 \quad (i, j = 1, \cdots, N) \quad \text{as } t \to +\infty, \tag{7}$$

or, in particular, if the system states with any given initial data possess the asymptotic synchronization in the pinning sense:

$$u^{(i)}(t, \cdot) - u(t, \cdot) \to 0 \quad (i = 1, \cdots, N) \quad \text{as } t \to +\infty, \tag{8}$$

where $u = u(t, \cdot)$ is called the asymptotically synchronizable state, which is a priori unknown? If the answer of this question is positive, these conclusions should be realized spontaneously on an infinite time interval $[0, +\infty)$, and are naturally developed results decided by the nature of the system itself.

In this lecture, we will study the long-time behaviour of a second order abstract linear evolution equation

$$U'' + \mathcal{L}U + AU + D\mathcal{G}U' = 0, \tag{9}$$

where A and D are symmetric and positive semi-definite matrices, $\mathcal{L}$ is the duality mapping from a Hilbert space V onto its dual space V', and $\mathcal{G}$ is a linear continuous operator from V into V'.

We will present two approaches for the asymptotic stability of evolution systems in §2 and §3, respectively.

According to the classic theory of semi-groups,[2,4,35] the asymptotic stability of system (9) is equivalent to the uniqueness of solution to the following elliptic system with $\beta \in \mathbb{R}$ and $\Phi = (\phi^{(1)}, \cdots, \phi^{(N)})^T$:

$$\mathcal{L}\Phi + A\Phi = \beta^2 \Phi \quad \text{with} \quad D\mathcal{G}\Phi = 0. \tag{10}$$

On the other hand, by LaSalle's invariance principle,[5,8] the asymptotic stability of system (9) is equivalent to the uniqueness of solution to the following evolutional system with $\Psi = (\psi^{(1)}, \cdots, \psi^{(N)})^T$:

$$\Psi'' + \mathcal{L}\Psi + A\Psi = 0 \quad \text{with} \quad D\mathcal{G}\Psi = 0. \tag{11}$$

No matter what method is adopted, the asymptotic stability is ultimately based on the uniqueness of the corresponding adjoint problem. So, the study on the uniqueness theorem is a necessary task in the stability theory.

We first observe that Kalman's rank condition

$$\text{rank}(D, AD, \cdots, A^{N-1}D) = N \tag{12}$$

is necessary for the uniqueness of solution to the over-determined system (10), or (11), therefore, necessary for the asymptotic stability of system (9). However, since the matrix D in (12) is not invertible in general, the D-observation cannot imply the nullity of the full observation $\mathcal{G}\Phi = 0$. So, only Kalman's rank condition is not sufficient and some additional conditions should be required.

Indeed, we will show that Kalman's rank condition is actually sufficient for the uniqueness of solution to the elliptic system (10) if (i) $\|A - aI\|$ is small for some $a \in \mathbb{R}$; (ii) there exists $c > 0$ independent of $\beta \in \mathbb{R}$ and $f \in H$, such that the following uniform estimate

$$\|\phi\|_H \leqslant c\|f\|_H \tag{13}$$

holds for any given solution $\phi \in V$ to the over-determined scalar problem

$$\beta^2 \phi - L\phi = f \quad \text{with} \quad g\phi = 0. \tag{14}$$

Thus, "*under Kalman's rank condition, the observability of a scalar problem implies the uniqueness of solution to a complex system.*" Using this method, we may provide a direct and efficient approach to solve a seemingly difficult problem of uniqueness of a complex system. However, since the observability inequality (13) is usually based on the multiplier method, the geometrical multiplier condition should be naturally required in applications.

For the evolutional system (11), the approach based on Ingham's inequality needs the gap condition of $\mathcal{L}$, therefore, it is essentially restricted to one-dimensional problems.[18],[27] In §3, we will propose a method, which is based on the non-harmonic Fourier's series at the infinite horizon and does not need any gap condition of $\mathcal{L}$.

Now let us outline the main idea in the study of asymptotic synchronization. Define the auxiliary variable $W = (w^{(1)}, \cdots, w^{(N-1)})^T$ by

$$w^{(i)} = u^{(i)} - u^{(i+1)}, \quad 1 \leqslant i \leqslant N - 1. \tag{15}$$

Assume that A satisfies the row-sum condition, respectively, D satisfies the null row-sum condition:

$$\sum_{j=1}^{N} a_{ij} = cst, \quad \text{respectively,} \quad \sum_{j=1}^{N} d_{ij} = 0, \quad 1 \leqslant i \leqslant N, \tag{16}$$

then there exist symmetric and positive semi-definite matrices A_1 and D_1 of order $(N-1)$, such that

$$W'' + \mathcal{L}W + A_1 W + D_1 \mathcal{G}W' = 0. \tag{17}$$

By this way, we transform the asymptotic synchronization of system (9) to the asymptotic stability of the reduced system (17).

The above approach is direct and efficient. However, the necessity of the row-sum conditions (16) is a delicate question. In fact, these conditions are usually imposed as physically reasonable hypotheses for systems of ordinary differential equations.

We will show that if system (9) is asymptotically synchronizable, then

$$\text{rank}(D, AD, \cdots, A^{N-1}D) \geqslant N - 1. \tag{18}$$

Moreover, in the case of equality, the conditions given by (16) hold. By this way, we clarify that the necessity of the row-sum conditions is in fact the consequence of the minimality of the rank of Kalman's matrix. This makes the theory of asymptotic synchronization more complete for systems of partial differential equations.

In §4, we will also examine the uniform synchronization in the consensus sense:

$$\|u^{(i)}(t) - u^{(j)}(t)\| \leqslant Me^{-\omega t} \quad (i, j = 1, \ldots, N) \tag{19}$$

or the uniform synchronization in the pinning sense:

$$\|u^{(i)}(t) - u(t)\| \leqslant Me^{-\omega t} \quad (i = 1, \cdots, N), \tag{20}$$

where M and ω are two positive constants. We first indicate that $\text{rank}(D) \geqslant N - 1$ is a necessary condition for the uniform synchronization. Then, under the minimum rank condition, we clarify the algebraic structure of the coupling matrices. We then establish the uniform synchronization by the method of compact perturbation and further give the dynamics of the asymptotic orbit.

The materials of the lecture are mainly selected from the recent works [27, 30–32].

2. Asymptotic synchronization by frequency domain approach

2.1. *Asymptotic stability*

Let H and V be two separated Hilbert spaces such that $V \subset H$ with dense and compact imbedding. Let L be the duality mapping from V onto its dual space V', such that

$$\langle L\phi, \psi \rangle_{V',V} = (\phi, \psi)_V, \quad \forall \phi, \psi \in V. \tag{1}$$

Let g be a linear continuous operator from V into V', such that

$$\langle g\phi, \psi \rangle_{V',V} = \langle g\psi, \phi \rangle_{V',V} \tag{2}$$

and

$$\langle g\phi, \phi \rangle_{V',V} \geqslant 0, \text{ and } \langle g\phi, \phi \rangle_{V',V} = 0 \text{ if and only if } g\phi = 0. \tag{3}$$

In what follows, we assume furthermore that the operator g is compact from V into V'.

Denote by $\mathcal{V}$ and $\mathcal{H}$ the product spaces:

$$\mathcal{V} = V^N, \quad \mathcal{H} = H^N. \tag{4}$$

Let $U = (u^{(1)}, \cdots, u^{(N)})^T$. Define the vector operators $\mathcal{L}$, respectively, $\mathcal{G}$ by

$$\mathcal{L}U = \begin{pmatrix} Lu^{(1)} \\ \vdots \\ Lu^{(N)} \end{pmatrix}, \quad \mathcal{G}U = \begin{pmatrix} gu^{(1)} \\ \vdots \\ gu^{(N)} \end{pmatrix}. \tag{5}$$

Let A and D be symmetric and positive semi-definite matrices. Consider the following second order evolution equations:

$$U'' + \mathcal{L}U + AU + D\mathcal{G}U' = 0. \tag{6}$$

Defining the linear operator $\mathcal{A}$ by

$$\mathcal{A}(U, \widehat{U}) = (\widehat{U}, -\mathcal{L}U - AU - D\mathcal{G}\widehat{U}) \tag{7}$$

with the domain

$$D(\mathcal{A}) = \{(U, \widehat{U}) \in \mathcal{V} \times \mathcal{V} : \mathcal{L}U + AU + D\mathcal{G}\widehat{U} \in \mathcal{V}\}, \tag{8}$$

we transform (6) into the following formulation:

$$(U, \widehat{U})' = \mathcal{A}(U, \widehat{U}). \tag{9}$$

Proposition 1. *The operator $\mathcal{A}$ defined in (7)–(8) generates a semi-group of contractions on the space $\mathcal{V} \times \mathcal{H}$. Moreover, $\mathcal{A}^{-1}$ is compact in $\mathcal{V} \times \mathcal{H}$.*

Proof. Using (1)–(3), we first check that $\mathcal{A}$ is dissipative:

$$(\mathcal{A}(U, \widehat{U}), (U, \widehat{U}))_{\mathcal{V} \times \mathcal{H}} = -\langle D\mathcal{G}\widehat{U}, \widehat{U} \rangle_{\mathcal{V}', \mathcal{V}} \leqslant 0.$$

We next show that $R(\mathcal{A}) = \mathcal{V} \times \mathcal{H}$. Then, by Hill-Yosida's Theorem,[35] it generates a semi-group of contractions on the space $\mathcal{V} \times \mathcal{H}$.

To this end, for any given $(F, \widehat{F}) \in \mathcal{V} \times \mathcal{H}$, we solve the equation $\mathcal{A}(U, \widehat{U}) = (F, \widehat{F})$, namely,

$$\widehat{U} = F, \quad \mathcal{L}U + AU = -(D\mathcal{G}F + \widehat{F}).$$

Noting (1) and the convention of pivot space, it follows that

$$(U, \Phi)_{\mathcal{V}} + (AU, \Phi)_{\mathcal{H}} = -\langle D\mathcal{G}F + \widehat{F}, \Phi \rangle_{\mathcal{V}', \mathcal{V}}, \quad \forall \Phi \in \mathcal{V},$$

which, due to Lax-Milgram's Lemma, admits a unique solution $U \in \mathcal{V}$ with the continuous dependence:

$$\|U\|_{\mathcal{V}} \leqslant c\|(D\mathcal{G}F + \widehat{F})\|_{\mathcal{V}'}.$$

Since the operator g is compact from V to V', so is the operator $\mathcal{G}$ from $\mathcal{V}$ to $\mathcal{V}'$. On the other hand, noting that the embeddings $V \subset H \subset V'$ are compact, the mapping

$$(F, \widehat{F}) \to U = -(\mathcal{L} + A)^{-1}(D\mathcal{G}F + \widehat{F})$$

is compact from $\mathcal{V} \times \mathcal{H}$ to $\mathcal{V}$. Therefore the mapping

$$(F, \widehat{F}) \to (U, \widehat{U}) = \left(-(\mathcal{L} + A)^{-1}(D\mathcal{G}F + \widehat{F}), F\right)$$

is compact from $\mathcal{V} \times \mathcal{H}$ into $\mathcal{V} \times \mathcal{H}$. Therefore, $\mathcal{A}^{-1}$ is compact in $\mathcal{V} \times \mathcal{H}$. □

Definition 1. System (6) is asymptotically (strongly) stable in $\mathcal{V} \times \mathcal{H}$ if for any given initial data $(U_0, U_1) \in \mathcal{V} \times \mathcal{H}$, the corresponding solution U satisfies

$$(U(t), U'(t)) \to (0, 0) \quad \text{in } \mathcal{V} \times \mathcal{H} \text{ as } t \to +\infty. \tag{10}$$

The following result, usually said the frequency domain approach, suggests to handle the asymptotic stability of the whole system (6) by means of the uniqueness of a scalar problem.

Theorem 1. *System* (6) *is asymptotically stable if and only if for any given* $\beta \in \mathbb{R}$, *the following eigen-system for the variable* $\Phi = (\phi^{(1)}, \cdots, \phi^{(N)})^T$:

$$\mathcal{L}\Phi + A\Phi = \beta^2 \Phi \tag{11}$$

associated with the D-observation

$$D\mathcal{G}\Phi = 0 \tag{12}$$

has only the trivial solution $\Phi \equiv 0$.

Proof. Noting that $\mathcal{A}^{-1}$ is compact in the space $\mathcal{V} \times \mathcal{H}$, by the classic theory of semi-groups,[2,4] the dissipative system (6) is asymptotically stable if and only if $\mathcal{A}$ has no pure imaginary eigenvalues.

Indeed, assume that $\mathcal{A}$ admits a pure imaginary eigenvalue, namely, there exist $\beta \in \mathbb{R}$ and a non-trivial $(\Phi, \Psi) \in \mathcal{V} \times \mathcal{H}$, such that

$$\mathcal{A}(\Phi, \Psi) = i\beta(\Phi, \Psi).$$

It follows that

$$\mathcal{L}\Phi + A\Phi + i\beta D\mathcal{G}\Phi = \beta^2 \Phi. \tag{13}$$

Since $\mathcal{L} + A$ is coercive, we have $\beta \neq 0$. Then, noting that $\mathcal{L}$ and $D\mathcal{G}$ are symmetric and positive semi-definite, we deduce that (13) is equivalent to system (11)–(12), which gives a contradiction. □

2.2. *Uniqueness theorem for elliptic systems*

Let us recall the following generalized rank condition of Kalman's type, which will play an important role in the study of uniqueness.

Proposition 2. *([27, Lemma 2.1]) Let $d \geqslant 0$ be an integer. Let A and D be two symmetric matrices. The Kalman's rank condition*

$$\text{rank}(D, AD, \cdots, A^{N-1}D) = N - d \tag{14}$$

holds if and only if d is the largest dimension of the subspaces which are invariant for A and contained in $\text{Ker}(D)$.

Proposition 3. *Assume that system (11)–(12) has only the trivial solution. Then the pair (A, D) necessarily satisfies Kalman's rank condition*

$$\text{rank}(D, AD, \cdots, A^{N-1}D) = N. \tag{15}$$

Proof. If (15) fails, by Proposition 2, there exist a number $\lambda \in \mathbb{R}$ and a non-trivial vector $E \in \mathbb{R}^N$, such that

$$AE = \lambda E \quad \text{and} \quad DE = 0.$$

Noting that L is self-adjoint and the imbedding from V into H is compact, by the spectral theory of compact self-adjoint operators, there exist $v \in V$ and $\alpha \in \mathbb{R}^+$ large enough, such that $Lv = \alpha v$. Defining

$$\beta^2 = \alpha + \lambda > 0 \quad \text{and} \quad \Phi = vE,$$

it is easy to check that Φ is a solution to (11). Moreover, noting that $\mathcal{G}$ is of diagonal form, we have

$$D\mathcal{G}\Phi = gvDE = 0.$$

Thus, we get a non-trivial solution of (11)–(12), which contradicts the assumption. $\qquad\square$

Since the matrix D in (15) is not necessarily invertible, Carleman's classic uniqueness theorem[10,44] is not applicable in thus case. Furthermore, it was shown that only Kalman's rank condition (14) is not sufficient for the uniqueness of solution to (11)–(12) in general.[27]

Definition 2. Let $\lambda_1, \cdots, \lambda_m$ denote the distinct eigenvalues of A. The matrix A satisfies the ϵ-closing condition if there exists a number a such that

$$\sup_{1 \leqslant l \leqslant m} |\lambda_l - a| \leqslant \epsilon. \tag{16}$$

Definition 3. The operator L satisfies the c-gap condition if there exists a positive number c, such that

$$|\alpha_n - \alpha_{n'}| \geqslant c \qquad (17)$$

holds true for all distinct eigenvalues $\alpha_n \neq \alpha_{n'}$ of L.

Definition 4. The operator L is g-observable if there exists a constant $c > 0$, independent of $\beta \in \mathbb{R}$ and $f \in H$, such that the following observability inequality

$$c\|\phi\|_H \leqslant \|f\|_H \qquad (18)$$

holds for any given solution $\phi \in V$ to the over-determined scalar problem

$$\beta^2 \phi - L\phi = f \quad \text{with} \quad g\phi = 0. \qquad (19)$$

Theorem 2. *Assume that the pair (A, D) satisfies Kalman's rank condition* (15). *Then, system* (11)–(12) *has only the trivial solution $\Phi = 0$ in any one of the following cases:*

(i) *the operator g is global, namely,*

$$g\phi = 0 \Longrightarrow \phi = 0; \qquad (20)$$

(ii) *the matrix A satisfies the ϵ-closing condition* (16) *with $\epsilon > 0$ small enough, and L satisfies the c-gap condition* (17) *and the over-determined scalar problem*

$$\beta^2 \phi = L\phi \quad \text{and} \quad g\phi = 0 \qquad (21)$$

has only the trivial solution $\phi \equiv 0$;

(iii) *the matrix A satisfies the ϵ-closing condition* (16) *with $\epsilon > 0$ small enough, and the operator L is g-observable.*

In order to clarify the main idea, we first make some algebraic preliminaries.

First, since Kalman's rank condition (15) is invariant under invertible linear transformation,[28] without loss of generality, A can be written as

$$A = \text{diag}(\overbrace{\lambda_1, \cdots, \lambda_1}^{\sigma_1}, \cdots\cdots, \overbrace{\lambda_m, \cdots, \lambda_m}^{\sigma_m}),$$

where λ_l is an eigenvalue of A with multiplicity σ_l $(l = 1, \cdots, m)$.

Accordingly, defining

$$\mu_r = \mu_{r-1} + \sigma_r, \quad r = 1, \cdots, m \quad \text{with} \quad \mu_0 = 0,$$

we regroup the components of Φ into

$$(\phi^{(1)}, \cdots, \phi^{(\mu_1)}), \ (\phi^{(\mu_1+1)}, \cdots, \phi^{(\mu_2)}), \cdots\cdots, (\phi^{(\mu_{m-1}+1)}, \cdots, \phi^{(\mu_m)}).$$

On the other hand, if we replace A by $A + bI$, and β^2 by $\beta^2 + b$ for any given $b > 0$ in (11), this will not modify anything in Theorem 2. So, without loss of generality, we may assume that $\lambda_l > 0$ for $1 \leqslant l \leqslant m$.

Denote by ϵ_i $(i = 1, \cdots, N)$ the canonical basis vectors in $\mathbb{R}^N$ and by d_i the i-th column-vector of the matrix D. Noting that $D\epsilon_i = d_i$ and the subspace $\mathrm{Span}\{\epsilon_{\mu_{l-1}+1}, \cdots, \epsilon_{\mu_l}\}$ is invariant for A, by Proposition 2, Kalman's rank condition (15) implies that

$$\mathrm{Span}\{\epsilon_{\mu_{l-1}+1}, \cdots, \epsilon_{\mu_l}\} \bigcap \mathrm{Ker}(D) = \{0\},$$

namely,

$$\sum_{i=\mu_{l-1}+1}^{\mu_l} \alpha_i d_i = 0 \text{ if and only if } \alpha_i = 0 \text{ for } i = \mu_{l-1}+1, \cdots, \mu_l.$$

Therefore, for any given l with $1 \leqslant l \leqslant m$, the vectors $d_{\mu_{l-1}+1}, \cdots, d_{\mu_l}$ are linearly independent.

Now we proceed the proof of Theorem 2.

Case (i). From (5) and (20), we have

$$D\mathcal{G}\Phi = \mathcal{G}(D\Phi) = 0 \Longrightarrow D\Phi \equiv 0.$$

Then, applying D to (11), it follows that

$$DA\Phi \equiv 0,$$

namely,

$$\sum_{l=1}^{m} \sum_{i=\mu_{l-1}+1}^{\mu_l} \lambda_l \phi^{(i)} d_i = 0.$$

We write (11) as

$$L\phi^{(i)} = (\beta^2 - \lambda_l)\phi^{(i)}, \quad \mu_{l-1}+1 \leqslant i \leqslant \mu_l, \quad 1 \leqslant l \leqslant m. \tag{22}$$

Since L is self-adjoint, the eigen-spaces

$$\mathrm{Span}\{\phi^{(\mu_{l-1}+1)}, \cdots, \phi^{(\mu_l)}\}, \quad 1 \leqslant l \leqslant m$$

are mutually orthogonal, then we have

$$\sum_{i=\mu_{l-1}+1}^{\mu_l} \lambda_l \phi^{(i)} d_i = 0, \quad 1 \leqslant l \leqslant m.$$

Noting $\lambda_l > 0$ and the linear independence of $d_{\mu_{l-1}+1}, \cdots, d_{\mu_l}$, it follows that

$$\phi^{(i)} = 0, \quad \mu_{l-1} + 1 \leqslant i \leqslant \mu_l, \quad 1 \leqslant l \leqslant m,$$

namely, $\Phi \equiv 0$.

Case (ii). Assume that there exist l and k with $l \neq k$, such that

$$\phi^{(i)} \neq 0, \quad \mu_{l-1} + 1 \leqslant i \leqslant \mu_l,$$

and

$$\phi^{(i')} \neq 0, \quad \mu_{k-1} + 1 \leqslant i' \leqslant \mu_k.$$

There exist α_n and $\alpha_{n'}$ such that

$$\beta^2 - \lambda_l = \alpha_n \quad \text{and} \quad \beta^2 - \lambda_k = \alpha_{n'},$$

then

$$\lambda_k - \lambda_l = \alpha_n - \alpha_{n'}.$$

However, because of the ϵ-closing condition (16) and the c-gap condition (17), the above equality cannot be satisfied for $\epsilon > 0$ small enough. Therefore, there exists at most an integer k with $1 \leqslant k \leqslant m$, such that

$$\phi^{(i)} = 0, \quad \mu_{l-1} + 1 \leqslant i \leqslant \mu_l, \quad l \neq k. \tag{23}$$

Then, (12) becomes

$$D\mathcal{G}\Phi = \sum_{l=1}^{m} \sum_{i=\mu_{l-1}+1}^{\mu_l} g\phi^{(i)} d_i = \sum_{i=\mu_{k-1}+1}^{\mu_k} g\phi^{(i)} d_i = 0.$$

Noting (22) and the linear independence of $d_{\mu_{k-1}+1}, \cdots, d_{\mu_k}$, it follows that

$$L\phi^{(i)} = (\beta^2 - \lambda_k)\phi^{(i)} \quad \text{with} \quad g\phi^{(i)} = 0 \tag{24}$$

for $\mu_{k-1} + 1 \leqslant i \leqslant \mu_k$. Then the uniqueness of solution to the scalar problem (21) implies that

$$\phi^{(i)} = 0, \quad \mu_{k-1} + 1 \leqslant i \leqslant \mu_k,$$

which, combining with (23), leads to

$$\phi^{(i)} = 0, \quad \mu_{l-1} + 1 \leqslant i \leqslant \mu_l, \quad 1 \leqslant l \leqslant m,$$

namely, $\Phi \equiv 0$.

Case (iii). Applying D to (11) and noting $W = D\Phi$, we get

$$(\beta^2 - a)W - \mathcal{L}W = DA\Phi - aW. \tag{25}$$

On the other hand, noting the diagonal form of $\mathcal{G}$ in (5), condition (12) leads to

$$\mathcal{G}W = \mathcal{G}D\Phi = D\mathcal{G}\Phi = 0. \tag{26}$$

For $1 \leqslant i \leqslant N, 1 \leqslant j \leqslant M$, setting

$$W = (w_j), \quad DA\Phi - aW = (f_j) \quad \text{and} \quad D = (d_{ij}),$$

we have

$$w_j = \sum_{i=1}^{m} d_{ji}\phi^{(i)} = \sum_{l=1}^{m} \sum_{i=\mu_{l-1}+1}^{\mu_l} d_{ji}\phi^{(i)}$$

and

$$f_j = \sum_{l=1}^{m}(\lambda_l - a) \sum_{i=\mu_{l-1}+1}^{\mu_l} d_{ji}\phi^{(i)}.$$

Then, taking the j-th component of (25) and (26), we get

$$(\beta^2 - a)w_j - Lw_j = f_j \tag{27}$$

associated with the D-observation

$$gw_j = 0. \tag{28}$$

If $\beta^2 - a \leqslant 0$, multiplying (27) by w_j, we get

$$-(\beta^2 - a)\|w_j\|_H^2 + \|w_j\|_V^2 = -(f_j, w_j)_H \leqslant \|f_j\|_H\|w_j\|_H.$$

It follows that

$$c\|w_j\|_H \leqslant \|f_j\|_H. \tag{29}$$

If $\beta^2 - a > 0$, then w_j satisfies the scalar problem (18). Since the operator L is g-observable, we get again (29).

On the other hand, noting that L is self-adjoint, we have

$$(\phi^{(i)}, \phi^{(i')})_H = 0, \quad \mu_{l-1} + 1 \leqslant i \leqslant \mu_l, \quad \mu_{k-1} + 1 \leqslant i' \leqslant \mu_k, \quad l \neq k.$$

Then it follows from (27) that

$$\|f_j\|_H^2 \leqslant \sup_{1 \leqslant l \leqslant m} |a - \lambda_l|^2 \sum_{l=1}^{m} \left\| \sum_{i=\mu_{l-1}+1}^{\mu_l} d_{ji}\phi^{(i)} \right\|_H^2$$

$$= \sup_{1 \leqslant l \leqslant m} |a - \lambda_l|^2 \|w_j\|_H^2,$$

hence, by the ϵ-closing condition (16) we get

$$\|f_j\|_H \leqslant \sup_{1 \leqslant l \leqslant m} |a - \lambda_l| \|w_j\|_H \leqslant \epsilon\|w_j\|_H.$$

Thus, it follows from (29) that

$$\sum_{i=\mu_{l-1}+1}^{\mu_l} d_{ij}\phi^{(i)} = 0, \quad 1 \leqslant j \leqslant M, \quad 1 \leqslant l \leqslant m,$$

provided that $\epsilon < c$, namely, we have

$$\sum_{i=\mu_{l-1}+1}^{\mu_l} d_i\phi^{(i)} = 0, \quad 1 \leqslant l \leqslant m.$$

Then, the linear independence of $d_{\mu_{l-1}+1}, \cdots, d_{\mu_l}$ implies that

$$\phi^{(i)} = 0, \quad \mu_{l-1}+1 \leqslant i \leqslant \mu_l, \quad 1 \leqslant l \leqslant m,$$

namely, $\Phi \equiv 0$. The proof is then complete.

Theorem 2 shows that a seemingly difficult uniqueness of solution to a complex system can be established through the observability of an over-determined scalar problem. This idea will be developed for studying the asymptotic stability and the asymptotic synchronization of second order evolution systems.

The first case (i) corresponds to the globally distributed observation. In this case, similarly to the finite dimension case, without any additional assumptions on the matrix A either on the operator L, only the Kalman's rank condition is sufficient for the uniqueness of solution to the over-determined system. In the second case (ii), the condition $g\phi = 0$ is not *a priori* assumed to yield $\phi = 0$. However, thanks to the c-gap condition (17), the uniqueness of solution to the scalar problem (21), which is weaker than the observability inequality, implies also the uniqueness of solution to the over-determined system. In the last case (iii), the observability inequality (18) does not come from the well-posedness, but rather from Carleman's estimate, and requires additional information on the differential operator L as well as on the geometric condition of the domain (see examples in §2.4).

Theorem 3. *Let the pair of matrices (A, D) satisfy Kalman's rank condition (15). Then, system (6) is asymptotically stable in any one of the situations described in Theorem 2.*

Proof. This is a direct consequence of Theorems 1 and 2.   □

2.3. *Asymptotic synchronization by groups*

By Proposition 3, when the pair of matrices (A, D) does not satisfy Kalman's rank condition (15), the dissipative system (6) is not asymptotically stable. Instead of asymptotic stability, we introduce the asymptotic synchronization by p-groups.

We return to consider the asymptotic synchronization by p-groups, a weakened notion of stability, described as follows.

Let $p \geqslant 1$ be an integer and

$$0 = n_0 < n_1 < n_2 < \cdots < n_p = N \tag{30}$$

be integers such that $n_r - n_{r-1} \geqslant 2$ for $1 \leqslant r \leqslant p$. We re-arrange the components of the state variable U into p groups

$$\left(u^{(1)}, \cdots, u^{(n_1)}\right), \left(u^{(n_1+1)}, \cdots, u^{(n_2)}\right), \cdots\cdots, \left(u^{(n_{p-1}+1)}, \cdots, u^{(n_p)}\right). \tag{31}$$

Definition 5. System (6) is asymptotically synchronizable by p-groups if for any given initial data $(U_0, U_1) \in \mathcal{V} \times \mathcal{H}$, the corresponding solution U to system (6) satisfies

$$\left(u^{(k)}(t) - u^{(l)}(t), (u^{(k)})'(t) - (u^{(l)})'(t)\right) \to (0, 0) \quad \text{in } V \times H \tag{32}$$

as $t \to +\infty$ for all $n_{r-1} + 1 \leqslant k, l \leqslant n_r$ and $1 \leqslant r \leqslant p$.

Now let S_r be the full row-rank matrix of order $(n_r - n_{r-1} - 1) \times (n_r - n_{r-1})$:

$$S_r = \begin{pmatrix} 1 & -1 & & & \\ & 1 & -1 & & \\ & & \ddots & \ddots & \\ & & & 1 & -1 \end{pmatrix}, \quad 1 \leqslant r \leqslant p. \tag{33}$$

Define the $(N - p) \times N$ matrix C_p of synchronization by p-groups as

$$C_p = \begin{pmatrix} S_1 & & & \\ & S_2 & & \\ & & \ddots & \\ & & & S_p \end{pmatrix}. \tag{34}$$

Let $\epsilon_1, \cdots, \epsilon_N$ be the vectors of the canonical basis of $\mathbb{R}^N$. Defining

$$e_r = \sum_{i=n_{r-1}+1}^{n_r} \epsilon_i, \quad 1 \leqslant r \leqslant p, \tag{35}$$

we have

$$\text{Ker}(C_p) = \text{Span}\{e_1, \cdots, e_p\}. \tag{36}$$

Then the asymptotic synchronization by p-groups (32) can be equivalently written as

$$C_p(U(t), U'(t)) \to (0, 0) \quad \text{in } (V \times H)^{N-p} \text{ as } t \to +\infty. \tag{37}$$

Before starting the study on the asymptotic synchronization by p-groups, we first give some algebraic preliminaries.

Definition 6. The matrix A satisfies the condition of C_p-compatibility if

$$A\mathrm{Ker}(C_p) \subseteq \mathrm{Ker}(C_p). \tag{38}$$

Proposition 4. *The condition of C_p-compatibility (40) is equivalent to the existence of a symmetric and positive semi-definite matrix A_p of order $(N - p)$, such that*

$$(C_p C_p^T)^{-1/2} C_p A = A_p (C_p C_p^T)^{-1/2} C_p. \tag{39}$$

Proof. It is sufficient to note that $\mathrm{Ker}((C_p C_p^T)^{-1/2} C_p) = \mathrm{Ker}(C_p)$. Then, the desired result can be obtained by virtue of [29, Proposition 2.15].

Moreover, the reduced matrix A_p can be explicitly given by

$$A_p = (C_p C_p^T)^{-1/2} C_p A C_p^T (C_p C_p^T)^{-1/2}, \tag{40}$$

which shows that the reduced matrix A_p is also symmetric and positive semi-definite. □

Definition 7. The matrix D satisfies the condition of strong C_p-compatibility if

$$\mathrm{Ker}(C_p) \subseteq \mathrm{Ker}(D). \tag{41}$$

Proposition 5. *The condition of strong C_p-compatibility (41) is equivalent to the existence of a symmetric and positive semi-definite matrix R of order $(N - p)$, such that*

$$D = C_p^T R C_p. \tag{42}$$

Proof. Noting that $\mathrm{Ker}(\sqrt{D}) = \mathrm{Ker}(D)$, it follows from (41) that

$$\mathrm{Im}(\sqrt{D}) = (\mathrm{Ker}(\sqrt{D}))^{\perp} = (\mathrm{Ker}(D))^{\perp} \subseteq \mathrm{Im}(C_p^T).$$

Thus, there exists a matrix $\widehat{R}$ of order $(N - p) \times N$, such that $\sqrt{D} = C_p^T \widehat{R}$, which gives the expression in (42) with $R = \widehat{R}\widehat{R}^T$.

Moreover, setting

$$D_p = (C_p C_p^T)^{1/2} R (C_p C_p^T)^{1/2}, \tag{43}$$

we easily check that D_p satisfies

$$(C_p C_p^T)^{-1/2} C_p D = D_p (C_p C_p^T)^{-1/2} C_p. \tag{44}$$

This justifies well the terminology of the condition of strong C_p-compatibility. □

Remark 1. Noting the special form of (36), the conditions of compatibility (38) and (41) are equivalent to the following row-sum conditions by blocks:

$$\sum_{j=n_{s-1}+1}^{n_s} a_{ij} = \alpha_{rs}, \quad \text{respectively,} \quad \sum_{j=n_{s-1}+1}^{n_s} d_{ij} = 0 \qquad (45)$$

for all $n_{r-1} + 1 \leqslant i \leqslant n_r$ and $1 \leqslant r, s \leqslant p$.

Proposition 6. *Assume that the matrix A satisfies the condition of C_p-compatibility* (38) *and D is given by* (42). *Then the asymptotic synchronization by p-groups of system* (6) *is equivalent to the asymptotic stability of the following reduced system with the variable $W = (w^{(1)}, \cdots, w^{(N-p)})^T$:*

$$W'' + \mathcal{L}W + A_p W + D_p \mathcal{G} W' = 0. \qquad (46)$$

Proof. Applying the matrix $(C_p C_p^T)^{-1/2} C_p$ to system (6), noting the conditions of C_p-compatibility (39) and (44) and setting

$$W = (C_p C_p^T)^{-1/2} C_p U, \qquad (47)$$

we get the self-closed reduced system (46). Moreover, since the reduced matrices A_p and D_p are still symmetric and positive semi-definite, the asymptotic stability of the reduced system (46) can be obtained by Theorem 1.   □

The necessity of the conditions of C_p-compatibility is a delicate question. Intuitively, we think that this question is intrinsically linked with the rank of Kalman's matrix. In what follows, we will give a lower bound on the rank of Kalman's matrix in Theorem 4. Next, in Theorem 5, we show that the necessity of the conditions of C_p-compatibility is a consequence of the minimality of the rank of Kalman's matrix. We establish the asymptotic synchronization by p-groups in Theorem 6. We clarify the two kinds of asymptotic synchronizations by p-groups in Theorem 7 at the end of §2.3.

Theorem 4. *Assume that system* (6) *is asymptotically synchronizable by p-groups. Then we necessarily have*

$$\text{rank}(D, AD, \cdots, A^{N-1}D) \geqslant N - p. \qquad (48)$$

Proof. Assume that

$$\text{rank}(D, AD, \cdots, A^{N-1}D) = N - q \quad \text{with } q > p.$$

Noting that A is symmetric, by Proposition 2, without loss of generality, we may assume that there exists a subspace $\text{Span}\{E_1, \cdots, E_q\}$ such that

$$AE_r = \lambda_r E_r \quad \text{and} \quad DE_r = 0, \quad r = 1, \cdots, q.$$

Since

$$\dim \mathrm{Im}(C_p^T) + \dim \mathrm{Span}\{E_1, \cdots, E_q\} = N - p + q > N,$$

there exists a unit vector E such that

$$E \in \mathrm{Span}\{E_1, \cdots, E_q\} \bigcap \mathrm{Im}(C_p^T).$$

Let $x \in \mathbb{R}^{N-p}$ such that $E = C_p^T x$. Setting $\phi = (E, U) = (x, C_p U)$, by (37) we have

$$(\phi, \phi') \to (0,0) \quad \text{in } V \times H \quad \text{as } t \to +\infty. \tag{49}$$

On the other hand, since $E \in \mathrm{Span}\{E_1, \cdots, E_q\}$, we can write $E = \sum_{r=1}^q a_r E_r$. Then, setting $\phi_r = (E_r, U)$ for $r = 1, \cdots, q$, we get

$$\phi = \sum_{r=1}^q a_r (E_r, U) = \sum_{r=1}^q a_r \phi_r.$$

Applying E_r to system (6), we get

$$\phi_r'' + L\phi_r + \lambda_r \phi_r = 0, \quad r = 1, \cdots, q \tag{50}$$

associated with the initial data:

$$t = 0: \quad \phi_r = v_r, \quad \phi_r' = 0, \quad r = 1, \cdots, q, \tag{51}$$

where the functions v_r are given by

$$Lv_r = \beta_r^2 v_r, \quad r = 1, \cdots, q. \tag{52}$$

Since the eigenfunctions $v_1, \cdots, v_q$ are mutually orthonormal, so are the solutions $\phi_1, \cdots, \phi_q$. Then we have

$$\|\phi\|_H^2 = \left\| \sum_{r=1}^q a_r \phi_r \right\|_H^2 = \sum_{r=1}^q |a_r|^2 \|\phi_r\|_H^2 = \sum_{r=1}^q |a_r|^2 \equiv 1.$$

This contradicts the convergence (49). □

Now we show that the necessity of the conditions of C_p-compatibility is in fact the consequence of the minimality of the rank of Kalman's matrix.

Theorem 5. *Assume that system* (6) *is asymptotically synchronizable by p-groups under the minimal rank condition*

$$\mathrm{rank}(D, AD, \cdots, A^{N-1}D) = N - p. \tag{53}$$

Then, A satisfies the condition of C_p-compatibility (38) *and D satisfies the condition of strong C_p-compatibility* (41).

Proof. By Proposition 2, the rank condition (53) implies the existence of p orthonormal vectors $E_1, \cdots, E_p$ such that

$$AE_r = \lambda_r E_r \quad \text{and} \quad DE_r = 0, \quad r = 1, \cdots, p. \tag{54}$$

Let β_r^2 be the eigenvalue of L associated with the eigenvector v_r, defined by (52). Setting

$$\beta_r^2 = \mu_r^2 + \lambda_r > 0,$$

we check easily that for each r with $1 \leqslant r \leqslant p$, the function $e^{i\beta_r t} v_r E_r$ is a solution to system (6). Then, by (37) we have

$$e^{i\beta_r t} v_r C_p E_r \to 0 \quad \text{in } V^{N-p} \text{ as } t \to +\infty.$$

It follows thus that

$$C_p E_r = 0, \quad r = 1, \cdots, p,$$

namely,

$$\mathrm{Ker}(C_p) = \mathrm{Span}\{E_1, \cdots, E_p\},$$

which together with (54) implies conditions (38) and (41). $\qquad\square$

Proposition 7. *Assume that the matrix A satisfies the condition of C_p-compatibility* (38) *and D is given by* (42). *Then,*

$$\mathrm{rank}(D_p, A_p D_p, \cdots, A_p^{N-p-1} D_p) \tag{55}$$
$$= \mathrm{rank}(D, AD, \cdots, A^{N-1} D).$$

Proof. First, using (40) and (43), we have

$$A_p D_p = (C_p C_p^T)^{-1/2} C_p A D C_p^T (C_p C_p^T)^{-1/2}.$$

Successively, noting (39), we have

$$A_p^2 D_p = A_p (A_p D_p) = (C_p C_p^T)^{-1/2} C_p A^2 D C_p^T (C_p C_p^T)^{-1/2}$$

and so on. Then by Cayley-Hamilton's theorem, we get

$$\mathrm{rank}(D_p, AD_p, \cdots, A_p^{N-p-1} D_p) \tag{56}$$
$$= \mathrm{rank}(C_p C_p^T)^{-1/2} C_p (DC_p^T, ADC_p^T, \cdots, A^{N-1} DC_p^T)$$
$$= \mathrm{rank} C_p (DC_p^T, ADC_p^T, \cdots, A^{N-1} DC_p^T).$$

Now, noting (40), it is easy to see that $A^T \{\mathrm{Ker}(C_p)\}^\perp \subseteq \{\mathrm{Ker}(C_p)\}^\perp$. Since $A^T = A$ and $\{\mathrm{Ker}(C_p)\}^\perp = \mathrm{Im}(C_p^T)$, we get $A\mathrm{Im}(C_p^T) \subseteq \mathrm{Im}(C_p^T)$. By condition (42), we have $\mathrm{Im}(D) \subset \mathrm{Im}(C_p^T)$. Then, we successively get

$$\mathrm{Im}(AD) \subseteq A\mathrm{Im}(C_p^T) \subseteq \mathrm{Im}(C_p^T),$$
$$\mathrm{Im}(A^2 D) \subseteq A\mathrm{Im}(C_p^T) \subseteq \mathrm{Im}(C_p^T)$$

and so on. It follows that

$$\mathrm{Ker}(C_p) \cap \mathrm{Im}(DC_p^T, ADC_p^T, \cdots, A^{N-1}DC_p^T)$$
$$\subseteq \mathrm{Ker}(C_p) \cap \mathrm{Im}(C_p^T) = \mathrm{Ker}(C_p) \cap \{\mathrm{Ker}(C_p)\}^{\perp} = \{0\}.$$

Then, by [29, Proposition 2.7], we get

$$\mathrm{rank}C_p(DC_p^T, ADC_p^T, \cdots, A^{N-1}DC_p^T) \qquad (57)$$
$$= \mathrm{rank}(DC_p^T, ADC_p^T, \cdots, A^{N-1}DC_p^T).$$

Next, we write the transposition of the matrix on the right-hand side of (57) as

$$\begin{pmatrix} C_pD \\ C_pDA \\ \vdots \\ C_pDA^{N-1} \end{pmatrix} = \begin{pmatrix} C_p & & & \\ & C_p & & \\ & & \ddots & \\ & & & C_p \end{pmatrix} \begin{pmatrix} D \\ DA \\ \vdots \\ DA^{N-1} \end{pmatrix}.$$

By (42), it is easy to see

$$\mathrm{Ker}\begin{pmatrix} C_p & & & \\ & C_p & & \\ & & \ddots & \\ & & & C_p \end{pmatrix} \cap \mathrm{Im}\begin{pmatrix} D \\ DA \\ \vdots \\ DA^{N-1} \end{pmatrix} = \{0\}.$$

Similarly as for (57), we get

$$\mathrm{rank}\begin{pmatrix} C_pD \\ C_pDA \\ \vdots \\ C_pDA^{N-1} \end{pmatrix} = \mathrm{rank}\begin{pmatrix} D \\ DA \\ \vdots \\ DA^{N-1} \end{pmatrix},$$

namely,

$$\mathrm{rank}(DC_p^T, ADC_p^T, \cdots, A^{N-1}DC_p^T) \qquad (58)$$
$$= \mathrm{rank}(D, AD, \cdots, A^{N-1}D).$$

Finally, combining (56)–(58), we obtain (55).   □

As for the asymptotic stability, the following theorem indicates that the asymptotic synchronization by p-groups of system (6) can be reduced to the uniqueness of solution to an over-determined scalar problem.

Theorem 6. *Let the pair (A, D) satisfy Kalman's rank condition (53). Assume that A satisfies the condition of C_p-compatibility (39) and D is given by (42). Then, system (6) is asymptotically synchronizable by p-groups in any one of the cases described in Theorem 2.*

Proof. By Proposition 6, it suffices to show the asymptotic stability of the reduced system (46). By Theorem 1, system (46) is asymptotically stable under the condition

$$\text{rank}(D_p, A_p D_p, \cdots, A_p^{N-p-1} D_p) = N - p, \tag{59}$$

which is true by Proposition 7 and condition (53). □

Theorem 7. *Assume that A satisfies the condition of C_p-compatibility (39) and D is given by (42). If system (6) is asymptotically synchronizable by p-groups, then for any given initial data $(U_0, U_1) \in V \times \mathcal{H}$, there exist some scalar functions $u_1, \cdots, u_p$, such that*

$$\left(u^{(k)}(t) - u_r(t), (u^{(k)})'(t) - u_r'(t)\right) \to (0, 0) \quad in \ V \times H \tag{60}$$

as $t \to +\infty$ for all $n_{r-1} + 1 \leqslant k \leqslant n_r$ and $1 \leqslant r \leqslant p$.

Proof. Noting (36), there exist some real numbers α_{rs} with $\alpha_{rs} = \alpha_{sr}$, such that

$$A e_r = \sum_{s=1}^{p} \alpha_{rs} e_s, \quad r = 1, \cdots, p.$$

Then, applying e_r to (6) and setting $v_r = (e_r, U)$ for $r = 1, \cdots p$, we get

$$v_r'' + L v_r + \sum_{s=1}^{p} \alpha_{rs} v_s = 0 \tag{61}$$

associated with the initial data

$$t = 0: \quad v_r = (U_0, e_r), \quad v_r' = (U_1, e_r). \tag{62}$$

Setting

$$S = \begin{pmatrix} e_1^T \\ \vdots \\ e_p^T \\ C_p \end{pmatrix},$$

by (37) we have

$$SU = \begin{pmatrix} (e_1, U) \\ \vdots \\ (e_p, U) \\ C_p U \end{pmatrix} \to \begin{pmatrix} v_1 \\ \vdots \\ v_p \\ 0 \end{pmatrix} \quad in \ V \times \mathcal{H} \ as \ t \to +\infty.$$

On the other hand, since

$$Se_r = \|e_r\|^2 \epsilon_r, \quad r = 1, \cdots, p,$$

setting $u_r = v_r/\|e_r\|^2$, it follows that

$$U \to S^{-1} \begin{pmatrix} v_1 \\ \vdots \\ v_p \\ 0 \end{pmatrix} = \sum_{r=1}^{p} v_r S^{-1} \epsilon_r = \sum_{r=1}^{p} \frac{v_r}{\|e_r\|^2} e_r = \sum_{r=1}^{p} u_r e_r$$

in $\mathcal{V} \times \mathcal{H}$ as $t \to +\infty$. The proof is thus complete. $\qquad\square$

Remark 2. The convergence (32) will be called the asymptotic synchronization by p-groups in the consensus sense, and the convergence given in (60) will be called in the pinning sense. Theorem 7 clarifies that the two notions are simply the same. However, the functions $u_1, \cdots, u_p$ given by Theorem 7 are not unique. In fact, any functions $\widehat{u}_1, \cdots, \widehat{u}_p$, such that for $r = 1, \cdots, p$,

$$(\widehat{u}_r - u_r, \widehat{u}_r' - u_r') \to (0,0) \quad \text{in } \mathcal{V} \times \mathcal{H} \quad \text{as } t \to +\infty \qquad (63)$$

satisfy also (60).

2.4. *Applications*

We will give some classical examples to illustrate possible applications of the developed abstract theory.

2.4.1. *Wave equations with boundary damping*

Let Ω be a bounded domain in $\mathbb{R}^n$ with the boundary $\Gamma = \Gamma_1 \cup \Gamma_0$, such that $\text{mes}(\Gamma_1) > 0$. Consider the following system of wave equations:

$$\begin{cases} U'' - \Delta U + AU = 0, & \text{in } \mathbb{R}^+ \times \Omega, \\ U = 0, & \text{on } \mathbb{R}^+ \times \Gamma_0, \\ \partial_\nu U + DU' = 0, & \text{on } \mathbb{R}^+ \times \Gamma_1, \end{cases} \qquad (64)$$

where ∂_ν denotes the outward normal derivative on the boundary.

Let $H^1_{\Gamma_0}(\Omega)$ denote the subspace of $H^1(\Omega)$, composed of functions with vanishing trace on Γ_0. Multiplying system (64) by a test function $\Phi \in (H^1_{\Gamma_0}(\Omega))^N$ and integrating by parts, we get the following variational formulation:

$$\int_\Omega ((U'', \Phi) + (\nabla U, \nabla \Phi) + (AU, \Phi)) dx + \int_{\Gamma_1} (DU', \Phi) d\Gamma = 0. \qquad (65)$$

Let L and g be defined by

$$\langle Lu, \phi \rangle = \int_{\Omega} \nabla u \cdot \nabla \phi \, dx \quad \text{and} \quad \langle gu, \phi \rangle = \int_{\Gamma_1} u \phi \, d\Gamma, \tag{66}$$

respectively. Setting $\mathcal{L}$ and $\mathcal{G}$ as in (5), the variational equation (65) can be rewritten as

$$U'' + \mathcal{L}U + AU + D\mathcal{G}U' = 0. \tag{67}$$

Obviously, the operators L and g defined by (66) satisfy conditions (1)–(3). Then by Proposition 1, system (67) generates a semi-group of contractions.

We assume that there exists $x_0 \in \mathbb{R}^n$, such that setting $m = x - x_0$, we have $(m \cdot \nu) \leqslant 0$ on Γ_0, where ν denotes the outward normal vector on the boundary.

Proposition 8. *The operator L defined by (66) is g-observable.*

Proof. By definition, we have to show that there exists a positive constant c, independent of $\beta \in \mathbb{R}$ and $f \in L^2(\Omega)$, such that the following uniform estimate

$$\int_{\Omega} |\phi|^2 dx \leqslant c \int_{\Omega} |f|^2 dx \tag{68}$$

holds for any given solution ϕ to the over-determined scalar problem:

$$\begin{cases} \beta^2 \phi + \Delta \phi = f, & \text{in } \Omega, \\ \phi = 0, & \text{on } \Gamma_0, \\ \partial_\nu \phi = 0, & \text{on } \Gamma_1 \end{cases} \tag{69}$$

with the observation

$$\phi = 0 \quad \text{on } \Gamma_1. \tag{70}$$

Let us first recall Rellich's identity

$$2 \int_{\Omega} \Delta \phi (m \cdot \nabla \phi) dx = (n-2) \int_{\Omega} |\nabla \phi|^2 dx \tag{71}$$

$$+ \int_{\Gamma} (2 \partial_\nu \phi (m \cdot \nabla \phi) - (m \cdot \nu) |\nabla \phi|^2) d\Gamma$$

for all $\phi \in H^2(\Omega)$, where $m = x - x_0$.

Then, multiplying the equation in (69) by $2m \cdot \nabla \phi + (n-1)\phi$ and integrating by parts, we have

$$\int_{\Omega} (|\beta \phi|^2 + |\nabla \phi|^2) dx - \int_{\Gamma_0} (m \cdot \nu) |\partial_\nu \phi|^2 d\Gamma = - \int_{\Omega} f(2m \cdot \nabla \phi + (n-1)\phi) dx.$$

Since $(m \cdot \nu) \leqslant 0$ on Γ_0, by Cauchy-Schwarz's inequality, we get

$$\int_\Omega (|\beta\phi|^2 + |\nabla\phi|^2)dx \leqslant c \int_\Omega |f|^2 dx. \tag{72}$$

Since $\phi \in H_0^1(\Omega)$, by Poincaré's inequality, (72) is a stronger version of (68). The proof is complete. $\qquad\square$

Theorem 8. *Assume that the pair (A, D) satisfies the rank condition (53). Let A satisfy the condition of C_p-compatibility (39) and D be given by (42). Assume furthermore that A satisfies the ϵ-closing condition (16) with $\epsilon > 0$ small enough. Then system (64) is asymptotically synchronizable by p-groups in $(H_{\Gamma_0}^1(\Omega) \times L^2(\Omega))^N$.*

Proof. Noting that the trace operator g is compact from $H_{\Gamma_0}^1(\Omega)$ into $L^2(\Gamma_1)$. Furthermore, by Proposition 8, the operator L is g-observable. Then, by Theorem 6 we get the conclusion. $\qquad\square$

2.4.2. *Wave equations with internal damping*

Let Ω be a bounded domain in $\mathbb{R}^n$ with smooth boundary Γ. Consider the following coupled system of wave equations:

$$\begin{cases} U'' - \Delta U + AU + DU' = 0, & \text{in } \mathbb{R}^+ \times \Omega, \\ U = 0, & \text{on } \mathbb{R}^+ \times \Gamma. \end{cases} \tag{73}$$

Multiplying system (73) by a test function $\Phi \in (H_0^1(\Omega))^N$ and integrating by parts, we get the following variational formulation:

$$\int_\Omega ((U'', \Phi) + (\nabla U, \nabla\Phi) + (AU, \Phi))dx + \int_\omega (DU', \Phi)d\Gamma = 0. \tag{74}$$

Let L and g be defined by

$$\langle Lu, \phi \rangle = \int_\Omega \nabla u \cdot \nabla\phi dx \quad \text{and} \quad \langle gu, \phi \rangle = \int_\Omega u\phi dx, \tag{75}$$

respectively. The variational problem (74) can be rewritten as

$$U'' + \mathcal{L}U + AU + D\mathcal{G}U' = 0. \tag{76}$$

Theorem 9. *Assume that the pair (A, D) satisfies the rank condition (53). Assume furthermore that A satisfies the condition of C_p-compatibility (39) and D is given by (42). Then system (73) is asymptotically synchronizable by p-groups in $(H_0^1(\Omega) \times L^2(\Omega))^N$.*

Proof. It is a direct consequence of Theorem 6. $\qquad\square$

2.4.3. *Kirchhoff plate equations with boundary damping*

In this section Ω is a star-shaped domain in $\mathbb{R}^2$, occupied by an elastic thin plate. Let μ with $0 < \mu < 1/2$ be the Poisson ratio. Let the shear force operator B_1 and the bending moment operator B_2 be defined by

$$\begin{cases} B_1\phi = 2\nu_1\nu_2\partial_{xy}\phi - \nu_1^2\partial_{yy}\phi - \nu_2^2\partial_{xx}\phi, \\ B_2\phi = (\nu_1^2 - \nu_2^2)\partial_{xy}\phi + \nu_1\nu_2(\partial_{xx}\phi - \partial_{yy}\phi). \end{cases} \tag{77}$$

Consider the following system[19,20] for more precise description:

$$\begin{cases} U'' + \Delta^2 U + AU = 0, & \text{in } \mathbb{R}^+ \times \Omega, \\ \Delta U + (1-\mu)B_1 U = 0, & \text{on } \mathbb{R}^+ \times \Gamma, \\ \partial_\nu \Delta U + (1-\mu)\partial_\tau B_2 U = DU', & \text{on } \mathbb{R}^+ \times \Gamma, \end{cases} \tag{78}$$

where ∂_τ denotes the tangential derivative on the boundary. For the sake of simplicity, the shear force damping is acted on the whole boundary Γ. We assume that there exists $x_0 \in \mathbb{R}^n$, such that setting $m = x - x_0$, we have

$$(m \cdot \nu) \geqslant 0, \quad \forall x \in \Gamma. \tag{79}$$

Define the symmetric bilinear form by

$$a(\phi, \psi) = \phi_{xx}\psi_{xx} + \phi_{yy}\psi_{yy} + \mu(\phi_{xx}\psi_{yy} + \phi_{yy}\psi_{xx}) + 2(1-\mu)\phi_{xy}\psi_{xy}. \tag{80}$$

Multiplying system (78) by a test function $\Phi \in (H^2(\Omega))^N$ and integrating by parts, by means of Green's formula ([19, (4.3.20)]):

$$\int_\Omega \Delta^2\phi \, \psi dxdy = \int_\Omega a(\phi, \psi)dxdy \tag{81}$$

$$+ \int_\Gamma (\partial_\nu\Delta\phi + (1-\mu)\partial_\tau B_2\phi)\psi d\Gamma - \int_\Gamma (\Delta\phi + (1-\mu)B_1\phi)\partial_\nu\psi d\Gamma,$$

we get the following variational formulation:

$$\int_\Omega ((U'', \Phi) + a(U, \Phi) + (AU, \Phi))dxdy + \int_\Gamma (DU', \Phi)d\Gamma = 0. \tag{82}$$

Let L and g be defined by

$$\langle Lu, \phi \rangle = \int_\Omega a(u, \phi)dxdy \quad \text{and} \quad \langle gu, \phi \rangle = \int_\Gamma u\phi d\Gamma. \tag{83}$$

(82) can be interpreted as

$$U'' + \mathcal{L}U + AU + D\mathcal{G}U' = 0, \tag{84}$$

which, by Proposition 1, is well-posed in the sense of semi-group of contractions on the space $(H^2(\Omega) \times L^2(\Omega))^N$.

Proposition 9. *The operator L defined by* (83) *is g-observable.*

Proof. We have to show that there exists a positive constant c, independent of $\beta \in \mathbb{R}$ and $f \in L^2(\Omega)$, such that the following uniform estimate

$$\int_\Omega |\phi|^2 dxdy \leqslant c \int_\Omega |f|^2 dxdy \qquad (85)$$

holds for any solution ϕ to the over-determined scalar problem:

$$\begin{cases} \beta^2 \phi - \Delta^2 \phi = f, & \text{in } \Omega, \\ \Delta\phi + (1-\mu)B_1\phi = 0, & \text{on } \Gamma, \\ \partial_\nu \Delta\phi + (1-\mu)\partial_\tau B_2\phi = 0, & \text{on } \Gamma \end{cases} \qquad (86)$$

with the observation

$$\phi = 0 \quad \text{on } \Gamma. \qquad (87)$$

Multiplying the equation in (86) by $m \cdot \nabla\phi$ and integrating by parts, by means of the following identity ([19, Lemma 4.5.1]):

$$\int_\Omega \Delta^2\phi(m \cdot \nabla\phi)dxdy = \int_\Omega a(\phi,\phi)dxdy + \frac{1}{2}\int_\Gamma (m \cdot \nu)a(\phi,\phi)d\Gamma$$
$$+ \int_\Gamma (\partial_\nu\Delta\phi + (1-\mu)\partial_\tau B_2\phi)(m \cdot \nabla\phi)d\Gamma$$
$$- \int_\Gamma (\Delta\phi + (1-\mu)B_1\phi)\partial_\nu(m \cdot \nabla\phi)d\Gamma,$$

we have

$$\int_\Omega (|\beta\phi|^2 d + a(\phi,\phi))dxdy + \frac{1}{2}\int_\Gamma (m \cdot \nu)a(\phi,\phi)d\Gamma = -\int_\Omega f(m \cdot \nabla\phi)dxdy.$$

Using Cauchy-Schwarz's inequality, we get

$$\int_\Omega (|\beta\phi|^2 + a(\phi,\phi))dxdy + \int_\Gamma (m \cdot \nu)a(\phi,\phi)d\Gamma \leqslant c \int_\Omega |f|^2 dxdy. \qquad (88)$$

Since $\phi \in H^2(\Omega) \cap H_0^1(\Omega)$, there exists a constant $c > 0$, such that

$$\int_\Omega a(\phi,\phi)dxdy \geqslant c\|\phi\|_{H^2(\Omega)}^2. \qquad (89)$$

Noting (79), estimate (88) is a stronger version of (85). The proof is complete. □

Theorem 10. *Assume that the pair* (A, D) *satisfies the rank condition* (53). *Let* A *satisfy the condition of* C_p-*compatibility* (39), *and* D *be given by* (42). *Assume furthermore that* A *satisfies the ϵ-closing condition* (16) *with* $\epsilon > 0$ *small enough. Then system* (78) *is asymptotically synchronizable by p-groups in* $(H^2(\Omega) \times L^2(\Omega))^N$.

Proof. Since the trace operator g is compact from $H^2(\Omega)$ into $L^2(\Gamma)$, and the operator L is g-observable, by Theorem 6 we get the conclusion. $\square$

2.4.4. *Kirchhoff plate equations with internal dampings*

Let A, D_1 and D_2 be symmetric and positive semi-definite matrices. Consider the following system

$$\begin{cases} U'' + \Delta^2 U + AU + D_1 U' - D_2 \Delta U' = 0, & \text{in } \mathbb{R}^+ \times \Omega, \\ U = \partial_\nu U = 0, & \text{on } \mathbb{R}^+ \times \Gamma. \end{cases} \tag{90}$$

We will investigate the asymptotic behaviour of system (90) under the joint action of the dampings given by the shear force $D_1 U'$ and the bending moment $D_2 \Delta U'$.

Theorem 11. *System* (90) *is asymptotically stable in* $(H_0^2(\Omega) \times L^2(\Omega))^N$ *if and only if* (A, D) *satisfies Kalman's rank condition* (15) *with* $D = (D_1, D_2)$.

Proof. The asymptotic stability of system (90) is equivalent to the uniqueness of solution to the eigen-problem with $\beta \in \mathbb{R}$:

$$\begin{cases} \Delta^2 \Phi + A\Phi = \beta^2 \Phi, & \text{in } \Omega, \\ \Phi = \partial_\nu \Phi = 0, & \text{on } \Gamma \end{cases} \tag{91}$$

associated with the mixed observations

$$D_1 \Phi = 0 \quad \text{and} \quad D_2 \Delta \Phi = 0 \quad \text{in } \Omega. \tag{92}$$

Obviously, Kalman's rank condition (15) with $D = (D_1, D_2)$ is still necessary for the uniqueness of solution to (91)–(92). We then examine the sufficiency. Applying D_1 to (91), it follows that

$$\Delta^2 D_1 \Phi + D_1 A\Phi = \beta^2 D_1 \Phi \quad \text{in } \Omega. \tag{93}$$

The first observation in (92) implies that $D_1 A\Phi = 0$ in Ω. A simple recurrence successively gives

$$D_1 \Phi = D_1 A\Phi = \cdots = D_1 A^{N-1} \Phi = 0. \tag{94}$$

On the other hand, the second observation in (92) implies that $D_2 \Phi = 0$ in Ω. Then, by the same procedure as for D_1, we get

$$D_2 \Phi = D_2 A\Phi = \cdots = D_2 A^{N-1} \Phi = 0. \tag{95}$$

By (94) and (95), it follows that

$$\Phi \in \mathrm{Ker} \begin{pmatrix} D \\ DA \\ \vdots \\ DA^{N-1} \end{pmatrix} = \{\mathrm{Im}(D, AD, \cdots, A^{N-1}D)\}^{\perp} = \{0\},$$

then $\Phi \equiv 0$. The proof is complete. $\square$

In this example, the feedbacks are global. This case is easy to be treated, but it gives the idea of generalizing Theorem 2 to the case of several observations. Once it is realized, we can consider the problems with mixed dampings, in particular, plate equations with distributed and boundary dampings, wave equations with memory and thermal sources, etc. We are waiting for a wide field of applications and discover new challenge in the future.

3. Asymptotic synchronization by LaSalle's invariant principle

3.1. *Asymptotic stability (continued)*

In the previous section, we have established a general theory on the asymptotic synchronization for linear dissipative systems by frequency domain approach. However, since the observability inequality (18) is based on the multiplier method, the geometrical multiplier condition should be naturally required in applications. In this section, we will propose the so-called time domain approach, based on LaSalle's invariance principle.

More precisely, let Ω be a rectangular domain:

$$\Omega = (0,1) \times (0,a) \tag{1}$$

with the configuration of the boundary

$$\begin{cases} \Gamma_0 = \{(0,y) \cup (1,y),\ 0 < y < a\}, \\ \Gamma_1 = \{(x,0) \cup (x,a),\ 0 < x < 1\}. \end{cases} \tag{2}$$

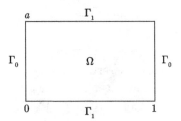

Fig. 1.   Rectangle with reflecting sides

Let A and D be symmetric and positive semi-definite matrices of order N. Consider the following coupled system of wave equations for the state variable $U = (u^{(1)}, \cdots, u^{(N)})^T$:

$$\begin{cases} U'' - \Delta U + AU = 0, & \text{in } \mathbb{R}^+ \times \Omega, \\ U = 0, & \text{on } \mathbb{R}^+ \times \Gamma_0, \\ \partial_\nu U + DU' = 0, & \text{on } \mathbb{R}^+ \times \Gamma_1, \\ t = 0 : U = U_0, \ U' = U_1, & \text{in } \Omega, \end{cases} \tag{3}$$

where ∂_ν denotes the outward normal derivative on the boundary Γ_1.

Since the horizontal rays of geometrical optics will be reflected between the two vertical boundaries Γ_0, they can not hit the non-diffractive points on the boundary Γ_1. So, Γ_1 can not be used to control the domain Ω in the sense of Bardos-Lebeau-Rauch in [3]. Therefore, the observability inequality (18) does not hold in the present case.

By Proposition 1, system (3) generates a C^0 semi-group of contractions on the space $(H^1_{\Gamma_0}(\Omega) \times L^2(\Omega))^N$. For the sake of completeness, we give a brief reminder here. We endow the space $(H^1_{\Gamma_0}(\Omega) \times L^2(\Omega))^N$ with the inner product:

$$((U, V), (\widetilde{U}, \widetilde{V})) = \int_\Omega \{ \langle \nabla U, \nabla \widetilde{U} \rangle + (AU, \widetilde{U}) + (V, \widetilde{V}) \} dx, \tag{4}$$

where $(\cdot, \cdot)$ denotes the inner product of $\mathbb{R}^N$, while $\langle \cdot, \cdot \rangle$ denotes the inner product of $\mathbb{M}^{N \times N}(\mathbb{R})$.

Let

$$D(\mathcal{A}) = \left\{ \begin{array}{l} (U, V) \in (H^1_{\Gamma_0}(\Omega) \times H^1_{\Gamma_0}(\Omega))^N \text{ such that} \\ \Delta U - AU \in (L^2(\Omega))^N \text{ and } \partial_\nu U + DV = 0 \text{ on } \Gamma_1 \end{array} \right\}.$$

We define the linear unbounded operator $\mathcal{A} : D(\mathcal{A}) \to (H^1_{\Gamma_0}(\Omega) \times L^2(\Omega))^N$ by

$$\mathcal{A}(U, V) = (V, \Delta U - AU).$$

Then system (3) can be formally transformed into the following abstract formulation:

$$(U, V)' = \mathcal{A}(U, V).$$

By Proposition 1, we can announce the following well-posedness result.

Proposition 10. *The operator $\mathcal{A}$ is m-accretive, therefore, generates a semigroup of contractions $e^{t\mathcal{A}}$ on the space $(H^1_{\Gamma_0}(\Omega) \times L^2(\Omega))^N$. Moreover, the resolvent of $\mathcal{A}$ is compact.*

Theorem 12. *System (3) is asymptotically stable if and only if for any given initial data $(\Phi_0, \Phi_1) \in (H^1_{\Gamma_0}(\Omega) \times L^2(\Omega))^N$, the following problem*

$$\begin{cases} \Phi'' - \Delta\Phi + A\Phi = 0, & \text{in } \mathbb{R}^+ \times \Omega, \\ \Phi = 0, & \text{on } \mathbb{R}^+ \times \Gamma_0, \\ \partial_\nu \Phi = 0, & \text{on } \mathbb{R}^+ \times \Gamma_1, \\ t = 0 : \Phi = \Phi_0, \ \Phi' = \Phi_1, & \text{in } \Omega \end{cases} \tag{5}$$

with the boundary D-observation at the infinite horizon

$$D\Phi = 0 \quad \text{on } \mathbb{R}^+ \times \Gamma_1 \tag{6}$$

has only the trivial solution $\Phi = 0$.

Proof. Since the resolvent of $\mathcal{A}$ is compact, for any given initial data $(U_0, U_1) \in D(\mathcal{A})$, the positive trajectory starting from (U_0, U_1) defined by

$$\mathcal{O}^+(U_0, U_1) = \bigcup_{t \geq 0} \{e^{t\mathcal{A}}(U_0, U_1)\}, \tag{7}$$

is bounded for the graph norm in $D(\mathcal{A})$, therefore, relatively compact in $(H_{\Gamma_0}^1(\Omega) \times L^2(\Omega))^N$. Then the ω-limit set $\omega(U_0, U_1) \subset D(\mathcal{A})$ defined by

$$\omega(U_0, U_1) = \{(\Psi_0, \Psi_1) : \exists t_n \to +\infty, \ e^{t_n \mathcal{A}}(U_0, U_1) \to (\Psi_0, \Psi_1)\} \tag{8}$$

is non-empty, compact and connex.

For any given initial data $(\Psi_0, \Psi_1) \in \omega(U_0, U_1) \subset D(\mathcal{A})$, denoted by $(\Psi(t), \Psi'(t)) = e^{t\mathcal{A}}(\Psi_0, \Psi_1)$ the corresponding trajectory, we have

$$\begin{cases} \Psi'' - \Delta\Psi + A\Psi = 0, & \text{in } \mathbb{R}^+ \times \Omega, \\ \Psi = 0, & \text{on } \mathbb{R}^+ \times \Gamma_0, \\ \partial_\nu \Psi + D\Psi' = 0, & \text{on } \mathbb{R}^+ \times \Gamma_1, \\ t = 0 : \Psi = \Psi_0, \ \Psi' = \Psi_1, & \text{in } \Omega. \end{cases} \tag{9}$$

By LaSalle's invariance principle (*cf.* [5, Theorem 9.2.3] or [8, Theorem 1], also [12, 37] for applications), $\omega(U_0, U_1)$ is invariant for $e^{t\mathcal{A}}$ and the restriction of $e^{t\mathcal{A}}$ on $\omega(U_0, U_1)$ is an isometry for all $t \geq 0$, therefore, the boundary damping should vanish:

$$\int_{\Gamma_1} (D\Psi', \Psi')_{\mathbb{R}^N} \, d\Gamma = 0, \quad \forall t \geq 0. \tag{10}$$

Since D is symmetric and positive semi-definite, we get thus the boundary dissipation

$$D\Psi' = 0 \quad \text{on } \mathbb{R}^+ \times \Gamma_1. \tag{11}$$

By the contraction of $e^{t\mathcal{A}}$, system (3) is asymptotically stable if and only if $\omega(U_0, U_1)$ is reduced to $(0, 0)$ for all the initial data $(U_0, U_1) \in D(\mathcal{A})$, therefore, if and only if the overdetermined system (9)–(11) with the initial data $(\Psi_0, \Psi_1) \in D(\mathcal{A})$ has only the trivial solution. However,[33] setting $\Phi = \Psi'$, we may check the equivalence between the uniqueness of solution to system (9)–(11) with the initial data $(\Psi_0, \Psi_1) \in D(\mathcal{A})$ and the uniqueness of solution to system (5)–(6) with the initial data $(\Phi_0, \Phi_1) \in (H_{\Gamma_0}^1(\Omega) \times L^2(\Omega))^N$. $\square$

3.2. *Uniqueness theorem for non-harmonic series*

Let Ω be the rectangular domain given by (1)–(2). Let $\lambda_n > 0$ be an eigenvalue of $-\Delta$ with the multiplicity s_n and $e_n^{(s)}$ with $1 \leqslant s \leqslant s_n$ be the associated eigen-functions:

$$\begin{cases} -\Delta e_n^{(s)} = \lambda_n e_n^{(s)}, & \text{in } \Omega, \\ e_n^{(s)} = 0, & \text{on } \Gamma_0, \\ \partial_\nu e_n^{(s)} = 0, & \text{on } \Gamma_1. \end{cases} \tag{12}$$

Let $\delta_l \geqslant 0$ be an eigenvalue of A with the multiplicity μ_l, associated with the eigenvectors $\omega_l^{(\mu)}$:

$$A\omega_l^{(\mu)} = \delta_l \omega_l^{(\mu)}, \quad 1 \leqslant l \leqslant d, \quad 1 \leqslant \mu \leqslant \mu_l.$$

Define the differential operator corresponding to problem (5) by

$$\mathcal{A}_0 \begin{pmatrix} \Phi \\ \Psi \end{pmatrix} = \begin{pmatrix} \Psi \\ \Delta\Phi - A\Phi \end{pmatrix} \tag{13}$$

with the domain of definition

$$D(\mathcal{A}_0) = \left\{ \begin{array}{l} (\Phi, \Psi) \in (H^1_{\Gamma_0}(\Omega) \times H^1_{\Gamma_0}(\Omega))^N \text{ such that} \\ \Delta\Phi - A\Phi \in (L^2(\Omega))^N \text{ and } \partial_\nu \Phi = 0 \text{ on } \Gamma_1 \end{array} \right\}. \tag{14}$$

To be clear, recall that $\mathbb{N}^*$ denotes the set of all the positive integers and $\mathbb{Z}^*$ the set of all the non-zero integers.

For any given $n \in \mathbb{N}^*$ and any given l with $1 \leqslant l \leqslant d$, we take

$$\|e_n^{(s)}\|_{L^2(\Omega)} = 1, \quad \|\omega_l^{(\mu)}\|_{\mathbb{R}^N} = 1 \tag{15}$$

and define

$$e_{-n}^{(s)} = e_n^{(s)}, \quad s_{-n} = s_n, \quad \beta_{\pm n, l} = \pm\sqrt{\lambda_n + \delta_l}. \tag{16}$$

Then $\beta_{n,l}$ is an eigenvalue of $\mathcal{A}_0$, associated with the eigenvectors

$$E_{n,l}^{(s,\mu)} = \begin{pmatrix} \dfrac{1}{i\beta_{n,l}} e_n^{(s)} \omega_l^{(\mu)} \\ e_n^{(s)} \omega_l^{(\mu)} \end{pmatrix}, \quad 1 \leqslant s \leqslant s_n, \quad 1 \leqslant \mu \leqslant \mu_l.$$

Since $\mathcal{A}_0$ is a skew-adjoint operator with the compact resolvent, the family of the eigenvectors $\{E_{n,l}^{(s,\mu)}\}$ for all $n \in \mathbb{Z}^*, 1 \leqslant l \leqslant d, 1 \leqslant s \leqslant s_n$ and $1 \leqslant \mu \leqslant \mu_l$ constitutes a Hilbert basis in the space $(H^1_{\Gamma_0}(\Omega) \times L^2(\Omega))^N$. Then, for any given initial data $(\Phi_0, \Phi_1) \in (H^1_{\Gamma_0}(\Omega) \times L^2(\Omega))^N$, we can decompose it as

$$\begin{pmatrix} \Phi_0 \\ \Phi_1 \end{pmatrix} = \sum_{n \in \mathbb{Z}^*} \sum_{1 \leqslant l \leqslant d} \left(\sum_{s=1}^{s_n} \sum_{\mu=1}^{\mu_l} a_{n,l}^{(s,\mu)} E_{n,l}^{(s,\mu)} \right)$$

with

$$\|\Phi_0\|_{(H^1(\Omega))^N}^2 + \|\Phi_1\|_{(L^2(\Omega))^N}^2 = \sum_{n\in\mathbb{Z}^*} \sum_{1\leqslant l\leqslant d} \left(\sum_{s=1}^{s_n}\sum_{\mu=1}^{\mu_l} |a_{n,l}^{(s,\mu)}|^2\right).$$

Accordingly, the solution to problem (5) is given by

$$\Phi = \sum_{n\in\mathbb{Z}^*} \sum_{1\leqslant l\leqslant d} \left(\sum_{s=1}^{s_n}\sum_{\mu=1}^{\mu_l} \frac{a_{n,l}^{(s,\mu)}}{i\beta_{n,l}} e_n^{(s)}\omega_l^{(\mu)}\right)e^{i\beta_{n,l}t} \quad \text{in } \mathbb{R}^+ \times \Omega,$$

and the D-observation (6) becomes

$$\sum_{n\in\mathbb{Z}^*} \sum_{1\leqslant l\leqslant d} \left(\sum_{s=1}^{s_n}\sum_{\mu=1}^{\mu_l} \frac{a_{n,l}^{(s,\mu)}}{i\beta_{n,l}} e_n^{(s)}|_{\Gamma_1} D\omega_l^{(\mu)}\right)e^{i\beta_{n,l}t} = 0 \quad \text{on } \mathbb{R}^+ \times \Gamma_1. \quad (17)$$

Noting (15) and the trace embedding, the above series (17) converges in $(C^0(0,+\infty;L^2(\Gamma_1)))^N$.

Theorem 13. *Let the pair (A, D) satisfy Kalman's rank condition:*

$$\operatorname{rank}(D, AD, \cdots, A^{N-1}D) = N. \quad (18)$$

Assume that for any given

$$\beta_{m,k} = \beta_{n,l} \text{ with } (m,k) \neq (n,l), \quad (19)$$

we have

$$\int_{\Gamma_1} e_m^{(r)}e_n^{(s)}d\Gamma = 0, \quad 1\leqslant r\leqslant s_m, \quad 1\leqslant s\leqslant s_n. \quad (20)$$

Then, the overdetermined system (5)–(6) has only the trivial solution.

Proof. For any given $\epsilon > 0$ small enough, let $N_\epsilon > 0$, such that

$$\sum_{|n|>N_\epsilon} \sum_{1\leqslant l\leqslant d} \sum_{s=1}^{s_n}\sum_{\mu=1}^{\mu_l} |a_{n,l}^{(s,\mu)}|^2 \leqslant \epsilon. \quad (21)$$

Let

$$\begin{pmatrix}\Phi_0^\epsilon\\\Phi_1^\epsilon\end{pmatrix} = \sum_{|n|\leqslant N_\epsilon} \sum_{1\leqslant l\leqslant d} \left(\sum_{s=1}^{s_n}\sum_{\mu=1}^{\mu_l} a_{n,l}^{(s,\mu)} E_{n,l}^{(s,\mu)}\right).$$

The corresponding solution Φ^ϵ to system (10) is given by

$$\Phi^\epsilon = \sum_{|n|\leqslant N_\epsilon} \sum_{1\leqslant l\leqslant d} \left(\sum_{s=1}^{s_n}\sum_{\mu=1}^{\mu_l} \frac{a_{n,l}^{(s,\mu)}}{i\beta_{n,l}} e_n^{(s)}\omega_l^{(\mu)}\right)e^{i\beta_{n,l}t} \quad \text{on } \mathbb{R}^+ \times \Omega,$$

and the corresponding observation (6) is written as

$$D\Phi^\epsilon = \sum_{|n|\leqslant N_\epsilon} \sum_{1\leqslant l\leqslant d} \left(\sum_{s=1}^{s_n} \sum_{\mu=1}^{\mu_l} \frac{a_{n,l}^{(s,\mu)}}{i\beta_{n,l}} e_n^{(s)}|_{\Gamma_1} D\omega_l^{(\mu)} \right) e^{i\beta_{n,l}t}. \tag{22}$$

By trace embedding, there exists a positive constant c_0 such that

$$\int_{\Gamma_1} |D(\Phi - \Phi^\epsilon)|^2 d\Gamma \leqslant c_0 (\|\Phi_0 - \Phi_0^\epsilon\|_{H^1(\Omega)}^2 + \|\Phi_1 - \Phi_1^\epsilon\|_{L(\Omega)}^2).$$

By (6) and (21), we have

$$\int_{\Gamma_1} |D\Phi^\epsilon|^2 d\Gamma \leqslant c_0 \sum_{|n|>N_\epsilon} \sum_{1\leqslant l\leqslant d} \sum_{s=1}^{s_n} \sum_{\mu=1}^{\mu_l} |a_{n,l}^{(s,\mu)}|^2 \leqslant c_0\epsilon. \tag{23}$$

On the other hand, we write (22) as

$$D\Phi^\epsilon = \sum_{|n|\leqslant N_\epsilon} \sum_{1\leqslant l\leqslant d} A_{n,l} e^{i\beta_{n,l}t}, \tag{24}$$

where

$$A_{n,l} = \sum_{s=1}^{s_n} \sum_{\mu=1}^{\mu_l} \frac{a_{n,l}^{(s,\mu)}}{i\beta_{n,l}} e_n^{(s)}|_{\Gamma_1} D\omega_l^{(\mu)}. \tag{25}$$

Then, it follows that

$$|D\Phi^\epsilon|^2 = \sum_{|n|\leqslant N_\epsilon, 1\leqslant l\leqslant d} |A_{n,l}|^2 + \sum_{\substack{|n|,|m|\leqslant N_\epsilon, 1\leqslant l,k\leqslant d \\ (m,k)\neq(n,l)}} A_{n,l}\bar{A}_{m,k} e^{i(\beta_{n,l}-\beta_{m,k})t}.$$

$$\tag{26}$$

But

$$\sum_{\substack{|n|,|m|\leqslant N_\epsilon, 1\leqslant l,k\leqslant d \\ (m,k)\neq(n,l)}} A_{n,l}\bar{A}_{m,k} e^{i(\beta_{n,l}-\beta_{m,k})t} \tag{27}$$

$$= \sum_{\substack{|n|,|m|\leqslant N_\epsilon, 1\leqslant l,k\leqslant d \\ (m,k)\neq(n,l),\beta_{m,k}=\beta_{n,l}}} A_{n,l}\bar{A}_{m,k} + \sum_{\substack{|n|,|m|\leqslant N_\epsilon, 1\leqslant l,k\leqslant d \\ (m,k)\neq(n,l),\beta_{m,k}\neq\beta_{n,l}}} A_{n,l}\bar{A}_{m,k} e^{i(\beta_{n,l}-\beta_{m,k})t}.$$

Noting (20) and (25), the integration on Γ_1 of the first term on the right-hand side of (27) vanishes. Then, it follows that

$$\int_0^T \int_{\Gamma_1} |D\Phi^\epsilon|^2 d\Gamma dt = \sum_{|n|\leqslant N_\epsilon} \sum_{1\leqslant l\leqslant d} T \int_{\Gamma_1} |A_{n,l}|^2 d\Gamma \tag{28}$$

$$+ \sum_{\substack{|n|,|m|\leqslant N_\epsilon, 1\leqslant l,k\leqslant d \\ (m,k)\neq(n,l),\beta_{m,k}\neq\beta_{n,l}}} \frac{e^{i(\beta_{n,l}-\beta_{m,k})T}-1}{i(\beta_{n,l}-\beta_{m,k})} \int_{\Gamma_1} A_{n,l}\bar{A}_{m,k} d\Gamma.$$

Let

$$C_\epsilon = \sum_{\substack{|n|,|m| \leqslant N_\epsilon, 1 \leqslant l, k \leqslant d \\ (m,k) \neq (n,l), \beta_{m,k} \neq \beta_{n,l}}} \frac{2}{|\beta_{n,l} - \beta_{m,k}|} \int_{\Gamma_1} |A_{n,l}\bar{A}_{m,k}| d\Gamma. \qquad (29)$$

It follows from (23) and (28) that

$$\sum_{|n| \leqslant N_\epsilon} \sum_{1 \leqslant l \leqslant d} \int_{\Gamma_1} |A_{n,l}|^2 d\Gamma \leqslant \frac{C_\epsilon}{T} + \frac{1}{T} \int_0^T \int_{\Gamma_1} |D\Phi^\epsilon|^2 d\Gamma dt$$

$$\leqslant \frac{C_\epsilon}{T} + c_0 \epsilon.$$

Taking $T \to +\infty$, we get

$$\sum_{|n| \leqslant N_\epsilon} \sum_{1 \leqslant l \leqslant d} \int_{\Gamma_1} |A_{n,l}|^2 d\Gamma \leqslant c_0 \epsilon$$

for any given $\epsilon > 0$ small enough. Then, noting (25), we get

$$\sum_{s=1}^{s_n} \left(\sum_{\mu=1}^{\mu_l} \frac{a_{n,l}^{(s,\mu)}}{i\beta_{n,l}} D\omega_l^{(\mu)} \right) e_n^{(s)}|_{\Gamma_1} = 0, \quad n \in \mathbb{Z}^*, \quad 1 \leqslant l \leqslant d.$$

Since $e_n^{(1)}, \cdots, e_n^{(s_n)}$ are the eigenfunctions associated with the same eigenvalue λ_n, by Carleman's uniqueness theorem,[10,44] their traces on Γ_1 are linearly independent, then we get

$$\sum_{\mu=1}^{\mu_l} \frac{a_{n,l}^{(s,\mu)}}{i\beta_{n,l}} D\omega_l^{(\mu)} = 0, \quad n \in \mathbb{Z}^*, \quad 1 \leqslant l \leqslant d, \quad 1 \leqslant s \leqslant s_n.$$

Since $\omega_l^{(1)}, \cdots, \omega_l^{(\mu_l)}$ are the eigenvectors associated with the same eigenvalue δ_l, by Proposition 2, Kalman's rank condition (18) implies that

$$\sum_{\mu=1}^{\mu_l} \frac{a_{n,l}^{(s,\mu)}}{i\beta_{n,l}} \omega_l^{(\mu)} = 0, \quad n \in \mathbb{Z}^*, \quad 1 \leqslant l \leqslant d, \quad 1 \leqslant s \leqslant s_n.$$

Then, by the linear independence of $\omega_l^{(1)}, \cdots, \omega_l^{(\mu_l)}$, we have

$$a_{n,l}^{(s,\mu)} = 0, \quad n \in \mathbb{Z}^*, \quad 1 \leqslant l \leqslant d, \quad 1 \leqslant s \leqslant s_n, \quad 1 \leqslant \mu \leqslant \mu_l.$$

Then $\Phi_0 \equiv \Phi_1 \equiv 0$. The proof is thus achieved. $\qquad \square$

Remark 3. If there exist positive constants C and γ, such that the sequence $\{\beta_{n,l}\}$ satisfies the following gap condition

$$|\beta_{n+1,l} - \beta_{n,l}| \geqslant \frac{C}{n^\gamma} \qquad (30)$$

for all $n \in \mathbb{Z}^*$ and $1 \leqslant l \leqslant d$, then, using the generalized Ingham's inequality in [18] and proceeding as in [27], we can show the uniqueness of solution to system (5) with the D-observation on a finite interval

$$D\Phi = 0 \quad \text{on } \Gamma_1 \times (0, T). \tag{31}$$

However, the gap condition (30) restricts the applicability of Ingham's type approach, essentially to one-dimensional problems. Fortunately, the D-observation (6) at the infinite horizon $(0, +\infty)$ does not require any gap condition on the sequence $\{\beta_{n,l}\}$. Moreover, even the case that $\beta_{m,k} = \beta_{n,l}$ for some $(m, k) \neq (n, l)$ is also tolerated under the orthogonality condition (20). This shows the main difference between the two methods.

Theorem 14. *Let Ω be the rectangular domain given by (1)–(2). Let the pair (A, D) satisfy Kalman's rank condition (18). Assume that there exists a constant $c > 0$, such that $\|A - cI\|$ is small enough. Then the overdetermined system (5)–(6) has only the trivial solution.*

Proof. We are now going to apply Theorem 13. First we consider the corresponding eigen-problem of system (12) as follows

$$\begin{cases} -\Delta\phi = \lambda\phi, & \text{in } \Omega, \\ \phi = 0, & \text{on } \Gamma_0, \\ \partial_\nu \phi = 0, & \text{on } \Gamma_1, \end{cases}$$

where Ω is the rectangular domain with reflecting sides given by (1)–(2). By a straightforward computation as in [7], we get the following eigenvalues and the associated eigenvectors:

$$\lambda_{m,n} = \left(m^2 + \frac{n^2}{a^2}\right)\pi^2, \quad \phi_{m,n} = \sin(m\pi x)\cos\left(\frac{n\pi y}{a}\right) \tag{32}$$

for all $(m, n) \in \mathbb{N}^* \times \mathbb{N}$. In particular, on the boundary Γ_1 we have

$$\phi_{m,n} = \begin{cases} \sin(m\pi x), & \text{on } (x, y) = (0, 1) \times \{0\}, \\ (-1)^n \sin(m\pi x), & \text{on } (x, y) = (0, 1) \times \{a\}. \end{cases} \tag{33}$$

According to (16), we put

$$\beta_{m,n,k}^2 = \lambda_{m,n} + \delta_k = \left(m^2 + \frac{n^2}{a^2}\right)\pi^2 + \delta_k, \tag{34}$$

where $\delta_k (1 \leqslant k \leqslant d)$ stand for the distinct eigenvalues of A.

We next check that the following condition

$$\lambda_{m,n} + \delta_k = \lambda_{p,q} + \delta_l \quad \text{with } (m, n, k) \neq (p, q, l)$$

implies that

$$\int_{\Gamma_1} \phi_{m,n}\phi_{p,q}d\Gamma = 0.$$

Noting (33)–(34), it is sufficient to check that the equality

$$\left(m^2 - p^2 + \frac{n^2 - q^2}{a^2}\right)\pi^2 = \delta_l - \delta_k \quad \text{with } (m,n,k) \neq (p,q,l) \tag{35}$$

implies that $m \neq p$.

When $k = l$, it is easy to see that equality (35) implies that $m \neq p$. So we consider only the case that $k \neq l$. Define the set

$$C_a = \left\{m^2 - p^2 + \frac{n^2 - q^2}{a^2}, \quad (m,n),(p,q) \in \mathbb{N}^* \times \mathbb{N}\right\}.$$

If a^2 is a rational number, then C_a is a discrete set and the elements in C_a have gaps. When $\|A - cI\|$ is small enough, equality (35) can not take place.

If a^2 is an irrational number, by Dirichlet's theorem on Diophantine approximation (Corollary 1B in [42]), the set C_a is dense in $\mathbb{R}$, so there is no positive gap for the elements in C_a. However, since $k \neq l$, as $\|A - cI\|$ is small enough, we have

$$0 < |\delta_k - \delta_l| < \frac{\pi^2}{a^2},$$

then, equality (35) implies $m \neq p$. The proof is thus achieved. $\square$

On the other hand, in what follows, we will remove the smallness assumption of $\|A - cI\|$ and consider A from a small perturbation of cI to an arbitrarily given matrix. This will be an interesting improvement from the point of view of perturbation.

Theorem 15. *Let Ω be the rectangular domain given by (1)–(2). Let the pair (A, D) satisfy Kalman's rank condition (18). Assume that the elements of A are rational numbers and a is an algebraic number. Then the overdetermined system (5)–(6) has only the trivial solution.*

Proof. When the elements of A are rational numbers, the characteristic polynomial of A has only rational coefficients, then the roots $\delta_k(1 \leq k \leq d)$ should be algebraic numbers. On the other hand, since a as well as $\frac{1}{a^2}$ are all algebraic numbers, then

$$\lambda_{m,n} - \lambda_{p,q} = \left(m^2 - p^2 + \frac{n^2 - q^2}{a^2}\right)\pi^2$$

is a transcendent number or 0. Then the equality

$$\lambda_{m,n} - \lambda_{p,q} = \delta_l - \delta_k$$

implies that $k = l$, and it is easy to see that $m \neq p$. The proof is complete.

$\square$

As immediate consequences of Proposition 10, Theorems 14 and 15, we have the following results.

Theorem 16. *Let Ω be the rectangular domain given by (1)–(2). Let the pair (A, D) satisfy Kalman's rank condition (18). Assume that there exists a constant $c > 0$, such that $\|A - cI\|$ is small enough. Then system (3) is asymptotically stable.*

Theorem 17. *Let Ω be the rectangular domain given by (1)–(2). Let the pair (A, D) satisfy Kalman's rank condition (18). Assume that the elements of A are rational numbers and the constant a is an algebraic number. Then system (3) is asymptotically stable.*

3.3. *Asymptotic synchronization by groups*

In this section, we will continue the study when Ω is a rectangular domain with reflecting sides. Under the conditions of C_p-compatibility (38) and (41), applying $(C_p C_p^T)^{-1/2} C_p$ to system (3) and setting $W = (C_p C_p^T)^{-1/2} C_p U$, we get the following reduced system

$$\begin{cases} W'' - \Delta W + A_p W = 0, & \text{in } \mathbb{R}^+ \times \Omega, \\ W = 0, & \text{on } \mathbb{R}^+ \times \Gamma_0, \\ \partial_\nu W + D_p W' = 0, & \text{on } \mathbb{R}^+ \times \Gamma_1, \end{cases} \tag{36}$$

where A_p and D_p are given by (40) and (43).

Obviously, the asymptotic synchronization by p-groups of system (3) is equivalent to the asymptotic stability of the reduced system (36). Moreover, we have

Theorem 18. *Let Ω be the rectangular domain given by (1)–(2). Let A satisfy the condition of C_p-compatibility (38), respectively, D satisfy the condition of strong C_p-compatibility (41). Assume that A satisfies the Kalman's rank condition*

$$\text{rank}(D, AD, \cdots, A^{N-1}D) = N - p, \tag{37}$$

and that there exists a constant $c > 0$, such that $\|A - cI\|$ is small enough. Then system (3) is asymptotically synchronizable by p-groups.

288   T. Li and B. Rao

Proof. Since the reduced matrices A_p and D_p are still symmetric and positive semi-definite, the asymptotic stability of the reduced system (36) follows from Theorem 18, provided that

$$\text{rank}(D_p, A_p D_p, \cdots, A_p^{N-p-1} D_p) = N - p,$$

which is true by Proposition 7 and the Kalman's rank condition (37). $\square$

The following result does not require the condition of C_p-compatibility for A and the condition of strong C_p-compatibility for D.

Theorem 19. *Let Ω be the rectangular domain given by (1)–(2). Assume that there exists a constant $c > 0$, such that $\|A - cI\|$ is small enough. If system (3) is asymptotically synchronizable by p-groups, then there exist functions $u_1, \cdots, u_p$ in $C^0(0, +\infty; H^1_{\Gamma_0}(\Omega)) \cap C^1(0, +\infty; L^2(\Omega))$, such that, setting*

$$u = \sum_{r=1}^{p} u_r e_r, \tag{38}$$

we have

$$(U - u, U' - u') \to (0, 0) \text{ in } (H^1_{\Gamma_0}(\Omega) \times L^2(\Omega))^N \quad \text{as } t \to +\infty. \tag{39}$$

Proof. By [30, Theorem 4.7], the asymptotic synchronization by p-groups of system (3) implies

$$\text{rank}(D, AD, \cdots, A^{N-1}D) = N - q \tag{40}$$

with $q \leqslant p$. Then, by Proposition 2, we have

$$V = \text{Span}\{\tilde{e}_1, \cdots, \tilde{e}_q\} \subseteq \text{Ker}(D),$$

where V is the largest invariant subspace of A, contained in $\text{Ker}(D)$. Let $\widetilde{C}_q$ be a full row-rank matrix of order $(N - q) \times N$, defined by

$$\text{Ker}(\widetilde{C}_q) = V.$$

A satisfies the corresponding condition of $\widetilde{C}_q$-compatibility (38), and D satisfies the corresponding condition of strong $\widetilde{C}_q$-compatibility (41). Then, by Theorem 18, we get

$$\widetilde{C}_q(U, U') \to (0, 0) \quad \text{in } (H^1_{\Gamma_0}(\Omega) \times L^2(\Omega))^{N-q} \quad \text{as } t \to +\infty. \tag{41}$$

Since the matrix A is symmetric, we may assume that $\tilde{e}_1, \cdots, \tilde{e}_q$ are the orthonormal eigenvectors of A. It follows that

$$U = \tilde{u} + \widetilde{C}_q^T (\widetilde{C}_q \widetilde{C}_q^T)^{-1} \widetilde{C}_q U,$$

where

$$\tilde{u} = \sum_{r=1}^{q} \tilde{u}_r \tilde{e}_r \quad \text{with } \tilde{u}_r = (U, \tilde{e}_r), \quad 1 \leqslant r \leqslant q.$$

Then, noting (41), we have

$$(U - \tilde{u}, \ U' - \tilde{u}') \to (0, 0) \quad \text{in } (H^1_{\Gamma_0}(\Omega) \times L^2(\Omega))^N \quad \text{as } t \to +\infty. \quad (42)$$

Next, multiplying (3) by $\tilde{e}_s$ for $s = 1, \cdots, q$, we get

$$\begin{cases} \tilde{u}''_s - \Delta \tilde{u}_s + \lambda_s \tilde{u}_s = 0, & \text{in } \mathbb{R}^+ \times \Omega, \\ \tilde{u}_s = 0, & \text{on } \mathbb{R}^+ \times \Gamma_0, \\ \partial_\nu \tilde{u}_s = 0, & \text{on } \mathbb{R}^+ \times \Gamma_1. \end{cases} \quad (43)$$

For any given $r = 1, \cdots, q$, taking

$$t = 0 : \|(\tilde{u}_r, \tilde{u}'_r)\|_{H^1(\Omega) \times L^2(\Omega)} = 1, \quad \tilde{u}_s = \tilde{u}'_s = 0, \quad s \neq r,$$

it follows that

$$\|C_p \tilde{e}_r\| = \|C_p \tilde{e}_r (\tilde{u}_r, \tilde{u}'_r)\|$$
$$\leqslant \|C_p (U - \tilde{u}_r \tilde{e}_r, U' - \tilde{u}'_r \tilde{e}_r)\| + \|C_p (U, U')\| \to 0 \quad \text{as } t \to +\infty.$$

We thus get $C_p \tilde{e}_r = 0$ for $r = 1, \cdots, q$, namely, $\text{Ker}(\widetilde{C}_q) \subseteq \text{Ker}(C_p)$. We can then write

$$\tilde{e}_r = \sum_{s=1}^{p} \alpha_{sr} e_s, \quad r = 1, \cdots, q.$$

Accordingly, we have

$$\tilde{u} = \sum_{r=1}^{q} \tilde{u}_r \tilde{e}_r = \sum_{r=1}^{q} \sum_{s=1}^{p} \tilde{u}_r \alpha_{sr} e_s = \sum_{s=1}^{p} \left(\sum_{r=1}^{q} \alpha_{sr} \tilde{u}_r \right) e_s.$$

Then, inserting the expression

$$u_s = \sum_{r=1}^{q} \alpha_{sr} \tilde{u}_r, \quad s = 1, \cdots, p \quad (44)$$

into (42), we get the convergence (39), called the asymptotic synchronization by p-groups in the pinning sense. However, when $q < p$, it follows from (44) that the functions $u_1, \cdots, u_p$ are linearly dependent. $\qquad \square$

Remark 4. Instead of the smallness assumption of $\|A - cI\|$, Theorems 18 and 19 remain true if A has rational elements and a is an algebraic number.

4. Uniform synchronization by method of compact perturbation

4.1. *Uniform stability*

We will study the uniform stability of the second order evolution system

$$U'' + \mathcal{L}U + AU + D\mathcal{G}U' = 0. \tag{1}$$

By Proposition 1, system (1) generates a C^0 semi-group of contractions on the space $(V \times H)^N$.

Definition 8. System (1) is uniformly (exponentially) stable in the space $(V \times H)^N$, if there exist constants $M \geqslant 1$ and $\omega > 0$, such that for any given initial data $(U_0, U_1) \in (V \times H)^N$, the corresponding solution U to system (1) satisfies

$$\|(U(t), U'(t))\|_{(V \times H)^N} \leqslant Me^{-\omega t}\|(U_0, U_1)\|_{(V \times H)^N}, \quad t \geqslant 0. \tag{2}$$

Proposition 11. *Let $\mathcal{R}$ be a linear compact mapping from V to $L^2(0, T; H)$. Then we can not find positive constants $M \geqslant 1$ and $\omega > 0$, such that for all $\theta \in V$, the solution to the following problem*

$$\begin{cases} u'' + Lu = \mathcal{R}\theta, \\ t = 0: \quad u = \theta, \ u' = 0 \end{cases} \tag{3}$$

satisfies

$$\|(u(t), u'(t))\|_{V \times H} \leqslant Me^{-\omega t}\|\theta\|_V, \quad t \geqslant 0. \tag{4}$$

Proof. Noting that problem (3) is time invertible, by well-posedness we have

$$\|\theta\|_V \leqslant \|u(T)\|_V + \|u'(T)\|_H + \int_0^T \|\mathcal{R}\theta\|_H dt. \tag{5}$$

Assume by contradiction that (4) holds for all $\theta \in V$, then we have

$$\|\theta\|_V \leqslant Me^{-\omega T}\|\theta\|_V + \int_0^T \|\mathcal{R}\theta\|_H dt. \tag{6}$$

When T is large enough, it follows that for all $\theta \in V$, we have

$$\|\theta\|_V \leqslant \frac{\sqrt{T}}{1 - Me^{-\omega T}}\|\mathcal{R}\theta\|_{L^2(0,T;H)}. \tag{7}$$

This contradicts the compactness of $\mathcal{R}$. The proof is complete. $\qquad\square$

Theorem 20. *Let C_q be a full row-rank matrix of order $(N-q) \times N$ with $0 \leqslant q < N$. Assume that there exist constants $M \geqslant 1$ and $\omega > 0$, such that for any given initial data $(U_0, U_1) \in (V \times H)^N$, the corresponding solution U to system (1) satisfies*

$$\|C_q(U(t), U'(t))\|_{(V \times H)^{N-q}} \leqslant M e^{-\omega t} \|C_q(U_0, U_1)\|_{(V \times H)^N} \qquad (8)$$

for all $t \geqslant 0$. Then

$$\mathrm{rank}(C_q D) \geqslant N - q. \qquad (9)$$

Proof. Assume by contradiction that the rank condition (9) fails. Then, we have

$$\mathrm{rank}(C_q D) = \mathrm{rank}(D C_q^T) < N - q = \mathrm{rank}(C_q^T). \qquad (10)$$

By [29, Proposition 2.11], we have

$$\mathrm{Im}(C_q^T) \cap \mathrm{Ker}(D) \neq \{0\}. \qquad (11)$$

Let $E \in \mathrm{Im}(C_q^T)$ be a unit vector such that $DE = 0$. Applying E to system (1) associated with the initial data

$$t = 0: \quad U = \theta E, \quad U' = 0 \qquad (12)$$

with $\theta \in V$, and setting $u = (E, U)$, we get

$$\begin{cases} u'' + Lu = -(E, AU), \\ t = 0 : u = \theta, \ u' = 0, \end{cases} \qquad (13)$$

here and hereafter $(\cdot, \cdot)$ denotes the inner product with the associated norm $\| \cdot \|$ in the euclidian space $\mathbb{R}^N$.

Now, we define the linear mapping

$$\mathcal{R} : \theta \to (E, AU). \qquad (14)$$

Since the matrices A and D are symmetric and positive semi-definite, by the dissipation of system (1) with the initial data (12), we have

$$\|\mathcal{R}\theta\|_{L^2(0,T;V)} + \|\mathcal{R}\theta\|_{H^1(0,T;H)} \leqslant c_T \|\theta\|_V, \qquad (15)$$

where c_T is a positive constant depending only on T.

Noting that the imbedding from $L^2(0,T;V) \cap H^1(0,T;H)$ into $L^2(0,T;H)$ is compact [33, Theorem 5.1], the mapping $\mathcal{R}$ is compact from V into $L^2(0,T;H)$.

On the other hand, noting $E = C_q^T x$, we have

$$u = (E, U) = (x, C_q U). \qquad (16)$$

Then, it follows from (8) that

$$\|(u(t), u'(t))\|_{V \times H} \leqslant c \|C_q(U(t), U'(t))\|_{V \times H} \leqslant c M e^{-\omega t} \|\theta\|_V \qquad (17)$$

for all $t \geqslant 0$ and all $\theta \in V$. This contradicts Proposition 3. $\qquad \square$

In particular, taking $C_q = I$ in Theorem 20, we get immediately

Corollary 1. *If system* (1) *is uniformly stable, then we have* $\mathrm{rank}(D) = N$.

Following the classical theory,[41,45] the uniform stability of a semi-group is robust by compact perturbation. This property was served for obtaining the uniform stability.[17,38]

Theorem 21. *Assume that the scalar equation*

$$u'' + Lu + gu' = 0 \tag{18}$$

is uniformly stable in the space $V \times H$. *If* $\mathrm{rank}(D) = N$, *then system* (1) *is uniformly stable in the space* $(V \times H)^N$.

Proof. Since the mapping $U \to AU$ is compact from V^N into H^N, the asymptotic stability of system (1) and the uniform stability of the following decoupled system

$$U'' + \mathcal{L}U + D\mathcal{G}U' = 0 \tag{19}$$

yield the uniform stability of system (1).

First, since $\mathrm{rank}(D) = N$, the decoupled system (19) can be decomposed into N scalar equations of the same type as that in (18), therefore, it is uniformly stable.

Next, we will check that system (11)–(12) has only the trivial solution. Then, by Theorem 1, the whole system (1) is asymptotically stable.

Otherwise, there exist $\beta \in \mathbb{R}$ and a non trivial function Φ such that

$$\mathcal{L}\Phi + A\Phi = \beta^2 \Phi, \quad \mathcal{G}\Phi = 0. \tag{20}$$

Let $E \in \mathbb{R}^N$ and $\lambda \geqslant 0$ be such that $AE = \lambda E$. Applying E to (20) and setting $\phi = (E, \Phi)$, we get

$$L\phi = (\beta^2 - \lambda)\phi, \quad g\phi = 0.$$

Since L is coercive, we have $\widehat{\beta^2} = \beta^2 - \lambda > 0$. Then, we check that $u = e^{i\widehat{\beta}t}\phi$ satisfies (18), which contradicts its uniform stability. □

Remark 5. Roughly speaking, Theorem 21 indicates that the uniform stability of system (1) can be obtained by means of the scalar equation (18). It provides thus a direct and efficient approach to solve a seemingly difficult problem of uniform stability of a complex system.

4.2. Uniform synchronization by groups

By Corollary 1, when $\text{rank}(D) < N$, system (1) is not uniformly stable. Similarly to the case of asymptotic stability in §2.4, we turn to consider the uniform synchronization by p-groups.

For any given integer $p \geqslant 1$, let

$$0 = n_0 < n_1 < n_2 < \cdots < n_p = N \tag{21}$$

be integers, such that $n_r - n_{r-1} \geqslant 2$ for all r with $1 \leqslant r \leqslant p$. We re-arrange the components of the state variable U into p groups

$$(u^{(1)}, \cdots, u^{(n_1)}), (u^{(n_1+1)}, \cdots, u^{(n_2)}), \cdots\cdots, (u^{(n_{p-1}+1)}, \cdots, u^{(n_p)}). \tag{22}$$

Definition 9. System (1) is uniformly (exponentially) synchronizable by p-groups, if there exist constants $M \geqslant 1$ and $\omega > 0$, such that for any given initial data $(U_0, U_1) \in (V \times H)^N$, the corresponding solution U to system (1) satisfies

$$\|(u^{(k)}(t) - u^{(l)}(t), u^{(k)'}(t) - u^{(l)'}(t))\|_{V \times H} \tag{23}$$

$$\leqslant M e^{-\omega t} \|(u_0^{(k)} - u_0^{(l)}, u_1^{(k)} - u_1^{(l)})\|_{V \times H}, \quad t \geqslant 0$$

for all k, l with $n_{r-1} + 1 \leqslant k, l \leqslant n_r$ and all r with $1 \leqslant r \leqslant p$.

Let C_p be the matrix given by (34). Then (23) can be equivalently rewritten as

$$\|C_p(U(t), U'(t))\|_{(V \times H)^{N-p}} \leqslant M e^{-\omega t} \|C_p(U_0, U_1)\|_{(V \times H)^{N-p}} \tag{24}$$

for all $t \geqslant 0$. Assume that the matrix A satisfies the condition of C_p-compatibility (38) and D the strong condition of C_p-compatibility (41).

Applying $(C_p C_p^T)^{-1/2} C_p$ to (7) and setting $W = (C_p C_p^T)^{-1/2} C_p U$, we get a self-closed reduced system

$$W'' + \mathcal{L}W + A_p W + D_p \mathcal{G} W' = 0, \tag{25}$$

where A_p and D_p are given by (40) and (43), respectively. Moreover, it is easy to check the following basic result.

Proposition 12. *Assume that the matrices A and D satisfy the condition of C_p-compatibility (38) and the condition of strong C_p-compatibility (41), respectively. The uniform synchronization by p-groups of system (1) in the space $(V \times H)^N$ is equivalent to the uniform stability of the reduced system (25) in the space $(V \times H)^{N-p}$.*

Since the reduced matrices A_p and D_p are still symmetric and positive semi-definite, the uniform stability of the reduced system (25) can be treated by Theorem 21. So, the uniform synchronization by p-groups is reduced to the uniform stability. However, the necessity of the condition of C_p-compatibility for A and that of the condition of strong C_p-compatibility for D are intrinsically linked with the rank of the matrix D.

Proposition 13. *If system* (1) *is uniformly synchronizable by p-groups, then we necessarily have*

$$\operatorname{rank}(C_pD) \geqslant N - p. \tag{26}$$

Proof. It is sufficient to take $C_q = C_p$ in Theorem 20. □

Proposition 14. *The following rank condition*

$$\operatorname{rank}(D) = \operatorname{rank}(C_pD) = N - p \tag{27}$$

holds if and only if $\operatorname{Ker}(D)$ *and* $\operatorname{Ker}(C_p)$ *are bi-orthonormal.*

Proof. By [29, Proposition 2.11], rank condition (27) is equivalent to

$$\operatorname{Ker}(D) \cap \operatorname{Im}(C_p^T) = \operatorname{Ker}(C_p) \cap \operatorname{Im}(D) = \{0\}, \tag{28}$$

namely,

$$\operatorname{Ker}(D) \cap \{\operatorname{Ker}(C_p)\}^\perp = \operatorname{Ker}(C_p) \cap \{\operatorname{Ker}(D)\}^\perp = \{0\}. \tag{29}$$

Hence by [29, Proposition 2.5], $\operatorname{Ker}(D)$ and $\operatorname{Ker}(C_p)$ are bi-orthogonal. □

Theorem 22. *Assume that system* (1) *is uniformly synchronizable by p-groups under the minimal rank conditions* (27). *Then A satisfies the condition of C_p-compatibility* (38) *and D satisfies the condition of strong C_p-compatibility* (41).

Proof. Let U be the solution to system (1) with the following initial data:

$$U_0 = \sum_{r=1}^p u_{0r}e_r, \quad U_1 = \sum_{r=1}^p u_{1r}e_r, \tag{30}$$

where $u_{0r} \in V$ and $u_{1r} \in H$ for $r = 1, \cdots, p$. Then, noting (36), we have

$$\|C_p(U(t), U'(t))\|_{(V \times H)^{N-p}} \tag{31}$$
$$\leqslant Me^{-\omega t}\|C_p(U_0, U_1)\|_{(V \times H)^{N-p}} = 0, \quad t \geqslant 0.$$

There exist some functions $u_1, \cdots, u_p$ in $C^0(\mathbb{R}^+, V) \cap C^1(\mathbb{R}^+, H)$, such that

$$U = \sum_{s=1}^p u_s e_s. \tag{32}$$

Then

$$\sum_{s=1}^{p} u_s'' e_s + \sum_{s=1}^{p} Lu_s e_s + \sum_{s=1}^{p} gu_s' De_s + \sum_{s=1}^{p} u_s A e_s = 0. \qquad (33)$$

Applying C_p to both sides of the above system, we get

$$\sum_{s=1}^{p} gu_s' C_p De_s + \sum_{s=1}^{p} u_s C_p A e_s = 0. \qquad (34)$$

In particular, by the continuity at $t = 0$, we have

$$\sum_{s=1}^{p} gu_s'(0) C_p De_s + \sum_{s=1}^{p} u_s(0) C_p A e_s = 0, \qquad (35)$$

then

$$C_p A e_s = 0, \quad C_p De_s = 0, \quad s = 1, \cdots, p. \qquad (36)$$

Thus A satisfies the condition of C_p-compatibility (38).

We show that D satisfies the condition of strong C_p-compatibility (41). In fact, for $s = 1, \cdots, p$, we have

$$(De_s, d) = (e_s, Dd) = 0, \quad d \in \mathrm{Ker}(D), \qquad (37)$$

then $De_s \in \mathrm{Ker}(D)^{\perp} \cap \mathrm{Ker}(C_p)$. By Proposition 14, $\mathrm{Ker}(D)$ is bi-orthogonal to $\mathrm{Ker}(C_p)$, so $\mathrm{Ker}(D)^{\perp} \cap \mathrm{Ker}(C_p) = \{0\}$. Then

$$De_s = 0, \quad s = 1, \cdots, p. \qquad (38)$$

We get thus the condition of strong C_p-compatibility (41) for the matrix D. $\qquad\square$

Theorem 23. *Let A satisfy* (38) *and D be given by* (42) *with* $\mathrm{rank}(R) = N - p$. *Assume that the scalar equation* (18) *is uniformly stable in the space $V \times H$. Then system* (1) *is uniformly synchronizable by p-groups in $(V \times H)^N$.*

Proof. By Proposition 12, it is sufficient to show the uniform stability of the reduced system (25). By (42), $\mathrm{rank}(D_p) = \mathrm{rank}(R) = N - p$. Then by Theorem 21, the reduced system (25) is uniformly stable. $\qquad\square$

Theorem 24. *Assume that system* (1) *is uniformly synchronizable by p-groups in $(V \times H)^N$, then there exist some functions $u_1, \cdots, u_p$ in $C^0(\mathbb{R}^+, V) \cap C^1(\mathbb{R}^+, H)$ and some positive constants $M \geqslant 1$ and $\omega > 0$, such that setting*

$$u = \sum_{r=1}^{p} u_r e_r / \|e_r\|, \qquad (39)$$

we have

$$\|(U(t) - u(t), U'(t) - u'(t))\|_{(V \times H)^N} \tag{40}$$
$$\leqslant M e^{-\omega t} \|C_p(U_0, U_1)\|_{(V \times H)^{N-p}}, \quad t \geqslant 0.$$

Assume furthermore that A satisfies the condition of C_p-compatibility (38) and D satisfies the condition of strong C_p-compatibility (41). Then u obeys a conservative system.

Proof. Let U be the solution to system (1) with any given initial data $(U_0, U_1) \in (V \times H)^N$. For $r = 1, \cdots, p$, let $u_r = (U, e_r)/\|e_r\|$. Noting that $\mathbb{R}^N = \text{Ker}(C_p) \oplus \text{Im}(C_p^T)$, we have

$$U = \sum_{r=1}^p u_r e_r / \|e_r\| + C_p^T (C_p C_p^T)^{-1} C_p U \tag{41}$$
$$= u + C_p^T (C_p C_p^T)^{-1} C_p U.$$

By (4), we get

$$\|(U(t) - u(t), U'(t) - u'(t))\|_{(V \times H)^N} \tag{42}$$
$$\leqslant \|C_p^T (C_p C_p^T)^{-1}\| \|C_p(U(t), U'(t))\|_{(V \times H)^{N-p}}$$
$$\leqslant M' e^{-\omega t} \|C_p(U_0, U_1)\|_{(V \times H)^{N-p}}, \quad t \geqslant 0$$

for some constant $M' \geqslant 1$.

Now we will precisely show the dynamics of the functions $u_1, \cdots, u_p$. First, recall that the condition of C_p-compatibility (38) implies

$$A e_r = \sum_{s=1}^p \beta_{rs} \frac{\|e_r\|}{\|e_s\|} e_s, \quad r = 1, \cdots, p. \tag{43}$$

Moreover, since A is symmetric, a straightforward computation shows that

$$(A e_r, e_s) = \sum_{q=1}^p \beta_{rq} \frac{\|e_r\|}{\|e_q\|} (e_q, e_s) = \beta_{rs} \|e_r\| \|e_s\| \tag{44}$$

and

$$(e_r, A e_s) = \sum_{q=1}^p \beta_{sq} \frac{\|e_s\|}{\|e_q\|} (e_r, e_q) = \beta_{sr} \|e_s\| \|e_r\|. \tag{45}$$

It follows that

$$\beta_{rs} = \beta_{sr}, \quad 1 \leqslant r, s \leqslant p.$$

On the other hand, the condition of strong C_p-compatibility (41) implies

$$D e_r = 0, \quad r = 1, \cdots, p. \tag{46}$$

Then, applying e_r to system (1), we get the following conservative system

$$\begin{cases} u_r'' + Lu_r + \displaystyle\sum_{s=1}^{p} \beta_{rs} u_s = 0, \\ t = 0 : u_r = (U_0, e_r)/\|e_r\|, \quad u_r' = (U_1, e_r)/\|e_r\| \end{cases} \tag{47}$$

for $r = 1, \cdots, p$. □

Remark 6. Classically, the convergence (23) or equivalently (24) is called the uniform synchronization by p-groups in the consensus sense, while the convergence (40) is in the pinning sense. Moreover, $(u_1, \cdots, u_p)^T$ is called the uniformly synchronizable state by p-groups. Theorem 24 indicates that the two notions are simply the same.

Moreover, setting the matrix $B = (\beta_{rs})$, we define the energy by

$$E(t) = \|u(t)\|_{V^p}^2 + (Bu(t), u(t))_{H^p} + \|u'(t)\|_{H^p}^2.$$

Since B is symmetric, we have

$$E(t) = E(0), \quad t \geqslant 0. \tag{48}$$

Then the orbit of u is lacalized on the sphere (48) which is uniquely determined by the projection of the initial data (U_0, U_1) to $\text{Ker}(C_p)$.

Remark 7. The condition of strong C_p-compatibility (41) implies that [29, Proposition 2.13]

$$\text{rank}(D, AD, \cdots, A^{N-1}D) = N - p.$$

Following [30, Theorem 4.7], there does not exist an extended matrix C_q $(q < p)$, such that

$$C_q(U(t), U'(t)) \to (0, 0) \quad \text{in } (V \times H)^N \text{ as } t \to +\infty.$$

Unlike the case of approximate boundary synchronization by p-groups, there is no possibility to get any induced synchronization in the present situation (see [29, Chaper 11]).

4.3. *Applications*

4.3.1. *Wave equations with boundary damping*

Let $\Omega \subset \mathbb{R}^n$ be a bounded domain with a smooth boundary $\Gamma = \Gamma_1 \cup \Gamma_0$, such that $\Gamma_1 \cap \Gamma_0 = \emptyset$ and $\text{mes}(\Gamma_1) > 0$. For fixing idea, we assume that $\text{mes}(\Gamma_0) > 0$.

Consider the following wave equation

$$\begin{cases} u'' - \Delta u = 0, & \text{in } \mathbb{R}^+ \times \Omega, \\ u = 0, & \text{on } \mathbb{R}^+ \times \Gamma_0, \\ \partial_\nu u + u' = 0, & \text{on } \mathbb{R}^+ \times \Gamma_1, \end{cases} \tag{49}$$

where ∂_ν denotes the outward normal derivative on the boundary. The uniform stability of (49) was abundantly studied by different approaches in literatures, we only quote [6, 21, 23] and the references therein.

Now, let A and D be symmetric and positive semi-definite matrices of order N. We consider the following system of wave equations:

$$\begin{cases} U'' - \Delta U + AU = 0, & \text{in } \mathbb{R}^+ \times \Omega, \\ U = 0, & \text{on } \mathbb{R}^+ \times \Gamma_0, \\ \partial_\nu U + DU' = 0, & \text{on } \mathbb{R}^+ \times \Gamma_1. \end{cases} \tag{50}$$

Multiplying (50) by $\Phi \in H^1_{\Gamma_0}(\Omega)$ and integrating by parts, we get the following variational formulation:

$$\int_\Omega \big((U'', \Phi) + (\nabla U, \nabla \Phi) + (AU, \Phi) \big) dx + \int_{\Gamma_1} (DU', \Phi) d\Gamma = 0. \tag{51}$$

Define

$$\langle Lu, \phi \rangle = \int_\Omega \nabla u \cdot \nabla \phi \, dx, \quad \langle gv, \phi \rangle = \int_{\Gamma_1} v \phi \, d\Gamma. \tag{52}$$

Then (51) can be rewritten as

$$U'' + \mathcal{L}U + AU + D\mathcal{G}U' = 0. \tag{53}$$

Moreover, since the scalar equation (49) is uniformly stable in $H^1_{\Gamma_0}(\Omega) \times L^2(\Omega)$, applying Theorem 23 and Theorem 24, we immediately obtain the following

Theorem 25. *Let A satisfy the condition of C_p-compatibility* (38) *and D be given by* (42) *with* $\text{rank}(R) = N - p$. *Then the system of wave equations* (50) *is uniformly synchronizable by p-groups in the space* $(H^1_{\Gamma_0}(\Omega) \times L^2(\Omega))^N$.

Moreover, for any given initial data $(U_0, U_1) \in (H^1_{\Gamma_0}(\Omega) \times L^2(\Omega))^N$, consider the problem

$$\begin{cases} u''_r - \Delta u_r + \sum_{s=1}^p \beta_{rs} u_s = 0, & \text{in } \mathbb{R}^+ \times \Omega, \\ u_r = 0, & \text{on } \mathbb{R}^+ \times \Gamma_0, \\ \partial_\nu u_r = 0, & \text{on } \mathbb{R}^+ \times \Gamma_1, \\ t = 0 : u_r = (U_0, e_r)/\|e_r\|, \ u'_r = (U_1, e_r)/\|e_r\|, & \text{in } \Omega \end{cases} \tag{54}$$

for $r = 1, \cdots, p$, *and the coefficients* β_{rs} *are given by* (43). *Then, setting* $u = \sum_{r=1}^{p} u_r e_r / \|e_r\|$, *the corresponding solution* U *to system* (50) *satisfies*

$$\|(U(t) - u(t), U'(t) - u'(t))\|_{(H^1_{\Gamma_0}(\Omega) \times L^2(\Omega))^N} \tag{55}$$

$$\leqslant M e^{-\omega t} \|C_p(U_0, U_1)\|_{(H^1_{\Gamma_0}(\Omega) \times L^2(\Omega))^{N-p}}, \quad t \geqslant 0.$$

4.3.2. *Wave equations with locally distributed damping*

Let $\Omega \subset \mathbb{R}^n$ denote a bounded domain with smooth boundary Γ. Let $\omega \subset \Omega$ denote the damped domain.

Let a be a smooth function such that

$$a(x) \geqslant 0, \quad \forall x \in \Omega \quad \text{and} \quad a(x) \geqslant a_0 > 0, \quad \forall x \in \omega. \tag{56}$$

Consider the uniform stability of the following locally damped scalar wave system

$$\begin{cases} u'' - \Delta u + a u' = 0, & \text{in } \mathbb{R}^+ \times \Omega, \\ u = 0, & \text{on } \mathbb{R}^+ \times \Gamma. \end{cases} \tag{57}$$

This is a very challenge and promising issue. There is a large amount of literatures that we will comment briefly. The uniform decay was first established by multipliers in [13] as ω is a neighbourhood of the boundary. Later, the result was generalized in [48] to semi-linear case. When Ω is a compact Riemann manifold without boundary and ω satisfies the geometric optic condition, the uniform stability was established by a micro-local approach in [40].

Now, we consider the following system of locally damped wave equations:

$$\begin{cases} U'' - \Delta U + A U + a D U' = 0, & \text{in } \mathbb{R}^+ \times \Omega, \\ U = 0, & \text{on } \mathbb{R}^+ \times \Gamma, \end{cases} \tag{58}$$

where A and D are symmetric and positive semi-definite matrices with constant elements. Multiplying system (58) by $\Phi \in (H^1_0(\Omega))^N$ and integrating by parts, we get the following variational formulation:

$$\int_\Omega ((U'', \Phi) + (\nabla U, \nabla \Phi) + (AU, \Phi)) dx + \int_\Omega a(DU', \Phi) dx = 0. \tag{59}$$

Let L and g be the linear continuous mappings from $H^1_0(\Omega)$ into $H^{-1}(\Omega)$, defined by

$$\langle Lu, \phi \rangle = \int_\Omega \nabla u \cdot \nabla \phi \, dx \quad \text{and} \quad \langle gv, \phi \rangle = \int_\Omega a v \phi \, dx, \tag{60}$$

respectively. Then the variational problem (59) can be rewritten as

$$U'' + \mathcal{L}U + AU + D\mathcal{G}U' = 0. \tag{61}$$

Then, applying Theorem 23 and Theorem 24, we have

Theorem 26. *Assume that the damped domain $\omega \subset \Omega$ contains a neighbourhood of the whole boundary Γ. Assume furthermore that A satisfies the condition of C_p-compatibility (38) and D is given by (42) with rank$(R) = N - p$. Then system (61) is uniformly synchronizable by p-groups in the space $(H_0^1(\Omega) \times L^2(\Omega))^N$.*

Moreover, for any given initial data $(U_0, U_1) \in (H_0^1(\Omega) \times L^2(\Omega))^N$, consider the problem

$$\begin{cases} u_r'' - \Delta u_r + \displaystyle\sum_{s=1}^{p} \beta_{rs} u_s = 0, & \text{in } \mathbb{R}^+ \times \Omega, \\[2mm] u_r = 0, & \text{on } \mathbb{R}^+ \times \Gamma, \\[2mm] t = 0 : u_r = (U_0, e_r)/\|e_r\|, \ u_r' = (U_1, e_r)/\|e_r\|, & \text{in } \Omega \end{cases} \tag{62}$$

for $r = 1, \cdots, p$, and the coefficients β_{rs} are given by (43). Then, setting $u = \sum_{r=1}^{p} u_r e_r / \|e_r\|$, the corresponding solution U to system (61) satisfies

$$\|(U(t) - u(t), U'(t) - u'(t))\|_{(H^1(\Omega) \times L^2(\Omega))^N} \tag{63}$$
$$\leqslant M e^{-\omega t} \|C_p(U_0, U_1)\|_{(H^1(\Omega) \times L^2(\Omega))^{N-p}}, \quad t \geqslant 0.$$

4.3.3. *Kirchhoff plate equations with locally distributed viscous damping*

In this section Ω is a bounded domain in $\mathbb{R}^2$, occupied by an elastic thin plate. We refer to [19] for the stabilization of linear models.

Let a be a smooth and non-negative function such that (56) holds. Assume that ω contains a neighbourhood of the whole boundary Γ. Then, the following system of plate equation

$$\begin{cases} u'' + \Delta^2 u + a u' = 0, & \text{in } \mathbb{R}^+ \times \Omega, \\ u = \partial_\nu u = 0, & \text{on } \mathbb{R}^+ \times \Gamma \end{cases} \tag{64}$$

is uniformly stable in $H_0^2(\Omega) \times L^2(\Omega)$.[19]

Consider the following system

$$\begin{cases} U'' + \Delta^2 U + AU + aDU' = 0, & \text{in } \mathbb{R}^+ \times \Omega, \\ U = \partial_\nu U = 0, & \text{on } \mathbb{R}^+ \times \Gamma, \end{cases} \tag{65}$$

where A and D are symmetric and positive semi-definite matrices with constant elements. Multiplying system (65) by $\Phi \in (H_0^2(\Omega))^N$ and integrating by parts, we get the following variational formulation

$$\int_\Omega ((U'', \Phi) + (\Delta U, \Delta \Phi) + (AU, \Phi))dx + \int_\Omega a(DU', \Phi)dx = 0. \qquad (66)$$

Let L and g be defined by

$$\langle Lu, \phi \rangle = \int_\Omega \Delta u \Delta \phi dx \quad \text{and} \quad \langle gv, \phi \rangle = \int_\Omega av\phi dx, \qquad (67)$$

respectively. The variational formulation (66) can be interpreted as

$$U'' + \mathcal{L}U + AU + D\mathcal{G}U' = 0. \qquad (68)$$

Then, applying Theorem 23 and Theorem 24, we have

Theorem 27. *Let A satisfy the condition of C_p-compatibility (38) and D be given by (42) with* $\text{rank}(R) = N - p$. *Then system (65) is uniformly synchronizable by p-groups in* $(H_0^2(\Omega) \times L^2(\Omega))^N$. *Moreover, for any given initial data* $(U_0, U_1) \in (H_0^2(\Omega) \times L^2(\Omega))^N$, *consider the problem*

$$\begin{cases} u_r'' + \Delta^2 u_r + \displaystyle\sum_{s=1}^p \beta_{rs} u_s = 0, & \text{in } \mathbb{R}^+ \times \Omega, \\ u_r = \partial_\nu u_r = 0, & \text{on } \mathbb{R}^+ \times \Gamma, \\ t = 0 : u_r = (U_0, e_r)/\|e_r\|, \ u_r' = (U_1, e_r)/\|e_r\|, & \text{in } \Omega \end{cases} \qquad (69)$$

for $r = 1, \cdots, p$, *and the coefficients β_{rs} are given by (43). Then, setting $u = \sum_{r=1}^p u_r e_r / \|e_r\|$, the corresponding solution U to system (65) satisfies*

$$\|(U(t) - u(t), U'(t) - u'(t))\|_{(H^2(\Omega) \times L^2(\Omega))^N} \qquad (70)$$
$$\leqslant Me^{-\omega t} \|C_p(U_0, U_1)\|_{(H^2(\Omega) \times L^2(\Omega))^{N-p}}, \quad t \geqslant 0.$$

Remark 8. The above classical examples illustrate the applications of the abstract theory. In fact, Theorems 3.8 and 3.9 are also applicable for many other models, such as system of wave equations with viscoelastic (Kelvin-Voigt) damping, system of Kirchhoff plate equations with shear force and bending moment damping etc.

5. Perspective comments

At the end of this lecture, we want to mention several possibilities to enlarge the research field on the topic.

(i) Up to now, we have started the work on a simplified model with only one damping. We may consider a system with several dampings of different types:

$$U'' + \mathcal{L}U + AU + D_1\mathcal{G}_1U' + D_2\mathcal{G}_2U' = 0, \tag{1}$$

where $\mathcal{G}_1$ and $\mathcal{G}_2$ can be internal and boundary dampings for wave equations, and bending moment and shear force dampings for plate equations, respectively. Many related questions can be asked, for example:

(a) Let $D = (D_1, D_2)$ be the composite damping matrix. Is Kalman's rank condition on (A, D) still sufficient for the asymptotic stability of the corresponding system?

(b) Is the condition $\mathrm{rank}(D) = N$ still sufficient for the uniform stability of the corresponding system?

The main difficulty comes from the interaction of the numerous matrices A, D_1 and D_2, somewhat like for coupled Robin problem in [24]. The key idea is to separate them by compactness/uniqueness argument, so more regularity seems to be necessary.

We do not have any answer yet for each question, but the first attempt already shows some interesting results for developing the research in these directions.

(ii) For the models considered in §2.4, the observability inequality is obtained by the multiplier method under the geometrical multiplier condition. It is stronger than the required observability inequality, for example, (72) is much stronger than (68) etc. We hope that this regularity should be served to establish a polynomial decay rate for the smooth initial data:

$$\|C_p(U(t), U'(t))\|_{(V \times \mathcal{H})^{N-p}} = O((1+t)^{-\delta}), \tag{2}$$

where the constant $\delta > 0$ is independent of the initial data. We refer to [12, 39] and the references therein for the recent progress on the polynomial stability of indirectly damped wave equations.

(iii) By the definition of uniform synchronization by p-groups:

$$\|C_p(U(t), U'(t))\|_{(V \times H)^{N-p}} \leqslant Me^{-\omega t}\|C_p(U_0, U_1)\|_{(V \times H)^{N-p}}, \quad t \geqslant 0, \tag{3}$$

if $C_p(U_0, U_1) = (0, 0)$, then $C_pU(t) \equiv 0$ for all $t \geqslant 0$. Thus, for any given synchronized initial data, the solution is always synchronized. This simplifies much the study on the necessity of the conditions of C_p-compatibility given in Theorem 22. A more natural definition of uniform synchronization by p-groups should be given by

$$\|C_p(U(t), U'(t))\|_{(V \times H)^{N-p}} \leqslant Me^{-\omega t}\|(U_0, U_1)\|_{(V \times H)^N}, \quad t \geqslant 0. \tag{4}$$

In this case, the solution is not automatically synchronized even for the synchronized initial data. The situation will be chaotic and presents certainly many interesting questions.

(iv) Under Kalman's rank condition (18), the following outside problem

$$\begin{cases} \Phi'' - \Delta\Phi + A\Phi = 0, & \text{in } (0, +\infty) \times \Omega^c, \\ \partial_\nu \Phi = 0, & \text{on } (0, +\infty) \times \Gamma \end{cases} \tag{5}$$

associated with an observation on any finite interval

$$D\Phi = 0 \quad \text{on } (-T, T) \times \Gamma$$

has always non-trivial solutions, but it has only the trivial solution with the observation at the infinite horizon

$$D\Phi = 0 \quad \text{on } (-\infty, +\infty) \times \Gamma. \tag{6}$$

We may ask if the inside system

$$\begin{cases} \Phi'' - \Delta\Phi + A\Phi = 0, & \text{in } (0, +\infty) \times \Omega, \\ \partial_\nu \Phi = 0, & \text{on } (0, +\infty) \times \Gamma \end{cases} \tag{7}$$

with the observation on any finite interval:

$$D\Phi = 0 \quad \text{on } [0, T] \times \Gamma_1 \tag{8}$$

may admit non-trivial solutions. This question is open and related to the approximate controllability at the infinite horizon.

(v) The approaches should be quite flexible and can be easily applied to other types equations, such as Timoshenko beam,[1,16] Bresse beam,[34] thermo-elastic plate[9,11] and Maxwell equations etc.[15]

References

[1] F. Ammar-Khodja; A. Benabdallah; J. E. Munoz Rivera; R. Racke, Energy decay for Timoshenko systems of memory type, J. Diff. Eqs., 194 (2003), 82–115.

[2] W. Arendt; C. J. Batty, Tauberian theorems and stability of one-parameter semi-groups, Trans. Amer. Math. Soc., 306 (1988), 837–852.

[3] C. Bardos; G. Lebeau; J. Rauch, Sharp sufficient conditions for the observation, control, and stabilization of waves from the boundary, SIAM J. Control Optim., 30 (1992), 1024–1064.

[4] C. D. Benchimol, A note on weak stabilization of contraction semi-groups, SIAM J. Control Optim., 16 (1978), 373–379.

[5] Th. Cazenave; A. Haraux, An Introduction to Semilinear Evolution Equations, Clarendon Press, Oxford, 1998.

[6] F. Conrad; B. Rao, Decay of solutions of the wave equation in a star shaped domain with non linear boundary feedback, J. Asymp. Anal., 7 (1993), 159–177.

[7] R. Courant; D. Hilbert, Methods of Mathematical Physics, Vol. I, Interscience Publishers, Inc., New York, N.Y., 1953.

[8] C. M. Dafermos; M. Slemrod, Asymptotic behavior of nonlinear contraction semigroups. J. Functional Analysis 13 (1973), 97–106.

[9] L. de Teresa; E. Zuazua, Controllability of the linear system of thermoelastic plates, Adv. Diff. Eqs., 1 (1996), 369–402.

[10] N. Garofalo; F. Lin, Uniqueness of solution for elliptic operators: A geometric-variational approach, Comm. Pure Appl. Math., 40 (1987), 347–366.

[11] S. W. Hansen; B. Zhang, Boundary control of a linear thermo-elastic beam, J. Math. Anal. Appl., 210 (1997), 182–205.

[12] J. Hao; B. Rao, Influence of the hidden regularity on the stability of partially damped systems of wave equations, J. Math. Pures Appl., 143 (2020), 257–286.

[13] A. Haraux, Une remarque sur la stabilisation de certains systèmes du deuxième ordre en temps, Portugal Math., 46 (1989), 245–258.

[14] Ch. Huygens, Oeuvres Complètes, Vol.15, Swets & Zeitlinger B.V., Amsterdam, 1967.

[15] B. V. Kapitonov, Stabilization and exact boundary controllability for Maxwells equations, SIAM J. Control Optim., 32 (1994), 408–420.

[16] J. U. Kim; Y. Renardy, Boundary control of the Timoshenko beam, SIAM J. Control Optim., 25 (1987), 1417–1429.

[17] V. Komornik; B. Rao, Boundary stabilization of compactly wave equations, Asymp. Anal., 14 (1997), 339–359.

[18] V. Komornik; P. Loreti, Fourier Series in Control Theory, Springer Monogr. Math., Springer-Verlag, New York, 2005.

[19] J. E. Lagnese, Boundary Stabilization of Thin Plates, SIAM, Study in applied mathematics, Philadelphia, 1989.

[20] J. E. Lagnese; J.-L. Lions, Modelling, Analysis and Control of Thin Plates, Recherches en Mathématiques Appliquées, Masson, Paris, 1988.

[21] I. Lasiecka; D. Tataru, Uniform boundary stabilization of semilinear wave equations with nonlinear boundary damping, Diff. Int. Equs., 6 (1993), 507–533.

[22] I. Lasiecka; R. Triggiani, Uniform stabilization of a shallow shell model with nonlinear boundary dampings, J. Math. Anal. Appl., 269 (2002), 642–688.

[23] G. Lebeau, Equation des ondes amorties, Math. Phys. Stud., 19 (1996), 73–109.

[24] T.-T. Li; X. Lu; B. Rao, Exact boundary controllability and exact boundary synchronization for a coupled system of wave equations with coupled Robin boundary controls, ESAIM: COCV, 27 (2021), S7, 29 pp., https://doi.org/10.1051/cocv/2020047.

[25] T.-T. Li; B. Rao, Synchronisation exacte d'un système couplé d'équations des ondes par des contrôles frontières de Dirichlet, C. R. Acad. Sci. Paris, Ser. 1, 350 (2012), 767–772.

[26] T.-T. Li; B. Rao, Exact synchronization for a coupled system of wave equations with Dirichlet boundary controls, Chin. Ann. Math., 34B (2013), 139–160.

[27] T.-T. Li; B. Rao, Criteria of Kalman's type to the approximate controllability and the approximate synchronization for a coupled system of wave equations with Dirichlet boundary controls, SIAM J. Control Optim., 54 (2016), 49–72.

[28] T.-T. Li; B. Rao, Kalman's criterion on the uniqueness of continuation for the nilpotent system of wave equations, C. R. Acad. Sci. Paris, Ser. I, 356 (2018), 1188–1194.

[29] T.-T. Li; B. Rao, Boundary Synchronization for Hyperbolic Systems, Progress in Non Linear Differential Equations and Their Applications, Subseries in Control, 94, Birkhäuser, 2019.

[30] T.-T. Li; B. Rao, Uniqueness of solution to systems of elliptic operators and application to asymptotic synchronization of linear dissipative systems, ESAIM: COCV 26 (2020) 117, https://doi.org/10.1051/cocv/2020062.

[31] T.-T. Li; B. Rao, Uniform synchronization of an abstract linear second order evolution system, to appear in SIAM J. Control Optim., 2021.

[32] T.-T. Li; B. Rao, Asymptotic synchronization of a coupled system of wave equations on a rectangular domain with reflecting sides, to appear in Asymptotic Analysis, 2021.

[33] J.-L. Lions, Contrôlabilité Exacte, Perturbations et Stabilisation de Systèmes Distribués, Vol. 1, Masson, Paris, 1988.

[34] Z. Liu; B. Rao, Energy decay rate of the thermoelastic Bresse system, Z. Angew. Math. Phys., 60 (2009), 54–69.

[35] A. Pazy, Semi-groups of Linear Operators and Applications to Partial Differential Equations, Applied Mathematical Sciences, Springer-Verlag, 1983.

[36] A. Pikovsky; M. Rosenblum; J. Kurths, Synchronization: A Universal Concept in Nonlinear Sciences, Cambridge University Press, 2001.

[37] B. Rao, Stabilization of Kirchhoff plate equation in star-shaped domain by nonlinear boundary feedback, Nonlinear Anal. 20 (1993), 605–626.

[38] B. Rao, Stabilization of elastic plates with dynamical boundary control, SIAM J. Control Optim., 36 (1998), 148–163.

[39] B. Rao, On the sensitivity of the transmission of boundary dissipation for strongly coupled and indirectly damped systems of wave equations, Z. Angew. Math. Phys., 70 (2019), Paper No. 75, 25 pp.

[40] J. Rauch; M. Taylor, Exponential decay of solutions to hyperbolic equations in bounded domains, Indiana University Mathematics Journal, 24 (1974), 79–86.

[41] D. L. Russell, Decay rates for weakly damped systems in Hilbert space obtained with control-theoretic methods, J. Diff. Equs., 19 (1975), 344–370.

[42] W. Schmidt, Diophantine approximation, Lecture Notes in Mathematics, 785, Springer, 1980.

[43] S. Strogatz, SYNC: The Emerging Science of Spontaneous Order, New York, THEIA, 2003.

[44] D. Tataru, Unique continuation for solutions to PDE's; between Hörmander's theorem and Holmgren's theorem, Comm. Partial Diff. Equs., 20 (1995), 855–884.

[45] R. Triggiani, Lack of uniform stabilization for noncontractive semigroups under compact perturbation, Proc. Amer. Math. Soc., 105 (1989), 375–383.

[46] L. Wang; Q. Yan, Optimal control problem for exact synchronization of parabolic system, Math. Control Relat. Fields 9 (2019), 411–424.

[47] N. Wiener, Cybernetics, or Control and Communication in the Animal and the Machine, 2nd ed., The MIT Press/John Wiley & Sons, Inc., Cambridge, Mass./New York, London,1961.

[48] E. Zuazua, Exponential decay for the semilinear wave equation with locally distributed damping, Comm. Part. Diff. Equs., 15 (1990), 205–235.